CENTERING ANIMALS

CENTERING ANIMALS

in Latin American History

Martha Few and **Zeb Tortorici**, editors

FOREWORD BY ERICA FUDGE

Duke University Press
Durham and London   2013

Designed by Heather Hensley

Typeset in Arno Pro by Keystone Typesetting

Library of Congress Cataloging-in-Publication Data

Centering animals in Latin American history / Martha Few
and Zeb Tortorici, eds. ; foreword by Erica Fudge.

p. cm

Includes bibliographical references and index.

ISBN 978-0-8223-5383-6 (cloth : alk. paper)

ISBN 978-0-8223-5397-3 (pbk. : alk. paper)

1. Animals—Latin America—History. 2. Human-animal
relationships—Latin America. 3. Animals—Symbolic
aspects—Latin America. I. Few, Martha, 1964–

II. Tortorici, Zeb, 1978–

QL223.C468 2013

591.98—dc23   2012048664

FOR NEIL L. WHITEHEAD (1956–2012)

CONTENTS

FOREWORD

A number of years ago I wrote about the need for a history of animals that would allow us to comprehend more fully how far we humans were "embedded within and reliant upon the natural order."[1] Since that essay was published, the field of the history of animals has grown exponentially. No longer regarded as marginal or perceived as eccentric or even semiserious, the history of animals has, in the past ten years, gained a status far beyond what might have been imagined back in 2002. As this volume's title proclaims, there has been a "centering" of animals in history during this decade, and the essays brought together here add to that movement.

Historical work on animals is invaluable for numerous reasons, as *Centering Animals in Latin American History* makes clear. It is leading us to new insights about human-animal relations, and thus, in this collection alone, we encounter the significance of colonial practices; indigenous beliefs; scientific and medical ideas; conceptions of blasphemy, pollution, and masculinity; and perceptions and representations of animals in Latin America from the sixteenth century to the twentieth. But what we also come to see—in this collection and elsewhere—goes beyond this.

Histories have shown us how far humans and animals were not simply cohabiting, whether deliberately or accidentally, but how *significant* animals were to human culture, to the extent that I now think of human culture as "so-called human culture." It is not simply that animals were (and are) used in the production of vellum, parchment, and glue, all things on which writing and publishing have relied, which historians have rarely acknowledged fully.[2] It is also that animals were present as foci for human attention in ways that changed human worlds—in films, zoos, and photography, for example, and in science, pet-keeping, and sport, to name but a few of the topics covered in monographs published in the past decade.[3] In engaging with such issues, and in recognizing the role that animals have had in them, historians have expanded our understanding of the places and the periods they research.

But histories of animals are also going further than this. Some works have begun to outline the leading roles that animals have played in the making of so-called human history. One implication of such works is that if the roles of animals were not so addressed, our historical understanding would remain limited: there would be aspects that we would never fully explore and, moreover, that we would never be *able* to fully explore. This is not to suggest that animals were knowingly engaged in transforming the worlds around them. Actor network theory, as outlined by Bruno Latour, and so usefully introduced into animal studies by Chris Philo and Chris Wilbert, offers us a way of thinking about animals not just as the passive recipients of human actions but as active presences in the world, yet without anthropomorphizing them.[4] On a beautifully prosaic level, and as an illustration of such a conception of animal agency, cow manure was vital to agricultural improvement and hence to urbanization and industrialization in England, a fact that puts cattle at the center of that country's historical development.[5]

Not only is a new acknowledgment of the role that animals have played in transforming the past becoming evident, but the focus on animals brings into view issues previously ignored by historians. Sometimes these issues concern the history of the emotional and domestic life, as in studies of pet-keeping, for example. But on other occasions historical work returns us to the familiar places of our pasts and finds them to be somewhat different once animals are recognized as central. Thus, Virginia DeJohn Anderson's study of the relationships between the indigenous peoples and the English settlers in Chesapeake and New England is a history of cattle farming, not only because it was an economic factor in people's lives at that time but also because, she argues persuasively, the relationships between the native people and the settlers were forged through misunderstandings and negotiations over livestock.[6] Cows, once again, were agents.

The work that has emerged in the field of the history of animals in the past decade has reached beyond my idea that we must understand more fully how far humans are embedded within and reliant on the natural order. It is beginning to show us that the human world is not only, or simply, human at all. The world—as *Centering Animals in Latin American History* shows so well—is full of other beings: locusts, cattle, dogs (baptized or otherwise), monkeys, seals, birds, and goats. What it also shows us is that it is with animals that history is made.

ERICA FUDGE
UNIVERSITY OF STRATHCLYDE

Notes

1. Erica Fudge, "A Left-Handed Blow: Writing the History of Animals," in *Representing Animals*, ed. Nigel Rothfels (Bloomington: Indiana University Press, 2002), 13.
2. For a study that does pay attention to the animal source of documents, see Sarah Kay, "Legible Skins: Animals and the Ethics of Medieval Reading," *Postmedieval: A Journal of Medieval Cultural Studies* 2.1 (2011): 13–32.
3. See, in order, Jonathan Burt, *Animals in Film* (London: Reaktion, 2002); Nigel Rothfels, *Savages and Beasts: The Birth of the Modern Zoo* (Baltimore: Johns Hopkins University Press, 2002); Matthew Brower, *Developing Animals: Wildlife and Early American Photography* (Minneapolis: University of Minnesota Press, 2010); Anita Guerrini, *Experimenting with Humans and Animals: From Galen to Animal Rights* (Baltimore: Johns Hopkins University Press, 2003); Katherine C. Greer, *Pets in America: A History* (Chapel Hill: University of North Carolina Press, 2006); Peter Edwards, *Horse and Man in Early Modern England* (London: Hambledon Continuum, 2007); Donna Landry, *Noble Brutes: How Eastern Horses Transformed English Culture* (Baltimore: Johns Hopkins University Press, 2009); Sandra Swart, *Riding High: Horses, Humans and History in South Africa* (Johannesburg: Wits University Press, 2010).
4. Bruno Latour, *We Have Never Been Modern*, trans. Catherine Porter (Cambridge: Harvard University Press, 2001 [1991]); Chris Philo and Chris Wilbert, "Animal Spaces, Beastly Places: An Introduction," in *Animal Spaces, Beastly Places: New Geographies of Human-Animal Relations*, ed. Chris Philo and Chris Wilbert (London: Routledge, 2000), 1–36, especially 17.
5. See David Stone, "Medieval Farm Management and Technological Mentalities: Hinderclay before the Black Death," *Economic History Review* 54.4 (2001): 612–38, especially 625; and Liam Brunt, "Where There's Muck, There's Brass: The Market for Manure in the Industrial Revolution," *Economic History Review* 60.2 (2007): 333–72.
6. Virginia DeJohn Anderson, *Creatures of Empire: How Domestic Animals Transformed Early America* (Oxford: Oxford University Press, 2004).

ACKNOWLEDGMENTS

The individuals who have contributed to this project in its various states are numerous. *Centering Animals in Latin American History* began as a conference panel titled "Animals, Colonialism, and the Atlantic World" for the 2006 annual meeting of the American Society for Ethnohistory, which took place in Williamsburg, Virginia. We are grateful to the participants and audience members for their questions and especially for their enthusiasm. As the edited volume slowly took shape over the years, a number of individuals influenced the directions the project took. We are especially grateful to Roger Gathman, for his perceptive comments and feedback, and to Erica Fudge, for graciously offering to write the foreword to the volume. In the broader fields of critical animal studies and sexuality studies, Zeb Tortorici and Martha Few want to thank the following individuals for their encouragement and support: Margo DeMello, Robin Derby, Georgina Dopico Black, Gabriel Giorgi, Susan Kellogg, Marcy Norton, Pete Sigal, and Neil L. Whitehead.

Research for Tortorici's chapter in this volume was made possible by generous funding from the Animals and Society Institute (http://www .animalsandsociety.org), where he was a visiting fellow at North Carolina State University in the summer of 2007. Ken Shapiro and Tom Regan were absolutely crucial to the success of this summer institute. For pushing him to think more critically (in both theory and practice) about the relationships between humans and other animals, and for sharing wonderful conversation over countless vegan meals, Zeb is especially grateful to Osvaldo Gómez, Joanne Lin, Scott Lucas, Jennifer Ly, Yuki Maeda, Tom and Nancy Regan, Frederico Santos Soares de Freitas, Su Anne Takeda, Danielle Terrazas Williams, Richard Twine, and Tom Tyler.

Research for Few's chapter in this volume was made possible with funding from the History Department and the Social and Behavioral Sciences

Research Institute at the University of Arizona. Few drafted her chapter while a Visiting Scholar at Harvard University's David Rockefeller Center for Latin America Studies, during spring 2009, where she benefited from ongoing conversations about insects, especially with Paul Scolieri and Edwin Ortiz. She would also like to thank Bert Barickman, Alison Futrell, Kevin Gosner, Keisuke Hirano, Katrina Jagodinsky, Steve Johnstone, Fabio Lanza, and Neil Prendergast for reading the chapter and for their helpful suggestions. Thanks also to Alex Hidalgo, co-translator of Reinaldo Funes Monzote's essay.

In our effort to "center animals" through scholarship, we have decided to donate the proceeds of this volume to animal-welfare organizations in Latin America. There are a number of nonprofit organizations and NGOs throughout Latin America—spaying and neutering programs, animal shelters, humane societies, animal sanctuaries, free veterinary clinics, anti-bullfighting and anti-circus campaigns, and more—that do important political work with animals. For a few examples in Brazil, Argentina, Colombia, and Mexico, see the following websites: http://www.uipa.org.br; http://www.arca brasil.org.br; http://www.greatapeproject.org/pt-BR; http://www.pea.org .br; http://www.ranchodosgnomos.org.br; http://www.adda.org.ar; http: //www.adacolombia.org; http://www.gepda.org; http://www.animanatur alis.org; and, http://www.amigosac.org. We want to thank Frederico Santos Soares de Freitas and Gerardo Tristan respectively for information on Brazilian and Mexican organizations. It should be noted that many such organizations espouse a discourse of "animals rights" (*derechos de los animales* in Spanish; *direitos dos animais* in Portuguese), and that despite one's affinity or revulsion to such terminology, these organizations unquestionably improve the living conditions of animals throughout Latin America. Please contact the editors in order to find out more information about the organizations to which we have donated, or to suggest other animal-welfare organizations in Latin America and the Caribbean.

Finally, Neil L. Whitehead, in particular, pushed us to think about the cyclical relationship between animality, humanity, and divinity in the early stages of this project. Neil's initial and sustained enthusiasm helped *Centering Animals in Latin American History* come to fruition, and his brilliant concluding essay to this volume, "Loving, Being, Killing Animals," is testament to the type of radically innovative scholarship to which he was dedicated. We dedicate this book to Neil L. Whitehead—a phenomenally creative, incisive, prolific, and supportive scholar and colleague, who is sorely missed.

Writing Animal Histories

ZEB TORTORICI AND MARTHA FEW

On 12 January 1563, Juan Canuc was walking with his wife to a nearby ranch for livestock grazing when they heard a number of chickens clucking on the hillside near a large cross. According to Canuc's testimony, which was translated from Yucatecan Maya to Spanish by a court-appointed interpreter, the couple found a young boy who "had his underwear loose and was sitting on the ground with a turkey [*gallina de la tierra*, a "chicken of this land"—it being native to Mesoamerica] in between his legs."[1] The boy, a fourteen-year-old Maya named Pedro Na, from a small town near Mérida in Mexico's largely indigenous Yucatán Peninsula, had been caught in flagrante delicto, committing the "unnatural" crime of bestiality. Canuc handed Na over to Spanish authorities, who threw him in prison and then tried him, apparently eager to make an example out of the boy. As was customary in bestiality cases, colonial authorities "deposited" the bird as evidence, but it died within a few days as a result of being injured by Na. Under questioning, the boy admitted, "It was true that yesterday afternoon in the said road, he came across some turkeys and chickens, and he took the said turkey and went to the hillside with it and, with the carnal agitation [*alteración carnal*] he felt ... he had carnal access with the turkey."[2]

Some weeks later, Pedro Na was sentenced to be taken from prison on horseback, with his hands and feet bound, accompanied by a town crier proclaiming his crime as he was brought to the central plaza, there to be publicly castrated and then permanently expelled from the province of Yucatán. Adding a macabre touch to the role of animals in spectacles and processions of public shame, on 14 February 1563 authorities pronounced, "[Because] the turkey with which the said Pedro Na committed the crime is dead and has been preserved, we order that as the sentence is carried out, the dead turkey shall be hung from Pedro Na's neck as he is paraded through the streets of this city. After the said sentence is executed, the turkey shall be burned in live flames and turned into ashes."[3] The turkey was surely a rotten corpse by this point; after all, it had died at least one month before the sentence was carried out. As a number of historians have shown, bestiality cases like this one supply a surprising amount of information about human societies, sexual desires, legal traditions, the urban/rural dichotomy, and human-animal interactions in history.[4] But these cases also provide us with information about particular animals that is otherwise lost in the historical record, where one nonhuman animal tends to blur into another. Pedro Na's case allows us to look, however fleetingly, at the fate of an individual animal as well as the culturally and theologically determined meanings and symbolism with which it was imbued at one particular moment in history.

What can the turkey with which Pedro Na violently fornicated tell us about the experiences and representations of animals in history? The turkey, had it not died from the injuries inflicted by the sexual act, surely would have been killed by authorities, who, following injunctions set forth in the Book of Leviticus, would have burned, hung, or beaten to death those animals implicated in bestiality cases.[5] How do cases like this elucidate colonial encounters among and between human and nonhuman animals? How can we take the intertwined histories of two animals, one of them human, gleaned from archival documents such that we may critically interrogate human uses of and interactions with other animals? The case, which reveals the semiotic space in which one dead animal (put on public display) attained a political, moralistic, religious, and cultural significance in early colonial Mexico, certainly has the potential to broaden our understanding of animal symbolism and human dominion over animals as it was understood in the context of the violent interface between different cultures and the enormous changes wrought by colonization. Ultimately, the case of Pedro Na provides an entrance into the world in which certain forms of theological and juridical regulation of relationships and boundaries between humans and animals prevailed; but it also sets the stage to examine larger changes in animal husbandry and the use of European domesticated animals in rural agrarian economies in this mundane moment. We notice, for instance, the presence of Castilian chickens among flocks of native Mesoamerican turkeys and the types of interspecies gazing that characterized the everyday enactment of human-animal interactions and agricultural industry in a colonial landscape.

In this vein, we want to pose the question of what Latin American histories look like if we choose as the standpoint of our observation a focus not only on the human-animal interaction but also on the experiences and histories of the nonhuman actor in the unfolding of history. Despite the methodological challenges inherent in writing histories of animals, by using documents created by and for humans, does this "centering" of animals— human and nonhuman alike—allow us to write more comprehensive and less anthropocentric cultural histories? Does the centering of animals—the transforming of nonhuman animals into *central* actors in the historical narrative—provide us with significantly different versions of the past than those historical works that solely present animals as visible and important factors in history? This is, as can be seen in the individual essays in this volume, a tension that permeates the entire collection. Some of the contributors are primarily invested in denaturalizing the human/animal binary, whereas others are more intent on demonstrating how the exploitative practices of humans in the past have wreaked havoc on particular species and on the natural environment. A few of the contributors work carefully through the methodological implications of "centering animals" within historical narratives, while others are skeptical of that very practice of centering animals, or they seek to include nonhuman animals as yet another social actor (alongside social classes, women, the state, the church, etc.) into the histories they write. After all, while animals play significant roles in all of the essays included in *Centering Animals*, they remain, at times, nearer to the margins of analysis—the perhaps necessary casualty of using the category of the "human" in order to examine the "animal." The project of centering animals is, in this sense, a necessarily tenuous one, especially given that the mere analytical and figurative presence of animals—their visibility—by no means automatically centers them.[6]

We believe, however, with Erica Fudge, that "the history of animals is a necessary part of our reconceptualization of ourselves as human."[7] We need therefore to critically interrogate the category of the "human" in our exploration of the multiple ways in which such species-categorical boundaries were challenged, broken down, rendered ambiguous, and sometimes reified in history. In other words, these essays explore what Marcy Norton has recently termed the "modes of interaction" between species.[8] This task is peculiarly urgent with regard to Latin American history, given the importance of the Columbian exchange in reconfiguring the natural history of the continent, and the fact that early modern Spain was, as Georgina Dopico Black notes, "particularly interested in what lies at the edges of the human: the beast and the sovereign, but also the monster, the machine, the hermaphrodite, the native, the slave, and the divine."[9] Lauren Derby, in a recent essay, shows how animals for the most part "remain invisible" in much of the literature on landscape change in Latin America, in spite of the popularity of environmental history.[10] Derby also asserts that "bringing the animals into the analysis might move us closer to local understandings of the natural world and syncretism on the ground between European, indigenous and creole views and practices, enabling new ways of thinking about environmental change."[11] It is no exaggeration to say that these changes, from the vectors created for microbes to the transformation of vast areas of land into pasturage, have played a determining role in Latin American history.

Macrohistory, however, depends on countless individual human-animal interactions. A perusal of historical and judicial archives in colonial Mexico and Guatemala, for example, brings to light a number of files in the colonial period involving all sorts of animals (as well as animal parts and commodities): bestiality, animal abuse, bullfights and cockfights, crop destruction by insects and roving domesticated animals, cochineal dye production, feral animals, *abigeato* (the stealing of livestock), shape-shifting, witchcraft, and, among other things, the persistent use of rigged scales to falsely weigh meat with the intent of charging the consumer higher prices. Yet, as colonial historians of Mexico and Guatemala who have based our work on the meticulous exploration of the historical archives concerning our target timeframe—the seventeenth, eighteenth, and early nineteenth centuries—we knew we were missing much of the picture of human-animal histories and interactions in Latin American history. The desire to see close readings of human-animal relationships from a variety of Latin American historical contexts—to theorize the types of looking and interaction that take place among and between different species—is largely what initially inspired us, first, to organize a conference panel on human-animal interactions in colonial Latin America, and subsequently to edit this volume.[12] In doing so, we are responding to an impulse that traverses some of our other work, in which the examination of animals has come to play an increasingly important role.

This edited volume therefore addresses the histories of animals as they relate politically, economically, culturally, and scientifically to colonial and postcolonial Latin America. The essays compiled here contribute to the growing body of historical writing on the multiple interactions between humans and other mammals, birds, insects, and fish that have been mediated in Latin America by colonial and capitalist economies, criminal-justice systems, medicinal practices, religious institutions, and colonial and postcolonial states. We are interested in writing histories of animals and of human-animal relationships as a way to complicate and go beyond recent work on the critical examinations of difference through the analytic lenses of race, ethnicity, gender, and sexuality. *Centering Animals* is ultimately an attempt to counteract the relative invisibility of animals in Latin American historiography, premised on the fact that animals were visible and prevalent historically in colonial and postcolonial Latin American societies, and that the interactions between humans and animals have significantly shaped the narratives of Latin American histories and cultures.

The themes of the volume are in dialogue with recent work on historical and contemporary debates over animal rights, changing moral attitudes toward human dominion over nature and the animal world, the relations of animality to humanity and divinity, and competing local forms of ecological knowledge. This work challenges the modern construction of dualisms such as the human-animal and wild-domestic. We seek to denaturalize and historicize Western conceptualizations of animals and nature. Current work on the cultural histories of animals, while particularly strong for Europe and North America, has for the most part not focused on Latin America. That human-animal studies in history is still a relatively nascent field helps explain why the cultural histories of animals in the historiography of Latin America have yet to appear on a larger scale. For the literature that does exist, we are in accordance with Julie Livingston and Jasbir K. Puar that "the most compelling work in the field challenges the instrumentalization of animals that is driven solely by efforts to deepen the singularity of humanness or to enhance the capacities of human exceptionalism."[13] Given, however, the environmental exchanges and consequences of European conquest and colonialism in Latin America, competing Mesoamerican, Andean, and other indigenous conceptions of animals in the natural world, and the construction of racial, ethnic, gender, and sexual hierarchies that mirrored and interacted with the reconceptualization of human-nonhuman animal hierarchies, animal histories in colonial and postcolonial Latin America are rich in critical moments that are especially relevant to the animal-rights and environmental turn of the present.

Regarding the colonial Americas as a whole, significant contributions addressing animals have been made by a number of scholars. Robert Cunninghame Graham's *Horses of the Conquest* and John Grier Varner and Jeannette Johnson Varner's *Dogs of the Conquest* are two important examples.[14] Alfred W. Crosby's *The Columbian Exchange: Biological and Cultural Consequences of 1492* (1973) created a framework for understanding the discovery and conquest of the New World in terms of the ecological changes that humans brought about.[15] Among the works by scholars following in Crosby's path, Elinor G. K. Melville's *A Plague of Sheep: Environmental Consequences of the Conquest of Mexico* was a notable, and much cited, study that surveyed a whole area of biological, environmental, and cultural changes that had, up until then, not been sufficiently conceptualized by historians.[16] From the beginning, the Spanish were intent on transposing the ecology of Europe into the American sphere. Christopher Columbus transported cattle, horses, pigs, sheep, and goats to the Caribbean on his second voyage in 1493. A number of animals—used as sources of food and, in the case of horses and dogs, as weapons in warfare—accompanied Hernán Cortés, Francisco Pizarro, and other Spanish conquistadors on their expeditions.[17] As Crosby asserts—in a statement subsequently echoed by a number of environmental historians—Europeans would not have made nearly as great a change as they did in the Americas had they not arrived with their animals. Virginia DeJohn Anderson's research on the New England and Chesapeake colonies, published nearly a decade after the equally important collection *New England's Creatures: 1400–1900*, edited by Peter Benes, also examines the biological and environmental changes wrought by the European introduction of livestock, dogs, and horses there.[18] Keith Thomas's pioneering work on the cultural understanding of nature in early modern England cites an example in colonial Virginia whereby colonists sought to convert indigenous lifestyles (and correspondingly mitigate the threats of the wild) by offering native peoples one cow for every eight wolves that they could kill.[19] Jon T. Coleman's important book *Vicious: Wolves and Men in America* is unsurpassed in its use of diverse historical documentation to trace the near extermination (and eventual protection) of the wolf in North America.[20] Of course, these intentional changes in the environment designed by the colonizers were destined not to remain for long in their hands, as they and the animals they brought with them in ships became unconscious vectors for organisms they could not even see.[21] This body of historical work sets the backdrop for our edited collection, showing how European domesticated animals were commonly used, sometimes inadvertently, as tools in the "civilizing" missions wreaked upon the natives and as agents to "tame" the land and, simultaneously, how colonists were invested economically and morally in exterminating certain species that they deemed particularly noxious to the colonial enterprise.

There is, of course, recent and important work dealing specifically with animals in Latin America, including Abel A. Alves's *The Animals of Spain: An Introduction to Imperial Perceptions and Human Interaction with Other Animals, 1492–1826* and Miguel de Asúa and Roger French's *A New World of Animals: Early Modern Europeans on the Creatures of Iberian America*, which examines travel accounts, ethnographic depictions of indigenous cultures, reports of flora and fauna, and natural histories to show how Europeans encountered and conceptualized previously unknown creatures in the Americas.[22] *Zoologia fantástica do Brasil (séculos XVI e XVII)* similarly looks at the way the European representations of the creatures encountered in sixteenth- and seventeenth-century Brazil were figured in a discourse of monsters and fantasies.[23] A majority of the studies on animals in Latin America, however, look not at animals as real factors in history, but at their representation in terms of mythology, animal symbolism, and anthropology, or at their straightforward natural history in zooarchaeology, paleontology, evolution, ecology, and environmental studies. Two important multidisciplinary edited collections published respectively in Mexico and Colombia— *Relaciones hombre-fauna: Una zona interdisciplinaria de estudio* and *Rostros culturales de la fauna: Las relaciones entre los humanos y los animales en el contexto colombiano*—serve as good examples of this strain of human-animal studies in Latin America.[24] Another strand of the work on animals in colonial Latin America evokes considerable sophistication in the analyses of hybrid human-animal deities, shape-shifting and diabolical animals, animal metaphors, and animal symbolism as represented in archival documentation and codices.[25] The animals themselves, however, have rarely been accorded a standpoint that would make them the primary focus of such works (unlike in recent anthropological works that call attention to "transspecies ethnography").[26] Thus, while much has been written on animals as symbols, representations, and phantasmagorical creatures, we situate the essays in this volume within a historiography of colonial and postcolonial Latin America that aims to make the interactions between human and nonhuman animals the primary focus of the analysis, so as to ensure that the "real animal" in question does not disappear from historical discourse.

The vast number of works dealing with indigenous peoples in the Americas attests to the astrological and cosmological significance animals had in many communities. From the centrality of Quetzalcoatl, the Mesoamerican plumed-serpent deity, and a number of other deities with animal qualities and characteristics in the Nahua pantheon to the Quechua recognition of llamas, birds, and a serpent in the constellations, we see how animals were imbued with all sorts of cosmological significance in indigenous societies.[27] The existence and quotidian use of domesticated dogs, turkeys, and stingless bees in Mesoamerica, domesticated llamas and guinea pigs in the Andes, and the frequent use of animals as tribute in Incan and Nahua political communities all attest to their agricultural and economic importance prior to the arrival of Europeans in the New World.[28] Indigenous cultures from the northern parts of what are now Alaska and Canada to Tierra del Fuego in the far south had their own specialized bodies of knowledge about animals, as some of the chapters here highlight. For the Nahua-speaking regions of Mesoamerica, for example, book 11 of *The Florentine Codex*, "Earthly Things," is a compendium of information—though overseen and at least partially mediated by the Spanish friar Bernardino de Sahagún in his desire to create a European taxonomy of Mesoamerican animals—on indigenous knowledge of four-footed animals, fish and other aquatic creatures, birds, serpents, insects, medicinal plants, trees, rocks, water, and similar topics.[29]

Thus, as evidenced by a number of works on European and indigenous interactions with new animals on both sides of the Atlantic, all human groups had conceptual, cosmological, and linguistic difficulties in making sense of the strange and exotic animals with which they came into contact. Europeans often framed animals native to the Americas as monstrous and fantastic. In January of 1493 Columbus, most likely gazing on manatees, recorded that he had seen three "mermaids" near the coast of Hispaniola. They were, however, "not so beautiful as they are depicted, for only after a fashion had human form in their faces."[30] Likewise, the opossum was referred to by Vincent Yáñez Pinzón, a commander on a ship traveling with Columbus on his first voyage, and later described by John Ogilby in his 1671 *America*, as a composite creature with "the foremost part resembling a Fox, the hinder a Monkey, the feet were like a Mans, with Ears like an Owl; under whose Belly hung a great Bag, in which they carry'd the Young, which they drop not, nor forsake till they can feed themselves."[31] The linguistic conceptualization and formulation of new creatures posed similar problems. A number of European vernacular names for South American mammals are in fact derived from such indigenous languages as Quechua (llama and puma), Araucanian, Carib (peccary), and Tupi-Guaraní (jaguar, capybara, and tapir). Indigenous peoples, at least in the early phases of contact, as shown by James Lockhart, often used extant indigenous words and terminology to designate recently arrived and previously unknown European animals. The Nahuas, for example, used the autochthonous word for cotton, *ichcatl*, to signify European sheep. The word *maçatl*, the Nahuatl word for deer, was used to refer to European horses. And cows and bulls were called *quaquauhe*, "one with horns."[32] These linguistic events, as well as other historical testimonies, have proven invaluable in helping us work toward a historiography of human-animal interactions in early Latin America.

Another one of our primary aims is to situate these essays within the burgeoning field of human-animal studies. Kenneth Shapiro has aptly described this interdisciplinary field as "primarily devoted to examining, understanding, and critically evaluating the myriad of complex and multidimensional relationships between humans and other animals."[33] While Shapiro rightly places the formal inception of human-animal studies alongside the inaugural issue of *Anthrozoos*, in 1987, the social sciences have been interested in the relationships between humans and animals going all the way back to the natural histories and popular zoologies that were written before and during the Enlightenment. The massive effect of the information that flooded into Europe as a result of European exploration, colonialism, and exploitation of distant places overturned the entire inherited understanding of people, plants, and animals. In contrast to the humanistic and often overtly anthropocentric nature of previous research and writing on animals, much of the human-animal studies scholarship penned in the last two decades locates itself not only within a posthumanist paradigm, but also, to use Erica Fudge's terminology, within the "anti-humanist" tradition, where the *human* can be theorized as "a category of difference, not substance."[34] It is equally important to note that much of the recent historical, sociological, anthropological, and philosophical work currently defined as human-animal studies could not have been done without the socially activist influences of Peter Singer's *Animal Liberation*, originally published in 1975, and Tom Regan's *The Case for Animal Rights* (1983).[35] Much of the research at the present time being done in the field of human-animal studies is linked politically with the modern animal-rights movement and the controversies surrounding factory farming, new biotechnologies, and endangered species. As Nigel Rothfels recently put it, this is a "scholarship that is also embedded in ethics and activism."[36] In an attempt to more meaningfully "center animals" through scholarship, we have decided to donate the proceeds of *Centering Animals in Latin American History* to animal-welfare organizations in Latin America. We thus envision our work as equally invested in both ethics and activism. The relatively recent advent of human-animal studies has also influenced and, to a large extent, determined the geographic imbalance of the chapters included in this collection. While it was far from our original plan to have three chapters that focus on Mexico and none that look at Bolivia, Colombia, Ecuador, Paraguay, Uruguay, Venezuela, and a number of other Latin American countries, our goal from the beginning was never encyclopedic, but rather provisional. We envision the essays collected here—with geographic concentrations on Mexico, Guatemala, Cuba, the Dominican Republic, Puerto Rico, Peru, Chile, Argentina, and Brazil—as multiple starting points that theorize animals as historical actors and that will, we hope, provoke future interdisciplinary research and writing on animals in Latin American history.[37] In a recent essay, "Does 'the Animal' Exist," the historian Susan J. Pearson and the anthropologist Mary Weismantel urge scholars to employ interdisciplinary approaches in order to "recover animals' physical presence in social life; to embed that social life within political economy; and, finally, to plot the spatial dimensions of human-animal relations."[38] The various essays of this edited collection, while largely employing methodologies most familiar to the discipline of history, contextualize animals and their physical and symbolic presence within "social life" (in a variety of Latin American historical moments), yet this volume is merely a beginning. Since animals have figured more prominently in the works of Latin American anthropology than in those of history, it is our hope not only that historians will increasingly "center" animals in Latin American environmental, political, economic, social, and cultural histories, but that—as Pearson and Weismantel suggest—scholars utilize the theoretical and methodological tools available to them from a variety of academic disciplines in order to write such histories. To do so requires historians to "move beyond the archive and draw upon techniques derived from ethnography, oral history, [and] literary studies."[39]

Naturally, any attempt at "centering animals" in history requires rethinking and reconceptualizing the domains in which the animal plays a role in human societies. As highlighted in some of the essays included here, the contributors are aware of the conceptual pitfalls inherent in writing animal histories, some of which have been cogently theorized by Erica Fudge in "A Left-Handed Blow: Writing the History of Animals." Neel Ahuja, John Soluri, and Neil L. Whitehead, in their respective contributions, call into question the political and methodological investments grounded in "centering animals" in history. In recognition of Nigel Rothfels's argument that "there is an inescapable difference between what an animal *is* and what people *think* an animal is," the contributors to this volume examine the histories of real animals and human conceptions of other animals, as framed within the field of multiple and different Latin American social practices.[40] Thus, while not covering all of Latin America, we have selected essays that provide the reader with a range of geographic and cultural contexts in Latin America, spanning the course of Latin America's colonial and postcolonial history, up until the late twentieth century.

We have organized the essays here thematically to highlight some of the major themes regarding the histories of human-animal relationships that this edited volume addresses, dividing the book into three sections: "Animals, Culture, and Colonialism"; "Animals and Medicine, Science and Public Health"; and "The Meanings and Politics of Postcolonial Animals," with provocatively posthumanist concluding comments offered by Neil L. Whitehead. The authors in part I (León García Garagarza, Martha Few, and Zeb Tortorici) each explore the symbolic and enacted roles of animals in key processes of colonization such as indigenous rebellion, religion, public spectacles of entertainment, and economic production. The essays in part II (by Adam Warren, Heather McCrea, and Neel Ahuja) examine the role of animals and animal parts in medical cultures, public-health policies, and scientific experimentation in the eighteenth, nineteenth, and twentieth centuries. It is in this section that the unstable human-animal-microbe triad becomes most evident. In part III the authors (Reinaldo Funes Monzote, John Soluri, Regina Horta Duarte, and Lauren Derby) focus on the politics of animals in postcolonial Latin America through hunting, the commodification of animals and animal parts, movements for the protection of animals and the environment, and political symbolism.[41] This final section's focus on "postcolonial animals" is more theoretical than temporal. This section, therefore, addresses the fact that the desiderata of "the fate of the nonhuman animal" has been too long neglected in the interdisciplinary field of postcolonial studies.[42] As Philip Armstrong notes, "In identifying the costs borne by non-European 'others' in the pursuit of Western cultures' sense of privileged entitlement, post-colonialists have concentrated upon 'other' humans, cultures, and territories but seldom upon animals."[43] Together, these contributions explore the ways in which human agents within different social contexts have perceived, used, and treated animals. The discussion aims at both finding broad connections among different culture areas and elucidating local constructions of the meanings of animals. Tracing the epistemological limits and ethical considerations that have variously shaped human-animal interactions is an important starting point of any attempt to write animals into Latin American history.

The specific types of relationships between humans and other animals and the histories discussed in these essays deserve some attention. That the economic relationships that humans have with animals cannot be understated raises a number of salient questions. How do animals become part of the different socioeconomic communities created by and ultimately for humans? What are some of the myriad ways that animals are commodified as food, clothing, medicinal remedies, agricultural tools, and forms of transportation? All of the essays here speak to the processes of commodification in one way or another. León García Garagarza deals specifically with the ways that Juan Teton, an indigenous spiritual leader in mid-sixteenth-century Mexico, preached against the imposition of Catholicism and the consumption of European domesticated animals. Framing such consumption within eschatological beliefs and identifying the Spaniards with the legendary *tzitzimimeh* deities at the end of time, Teton warned Nahua commoners, for example, that "those who eat the flesh of [European] cows will be transformed into that." In his analysis, García Garagarza reconstructs the world of colonial conflict, spiritual conquest, and ontological fluidity through the vast changes brought about when European animals and animal symbolism were imposed on peoples who were unfamiliar with European animals.[44] Teton's brief moment came as Nahua speakers were still translating the advent of these animals and the introduction of a pastoral economy in early Mexico into the terms of earlier cultural constructs. From the European perspective, Rebecca Earle rightly asserts that in the Spanish Americas "diet was in fact central to the colonial endeavor," and here García Garagarza shows us that native diet was similarly central to indigenous religious beliefs and particular modes of anticolonial discourse.[45]

Adam Warren, in his essay, turns to another moment in the colonial interface between Europeans and American peoples by focusing on material taken from the animal-based medicinal remedies central to human healing and Andean medicine. As an example of how the latter operated, he uncovers the medical practices among one indigenous group, the Kallawaya, concerning the influence of animal remedies on human bodily humors. Then he shows how similar beliefs were deployed in eighteenth-century home-remedy guides, known as *recetarios*, and natural histories of the Andes, much in the same way that Miruna Achim has shown regarding the use of lizards in colonial Mexican medicine.[46] Using the example of remedies made from animal parts, including llama fat and condor skin, Warren demonstrates a sort of medicinal mestizaje through which Martín Delgar, one recetario author, translated indigenous healing practices involving animals for an urban and ethnically mixed colonial audience. Both García Garagarza and Warren deal with the convergence and conflicts of indigenous and Spanish knowledge of animals at specific historical moments, accessing the experiences and symbolic worlds of humans and other animals in Mesoamerica (as the old order collapsed) and the Andes (as the new colonial order emerged).

In a very different context, that of twentieth-century Puerto Rico, Neel Ahuja's essay also looks at the use of animals in the promotion of human health, specifically in reference to primate experimentation at U.S.-funded research institutions that were established in Puerto Rico in the mid-twentieth century under the aegis of finding vaccines and cures for polio and other diseases. Devoting part of his essay to a critique of the ways monkeys were envisioned as "'raw materials' for the production of scientific knowledges and pharmaceutical commodities," Ahuja devotes another part of the essay to the fascinating adaptation (and feralization) of escaped rhesus and African patas monkeys intended solely for research. Ahuja merges the methodologies of transnational history with a critical look at human-animal studies to offer a fascinating glimpse of the colonial and postcolonial politics at play in the global primate trade, perceptions of public health, and the unequal production and exchange of medical knowledge.

While Ahuja and most of the other contributors in this volume perhaps rightly avoid invoking any notion of "animal agency," it is a problematic yet worthwhile concept to bring up, especially within the context of the posthumanist turn in the humanities and social sciences. Cary Wolfe writes that posthumanism "names a historical moment in which the decentering of the human by its imbrications in technical, medical, informatic, and economic networks is increasingly impossible to ignore, a historical development that points toward the necessity of new theoretical paradigms (but also thrusts them on us), a new mode of thought that comes after the cultural repressions and fantasies, the philosophical protocols and evasions, of humanism as a historically specific phenomenon."[47] Posthumanist notions of subjectivity then seek to problematize earlier humanist paradigms of agency, which were plagued by the romantic assumption "of the essential and a priori distinction between humans and all other kinds of entities that exists."[48] In spite of such shifts, Chris Wilbert is justified in his recent assertion that agency has, "at a general analytical level in the social sciences, been seen anthropocentrically as a purely human property."[49]

If we return to the topic of bestiality that was discussed at the outset of this introduction, we easily find proof of animals that struggled against and sometimes resisted the sexual violence inflicted on them by humans. The large corpus of bestiality cases from colonial Mexico and Guatemala shows us that the nonhuman victims of bestiality tended to be the large European domesticated animals—donkeys, mares, mules, cows, goats, and sheep—that were ubiquitous in postconquest rural spaces and farming communities. Archival documentation demonstrates that one of the primary signs through which judicial authorities inferred that bestiality had either been intended or consummated was whether or not the suspect had bound the hind legs of the animal with rope or cloth. When faced with the invasive sexual acts of humans, many animals attempted to flee or kick their would-be-assailants. In an 1818 bestiality case from the Mexican town of Teocaltiche, for example, Tomás Amador, who had tied up a mule and covered its eyes, confessed that although his intention was to fornicate with the animal, "he got to unbuttoning his underwear, but he did not sin [with the animal] because the mule resisted [*se resistió*], kicking him in the shin."[50] This is one rare instance in which an animal, if the story is true, succeeded in "resisting" human sexual advances.

The question of whether or not to frame such acts as demonstrative of animal "agency," however, may be misguided, especially if we seriously consider the posthumanist goal of decentering of the human. Might the imposition of the category of agency on nonhuman animals be one of Cary Wolfe's "cultural repressions and fantasies" of humanism—one that fails to demonstrate how many animals, when faced with imminent danger, reacted instinctually and in ways that sometimes thwarted the plans of their human or nonhuman aggressors? Could the same be said for bulls in a bullring, rhesus monkeys escaping from animal institutions, birds that fled their human hunters, and seals that protected their young in the face of the human hunt for pelts? To simply declare that such animals have "agency" might be to approach the issues of animal sentience and instinct in an anthropocentric way, especially given that the notion of agency arises in a secular Western culture and is itself saturated with Western humanist beliefs. While agency, in this sense, can be a category bereft of meaning, Dorothee Brantz notes that historically "even though people may have wanted to treat animals as mere resources, this desire to instrumentalize the nonhuman has been repeatedly challenged, and not least by the animals themselves when they refused to 'play along.'"[51] While we will never be able to know what the animals were thinking, we can look at nonhuman animal reactions to invasive human cultural practices as evidence of their sentience, their cognition, and their instincts to circumvent what they perceived to be particularly threatening, dangerous, or painful situations.[52] On the other hand, it should be noted that historians regularly attribute agency (in a more metaphoric and collective sense) to a variety of nonhuman social entities including nation-states, market forces, the church, governments, classes, and genders. In this sense, the attribution of "agency" to nonhuman actors and entities is largely a means of making sense of the past through narrative mechanisms. There are, in essence, many types of "agency" to be found in these essays— some of them subtly anthropomorphizing, others less so.[53]

The human consumption of animals and animal parts is another theme that runs through these essays. Animal consumption and animal labor are particularly central to the essays by Reinaldo Funes Monzote and Regina Horta Duarte, both of which have been significantly revised, updated, and translated into English for inclusion in this volume.[54] John Soluri's contribution, focusing on competing local and international rights over sealing in Tierra del Fuego from the mid-nineteenth century to the early twentieth, also analyzes the human consumption of animals and animal products— seals and pelts—which were valued for both their use value (they were an important source of food for local populations) and their exchange value (the pelts were sold internationally at a profit by large-scale sealing expeditions). Here, however, histories of consumption and commodification are mediated by protectionist and environmentalist efforts to curtail the abuses of draft animals and trends toward the extinction of endangered avian and maritime species. In the Cuban context Funes Monzote examines the roles of animals—namely, oxen, horses, and mules—in the production of sugar throughout the nineteenth century. Emphasizing the brutal treatment that many working animals received, he links the growing cultural discomfort with the mistreatment of animals to the founding of the Sociedad Cubana Protectora de Animales y Plantas (Cuban Society for the Protection of Animals and Plants) in Havana in 1882. Inspired by the creation of Society for the Prevention of Cruelty to Animals in England in 1824 and similar organizations worldwide, the Cuban movement for the protection of animals successfully limited the "barbaric and cruel" mistreatment of beasts of burden, bullfighting, and cockfighting in Cuba—invoking terms that implicated both prejudices of class and race.[55] On numerous levels, the story that Funes Monzote tells testifies to the "pervasive violence enacted on animal bodies in public spaces that was ubiquitous . . . throughout much of the world in the nineteenth century."[56]

Both Duarte and Soluri look at competing hunting rights and the ways in which particular species of birds and seals, respectively, were intensively hunted when, due to shifts in fashion, the demand for particular animal products—plumes and pelts—would rise in American and European markets. Duarte surveys diverse sources including late nineteenth- and early twentieth-century hunting manuals, fashion magazines, ornithological texts, and contemporary scientific research on birds to examine how the international demand for exotic plumes negatively affected Brazil's avifauna and was of collateral importance in the destruction of the natural environment in the early part of the twentieth century.[57] Invasive hunting and environmental practices were partly curtailed by the rise in international legislation aimed at the protection of birds and their habitats, but the disjuncture between legal measures and the social realities in Brazil persists. Duarte also highlights another factor affecting both the production and consumption of animal products, namely, the racial and racist components of Brazil's modernizing projects, and in so doing raises the question of difference as determined by race and social class in conjunction with species alterity.

Soluri also uses ethnic and national identities as factors in the clashes over indigenous and international rights to hunt fur seals off the Patagonian coast. Using scientific texts, travel accounts, ships' logs, and archival documents, Soluri shows that these sources seldom describe seals as living organisms or as dynamic components of an ecosystem but rather as mere commodities whose histories become methodologically difficult (and perhaps even undesirable) to "center." The author offers an extended analysis of the conflicts of interest between local hunters (largely comprised of populations made up of the ancestors of the Chonos, Kawéskar, and Yámana) and international sealing expeditions, between national and international interests and conservationist policies, and between global markets and local ecologies. And then there are the interests of the seals themselves. Demonstrating the pervasive and deleterious ecological effects of large-scale hunting on populations of South American fur seals, Soluri's histories exemplify the following points made by the Animal Studies Group: "The killing of animals is a structural feature of all human-animal relations. It reflects human power over animals at its most extreme and yet also at its most commonplace. From a historical point of view, there is nothing new about this killing; what has changed is the scope of the technology involved and the intensity of its global impact on animal species."[58]

It is this troubling specter of animal death—and the ever-present fears of human death in societies that were populated by feral animals, epidemic contagion, insect infestations, disease transmissions, and environmental degradation—that looms largest in all of the chapters. Given that killing is a key element of most human-animal relationships, this perhaps comes as no surprise.[59] Two of the chapters in this volume focus specifically on the killing of "pests" and the effects of stray animals and potentially noxious insects like locusts and mosquitoes on (post)colonial economies and public health. Martha Few's contribution examines human perceptions of the periodic swarming of the *chapulín* (locust) in colonial Guatemala in relation to the tactics taken by colonial officials and the population at large to combat infestations and crop destruction. Using accounts written by political officials, Indian elites, farmers, priests, and European travelers, Few examines how potentially catastrophic swarms of locusts "forced periodic diversions of Indian tribute labor and other colonial wealth away from agricultural production," thus suspending the colonial order. Theorizing locusts as ambiguous "social agents" that have been imbued with economic, political, medical, and religious significance, Few examines the meanings behind insect exorcisms and state-sponsored killing campaigns.

While insects—a comparatively overlooked category in historical human-animal studies—are historical actors in their own right, their reception and treatment by humans are highly contingent on economic context and scientific knowledge of disease transmission.[60] Moving us into the realm of debates over postcolonial public health in nineteenth- and early twentieth-century Mexico, Heather McCrea analyzes the transformation of stray dogs, vagrant swine, mosquitoes, louses, ticks, and other creatures from the conceptual category of "pests" to that of "vectors of disease," in the latter role becoming impediments to social progress. This movement toward an analysis of insects, pests, viruses, and vectors is significant here and, in the words of Livingston and Puar, "complicate[s] any easy narrative about 'the animal turn'—the rise in interest in animal studies—that privileges certain sites, disciplines, and species."[61] Cholera epidemics and the spread of filth and disease through urban cemeteries and slaughterhouses provide only part of the backdrop to an era in which the civil effort was initiated to segregate humans, animals, and insects in the name of public health. McCrea also traces the racial implications and social applications of these categories in relation to the fact that the governing class and the Mexican press customarily referred to the country's indigenous Maya as "barbarous savages" that impeded national progress. The racial implications of such narratives are clear: "It was easy to subsume animals, insects, and the indigenous Maya into negative categories all associated with filth, pestilence, and backwardness." The elimination of Mexico's indigenous population that was often desired and unambiguously voiced by the country's elite—a narrative that repeats itself in the histories of all Latin American countries—fed into human beliefs and prejudices about animality and the perceived origins of social disorder.

The chapters outlined here have thus far broached the multiple meanings of animal uses and deaths, in the names of consumption, commodification, science, public health, fashion, and entertainment, as enacted in specific Latin American historical contexts. The chapters contributed by Zeb Tortorici and Lauren Derby also look at everyday social realities and unique cultural practices involving animals, but do so specifically from the vantage point of public spectacles centered on animals (respectively, dogs and goats) and their symbolic representations. Tortorici's contribution broaches the relationship between dogs and human (affective) pleasure by examining the unorthodox occurrences of canine weddings, baptisms, and funerals in eighteenth-century Mexico.[62] Focusing his analysis first on the 1770 Inquisition trial of one priest who jokingly wed two canines "in the name of the Father and the Mother of all dogs" at a local party in Mexico City, Tortorici examines the nature of animal spectacles and the disjunctures coding the popular and official beliefs surrounding privileged species, sacrosanct rituals, the meaning of mockery and desecration, and human-animal relationships. Juxtaposing carnivalesque spectacles and sardonic literary texts involving dogs with the Mexican Inquisition's concerns about heretical doctrinal infractions, Tortorici asserts that the social world that was built up around the domestic dog in Bourbon Mexico was indicative of a change in the sensibilities regarding animals and affective ties with pets. Such historical and literary representations of dogs provide us ultimately with a window into pet-keeping as a cultural practice "through which we can glean the colonizing logic and the cultural binaries that governed the ideas surrounding pets and affections bestowed on them: domestic/savage, pet/stray, purebred/mixed, leisurely/utilitarian, and private/public." Tortorici's chapter advances the project of "centering animals" in history by shifting from the discourses on and about animals to the actual histories of those animals, in order to better understand their mutable relationships with the humans around them.

Derby takes a different and innovative approach by looking at the way animals can figure as political symbols in modernity (though real animals remain present throughout the essay as well). Her study specifically looks at why Rafael Trujillo, the dictator who ruled the Dominican Republic from 1930 until 1961, was nicknamed the "goat" by his enemies, and why popular festivities performed during "the feast of the goat" exploited that nickname.[63] After his death, Trujillo was remembered and mocked through carnivalesque acts and widespread associations of the dictator with "the goat" in popular culture. To elucidate the cultural significance of goats in the Dominican Republic, Derby addresses lore about the sexual behavior of goats, popular animal metaphors and puns, and tales of shape-shifting and animal sacrifice. The choice of the goat as a symbol for the dead Trujillo can be explained only by going into the layered meanings of the goat in colonial and postcolonial Dominican history. In burning an image of a goat with Trujillo's face and consuming goat meat en masse on the first anniversary of Trujillo's death, Dominicans effected the ritual reestablishment of their dignity from an epoch of national humiliation—for, according to Derby, "for once they were the ones eating the goat, rather than the reverse." This chapter shows how humans imbued goats with meaning in one particular historical and cultural context.

Erica Fudge has noted that animals appear in historical archives and libraries "as absent presences: there, but not speaking."[64] This "absent-presence of animals" pervades the records on which we base our histories, laying bare the variety of anthropocentric attitudes embodied in these sources. The authors of these essays, however, find traces of animal histories—often mere bits and pieces, as of that unnamed turkey whose corpse was strung around the neck of Pedro Na while he underwent castration at the hands of colonial Spanish authorities in 1563—that provoke us into reading further into the hunted, killed, commodified, commercialized, dissected, cooked, consumed, and endangered animals that appear (and disappear) in the Latin American historical record. References to animals are scattered throughout indigenous cosmologies and codices, European accounts of exploration and colonization, natural histories, criminal cases and Inquisition trials, religious treatises, medical manuals and scientific experiments, ethnographies, public-health measures, artistic representations, and political campaigns for animal welfare. In these texts, real animals figure alongside the metaphorical animals that were used to symbolically represent the limits of humanity and thus sometimes occluded the real, sentient animal. Animals appear to be both everywhere and nowhere at the same time. The essays in this volume speak to the myriad and protean historical relationships between humans and other animals in Latin American history, a relationship as important as that between humans with each other. This focus essentially foregrounds the critical analysis of the uses to which so many nonhuman animals have been put to in the past. It is our hope that the essays in this collection meaningfully "center animals" historically, politically, and ethically. Perhaps not unlike the sixteenth-century human by-standers observing the castration of Pedro Na for the crime of bestiality and the burning of the turkey's corpse, we are witnesses to both tangible and textual fragments of animals whose multiple meanings and experiences we can only begin to understand.

Notes

1. Archivo General de las Indias (AGI), Justicia, leg. 248 (microfilm reel #191), f. 2: "Al dho Pedro Na yndio que tenya los cazaguelles quitados y sentado en el suelo y entre las piernas una gallina de la tierra y este testigo le hechó la mano e vido como a la dha gallina le salia e corria sangre del sieso y del dho Pedro Na tenya los caçaguelles desatados e descubierto el myembro engendratibo." For a closer analytical look at this case in the context of other bestiality cases, see Zeb Tortorici, "Contra Natura: Sin, Crime, and Unnatural Sexuality in Colonial Mexico 1530–1821" (PhD diss., University of California, Los Angeles, 2010). For evidence that the Spanish term *gallina de la tierra* was used to refer specifically to turkeys found for the first time in the Americas, see Rafael Heliodoro Valle, "El Español de la América Española," *Hispania* 36.1 (1953): 52–57; and Lawrence B. Kiddle, "Los nombres del pavo en el dialecto nuevomejicano," *Hispania* 24.2 (1941): 213–16.

2. AGI, Justicia, leg. 248 (microfilm reel #191), f. 3: "Que es verdad que ayer tarde en el dho camyno topo a estas gallinas con unos pollos e tomo de ellas la gallina que les mostraba e se entro con ella en el monte e se debaxo los caçaguelles e con la alteracion carnal que tubo tomo el myembro engendratibo en la mano e se lo metio por el sieso a la dha gallina e tubo aceso carnal con ella e della le hizo correr sangre por el sieso a la dha gallina e que en esto llegaron un yndio e una yndio."

3. AGI, Justicia, leg. 248 (microfilm reel #191), f. MVCXIX: "La gallina con que delinquyo y cometio el dho pedro na el dho delito esta muerta e se ha tenydo guardada mandaba e mando que para executar la dha sentencia se la cuelgen del pescueço al dho pedro na e sea traydo con ella por las calles acostumbradas desta zbdad e despues de executada la dha sentencia el susodho mando se quemase la dha gallina en llamas bibas e fuese hecha polbos."

4. For the historiography of bestiality in the early modern period, see Jonas Lilie-quist, "Peasants against Nature: Crossing the Boundaries between Man and Animal in Seventeenth- and Eighteenth-Century Sweden," *Journal of the History of Sexuality* 1.3 (1991): 393–423; P. G. Maxwell-Stuart, "'Wild, Filthie, Execrabill, Detestabill, and Unnatural Sin': Bestiality in Early Modern Scotland," in *Sodomy in Early Modern Europe*, ed. Tom Betteridge (Manchester: Manchester University Press, 2002), 82–93; William E. Monter, "Sodomy and Heresy in Early Modern Switzerland," in *The Gay Past: A Collection of Historical Documents*, ed. Salvatore J. Licata and Robert P. Petersen (New York: Routledge, 1986), 41–55; John M. Murrin, "'Things Fearful to Name': Bestiality in Early America," in *The Animal/Human Boundary: Historical Perspectives*, ed. Angela N. H. Creager and William

Chester Jordan (Rochester: University of Rochester Press, 2002), 115–56; Helmut Puff, "Nature on Trial: Acts 'Against Nature' in the Law Courts of Early Modern Germany and Switzerland," in *The Moral Authority of Nature*, ed. Lorraine Daston and Fernando Vidal (Chicago: University of Chicago Press, 2004), 232–53; and Laura Stokes, *Demons of Urban Reform: Early European Witch Trials and Criminal Justice, 1430–1530* (Houndmills, U.K.: Palgrave Macmillan, 2011). On the modern period, see Jens Rydström, *Sinners and Citizens: Bestiality and Homosexuality in Sweden, 1880–1950* (Chicago: University of Chicago Press, 2003). For work that broaches both early modern and contemporary issues, see Piers Beirne, *Confronting Animal Abuse: Law, Criminology, and Human-Animal Relationships* (Lanham: Rowman and Littlefield, 2009), especially chapter 3, "Toward a Sociology of Animal Sexual Assault." For the only historical treatments of bestiality in colonial Mexico, see Lee Penyak, "Criminal Sexuality in Central Mexico, 1750–1850" (PhD diss., University of Connecticut, 1993); Mílada Bazant, "Bestialismo, el delito nefando, 1800–1856," in *Documentos de Investigación* (Mexico City: El Colegio Mexiquense, 2000), 1–22; Zeb Tortorici, "Against Nature: Sodomy and Homosexuality in Colonial Latin America," *History Compass* 10.2 (2012): 161–78; and Tortorici, "Contra Natura."

5. One of the rules of the law in Leviticus is this: "If a man lie with a beast, he shall surely be put to death; and ye shall slay the beast. And if a woman approach unto any beast, and lie down thereto, thou shalt kill the woman and the beast; they shall surely be put to death; their blood shall be upon them" (Leviticus 20:13).

6. We are particularly grateful to Frederico Santos Soares de Freitas for pushing us to think more about the many ambiguities inherent in the project of centering animals (or other social actors) in the analytic sphere.

7. Erica Fudge, "A Left-Handed Blow: Writing the History of Animals," in *Representing Animals*, ed. Nigel Rothfels (Bloomington: Indiana University Press, 2002), 5.

8. Marcy Norton, "Going to the Birds: Birds as Things and Beings in Early Modernity," in *Early Modern Things: Objects and Their Histories, 1500–1800*, ed. Paula Findlen (London: Routledge, 2013), 53–83.

9. Georgina Dopico Black, "The Ban and the Bull: Cultural Studies, Animal Studies, and Spain," *Journal of Spanish Cultural Studies* 11.3–4 (2010): 237.

10. This, however, is changing for the better. See, for example, the recent scholarship by Marcy Norton: "Animals in Spain and Spanish America," in *Lexikon of the Hispanic Baroque: Technologies of a Transatlantic Culture*, ed. Ken Mills and Evonne Levy (Toronto: University of Toronto Press, forthcoming 2013); "Adoption and Predation: Human-Animal Relationships in the Caribbean and South America," invited seminar at the McNeil Center for Early American Studies, University of Pennsylvania, 13 January 2012; and "Beyond Anthropocentrism and Antianthropocentrism: Elite Hunting as a Mode of Interaction," invited paper for European Colloquium, History Department, Cornell University, 3 October 2011.

11. Lauren Derby, "Bringing the Animals Back In: Writing Quadrupeds into Caribbean History," *History Compass* 9.8 (2011): 602–21.

12. We originally organized a panel titled "Animals, Colonialism, and the Atlantic World" for the 2006 annual meeting of the American Society for Ethnohistory, which took place in Williamsburg, Virginia.

13. Julie Livingston and Jasbir K. Puar, "Interspecies," *Social Text* 29.1-106 (2011): 3.

14. See Robert Cunninghame Graham, *Horses of the Conquest: A Study of the Steeds of the Spanish Conquest* (Norman: University of Oklahoma Press, 1949); and John Grier Varner and Jeannette Johnson Varner, *Dogs of the Conquest* (Norman: University of Oklahoma Press, 1983).

15. For a good general discussion, see Alfred W. Crosby, *The Columbian Exchange: Biological and Cultural Consequences of 1492* (Westport: Greenwood Press, 1972), especially chapter 3, "Old World Plants and Animals in the New World"; and Alfred W. Crosby, *Ecological Imperialism: The Biological Expansion of Europe, 900–1900* (New York: Cambridge University Press, 1986). For European perceptions of animals native to the Americas, see Antonello Gerbi, *Nature in the New World: From Christopher Columbus to Gonzalo Fernández de Oviedo*, trans. Jeremy Moyle (Pittsburgh: University of Pittsburgh Press, 1975); and Kathleen Ann Myers, *Fernández de Oviedo's Chronicle of America: A New History for a New World*, trans. Nina M. Scott (Austin: University of Texas Press, 2007). On how Mesoamerican ideas about hummingbirds were received in Europe, see Iris Montero Sobrevilla, "Transatlantic Hum: Natural History and the Torpid Hummingbird, c. 1500–1800" (PhD diss., University of Cambridge, 2012).

16. See Elinor G. K. Melville, *A Plague of Sheep: Environmental Consequences of the Conquest of Mexico* (Cambridge: Cambridge University Press, 1994), for a discussion of the pastoralization of the Mexican landscape that proceeded hand-in-hand with evangelization. See also Andrew Sluyter, "The Ecological Origins and Consequences of Cattle Ranching in Sixteenth-Century New Spain," *Geographical Review* 86.2 (1996): 161–77; Renée González-Montagut, "Factors That Contributed to the Expansion of Cattle Ranching in Veracruz, Mexico," *Mexican Studies / Estudios Mexicanos* 15.1 (1999): 101–30; and David E. Vassberg, "Concerning Pigs, the Pizarros, and the Agro-pastoral Background of the Conquerors of Peru," *Latin American Research Review* 13.3 (1978): 47–61.

17. Aside from Varner and Varner, *Dogs of the Conquest*, see also Sara E. Johnson, "'You Should Give Them Blacks to Eat': Waging Inter-American Wars of Torture and Terror," *American Quarterly* 61.1 (2009): 65–92, for a discussion of the use of dogs as tools of torture in colonial Haiti and other historical contexts.

18. Virginia DeJohn Anderson, *Creatures of Empire: How Domestic Animals Transformed Early America* (Oxford: Oxford University Press, 2004). The edited collection by Peter Benes, *New England's Creatures: 1400–1900* (Boston: Boston University Press, 1995), was published well before the current trend in historical scholarship toward human-animal interactions.

19. Keith Thomas, *Man and the Natural World: Changing Attitudes in England 1500–1800* (New York: Penguin, 1983), 30.

20. Jon T. Coleman, *Vicious: Wolves and Men in America* (New Haven: Yale University Press, 2004).

21. For example, the *Aedes aegypti* mosquito, implicated in the spread of yellow fever, first appeared in America in the larvae carried in the water casks of slave traders from Africa. In turn, the yellow fever microorganism was carried in the mosquito. See Ralph T. Bryan, "Alien Species and Emerging Infectious Diseases: Past Lessons and Future Implications," in *Invasive Species and Biodiversity Management*, ed. Odd Terje Sandlund, Peter Johan Schei, and Slaug Viken (Dordrecht: Springer, 1999), 163–76. We thank Roger Gathman for this observation and reference.

22. Abel A. Alves, *The Animals of Spain: An Introduction to Imperial Perceptions and Human Interaction with Other Animals, 1492–1826* (Leiden: Brill, 2011); and Miguel de Asúa and Roger French, *A New World of Animals: Early Modern Europeans on the Creatures of Iberian America* (Burlington: Ashgate, 2005).

23. Afonso de Escragnolle Taunay and Odilon Nogueira de Matos, *Zoologia fantástica do Brasil (séculos XVI e XVII)* (São Paulo: Edusp, Museu Paulista Universidade de São Paulo, 1999).

24. Eduardo Corona M. and Joaquín Arroyo Cabrales, eds., *Relaciones hombre-fauna: Una zona interdisciplinaria de estudio* (Mexico City: Instituto Nacional de Antropología e Historia, 2002); and Astrid Ulloa and Luis Guillermo Baptiste-Ballera, eds., *Rostros culturales de la fauna: Las relaciones entre los humanos y los animales en el contexto colombiano* (Bogotá: Instituto Colombiano de Antropología e Historia, 2002). See also Alberto G. Flórez Malagón, ed., *El poder de la carne: Historias de ganaderías en la primera mitad del siglo XX en Colombia* (Bogota: Editorial Pontificia Universidad Javeriana, 2008), for a recent collection of essays on cattle raising in Colombia.

25. The works that fall into this camp are numerous. For a few examples, see Louise M. Burkhart, *The Slippery Earth: Nahua-Christian Moral Dialogue in Sixteenth-Century Mexico* (Tucson: University of Arizona Press, 1989); Alfredo López Austin, *Cuerpo humano e ideología: Las concepciones de los antiguos nahuas* (Mexico City: Universidad Nacional Autónoma de México, 1980); Elizabeth P. Benson, *Birds and Beasts of Ancient Latin America* (Gainesville: University of Florida Press, 1997); Serge Gruzinski, *Man-Gods of the Mexican Highlands: Indian Power and Colonial Society, 1520–1800* (Stanford: Stanford University Press, 1989); Lisa Sousa, "The Devil and Deviance in Native Criminal Narratives from Early Mexico," *The Americas* 59.2 (2002): 161–79; and Laura de Mello e Souza, *The Devil and the Land of the Holy Cross: Witchcraft, Slavery, and Popular Religion in Colonial Brazil* (Austin: University of Texas Press, 2004). For recent anthropological work on shape-shifting in the contemporary Amazonia, see Carlos Fausto, "A Blend of Blood and Tobacco: Shamans and Jaguars among the Parakanã of Eastern Amazonia," and Márnio Teixeira-Pinto, "Being Alone amid Others: Sorcery and Morality among the Arara, Carib, Brazil," both in *In Darkness and Secrecy: The Anthropology of Assault Sorcery and Witchcraft in Amazonia*, ed. Neil Whitehead and Robin Wright (Durham: Duke University Press, 2004).

26. Works that do focus centrally on animals are often anthropological. See, for example, Edmundo Morales, *The Guinea Pig: Healing, Food, and Ritual in the Andes* (Tucson: University of Arizona Press, 1995); Eduardo P. Archetti, *Guinea-Pigs: Food, Symbol and Conflict of Knowledge in Ecuador* (New York: Berg, 1997); Marion Schwartz, *A History of Dogs in the Early Americas* (New Haven: Yale University Press, 1997); Lynn Hirschkind, "Sal/Manteca/Panela: Ethnoveterinary Practice in Highland Ecuador," *American Anthropologist* 102.2 (2000): 290–302; and Eduardo Kohn, "How Dogs Dream: Amazonian Natures and the Politics of Transspecies Engagement," *American Ethnologist* 34.1 (2007): 3–24.

27. On cosmological animals in the Quechua world, see Gary Urton, "Animals and Astronomy in the Quechua Universe," *Proceedings of the American Philosophical Society* 125.2 (1981): 110–27.

28. On the eating of domesticated guinea pigs in the pre-Hispanic Andes and of domesticated turkeys and dogs in Mesoamerica, see Sophie D. Coe, *America's First Cuisines* (Austin: University of Texas Press, 1994).

29. Bernardino de Sahagún, *Florentine Codex: General History of the Things of New Spain*, trans. Arthur J. O. Anderson and Charles E. Dibble (Santa Fe: School of American Research; Salt Lake City: University of Utah Press, 1950–82).

30. Cited in Asúa and French, *A World of New Animals*, 4.

31. Cited in Susan Scott Parrish, "The Female Opossum and the Nature of the New World," *William and Mary Quarterly* 54.3 (1997): 475–514. Parrish offers a fascinating glimpse at the debates surrounding the sexuality and social meanings constructed around this particularly peculiar creature found in the Americas.

32. James Lockhart, *The Nahuas after the Conquest: A Social and Cultural History of the Indians of Central Mexico, Sixteenth through Eighteenth Centuries* (Stanford: Stanford University Press, 1992), 280. Lockhart writes that descriptive terms also abound: "*tentzone,* 'bearded one,' for goat; and *quanaca,* 'head flesh,' in reference to the rooster's comb and hence to the whole animal, then including the chicken as well" (Lockhart, *The Nahuas after the Conquest,* 279).

33. Kenneth Shapiro, *Human-Animal Studies: Growing the Field, Applying the Field* (Ann Arbor: Animals and Society Institute, 2008), 5.

34. Fudge, "A Left-Handed Blow," 15.

35. See Peter Singer, *Animal Liberation* (New York: Random House, 1975); and Tom Regan, *The Case for Animal Rights* (Berkeley: University of California Press, 2004).

36. Nigel Rothfels, "Foreword," in *Other Animals: Beyond the Human in Russian Culture and History*, ed. Jane Costlow and Amy Nelson (Pittsburgh: University of Pittsburgh Press, 2010), 1–16.

37. For some of the invaluable edited collections on animals in history, which we have used as partial models for this collection, see Angela N. H. Creager and William Chester Jordan, eds., *The Animal/Human Boundary: Historical Perspectives* (Rochester: University of Rochester Press, 2002); Gregory M. Pflugfelder and Brett L. Walker, eds., *JAPANimals: History and Culture in Japan's Animal Life* (Ann Arbor:

Center for Japanese Studies, University of Michigan, 2005); and Jane Costlow and Amy Nelson, eds., *Other Animals: Beyond the Human in Russian Culture and History* (Pittsburgh: University of Pittsburgh Press, 2010).

38. Susan J. Pearson and Mary Weismantel, "Does 'the Animal' Exist? Toward a Theory of Social Life with Animals," in *Beastly Natures: Animals, Humans, and the Study of History*, ed. Dorothee Brantz (Charlottesville: University of Virginia Press, 2010), 22.

39. Derby, "Bringing the Animals Back In," 603.

40. Nigel Rothfels, *Savages and Beasts: The Birth of the Modern Zoo* (Baltimore: Johns Hopkins University Press, 2002), 5.

41. For more on the commodification of animals and animal products in Spain and Latin America, see, for example, Carla Rahn Phillips and William D. Phillips Jr., *Spain's Golden Fleece: Wool Production and the Wool Trade from the Middle Ages to the Nineteenth Century* (Baltimore: Johns Hopkins University Press, 1997); Jeffrey M. Pilcher, *The Sausage Rebellion: Public Health, Private Enterprise, and Meat in Mexico City, 1890–1917* (Albuquerque: University of New Mexico Press, 2006); and Sandra Aguilar, "Nutrition and Modernity: Milk Consumption in 1940s and 1950s Mexico," *Radical History Review* 110 (2011): 36–58.

42. Philip Armstrong, "The Postcolonial Animal," *Society and Animals* 10.4 (2002): 413.

43. Armstrong, "The Postcolonial Animal," 413.

44. For a comparative look at ontological fluidity and porous human-animal boundaries among the Tupi and other indigenous peoples of the South American lowlands (through the concept of multinaturalism), see Eduardo Viveiros de Castro, *A inconstância da alma selvagem e outros ensaios de antropologia* (São Paulo: Cosac and Naify, 2002); and Eduardo Viveiros de Castro, "Cosmological Deixis and Amerindian Perspectivism," *Journal of the Royal Anthropological Institute* 4.3 (September 1998): 469–88. On diet in early colonial Spanish America, see Rebecca Earle, "'If You Eat Their Food . . .': Diets and Bodies in Early Colonial Spanish America," *American Historical Review* 115.3 (2010): 688–713.

45. Earle, "'If You Eat Their Food . . . ,'" 688.

46. Miruna Achim, *Lagartijas medicinales: Remedios americanos y debates científicos en la Ilustración* (Mexico City: Conaculta / UAM Cuajimalpa, 2008).

47. Cary Wolfe, *What Is Posthumanism?* (Minneapolis: University of Minnesota Press, 2010), xv.

48. Malcolm Ashmore, Robin Wooffitt, and Stella Harding, "Humans and Others, Agents and Things," *American Behavioral Scientist* 37 (1994): 734.

49. Chris Wilbert, "What Is Doing the Killing? Animal Attacks, Man-Eaters, and Shifting Boundaries and Flows of Human-Animal Relations," in *Killing Animals*, Animal Studies Group (Champaign: University of Illinois Press, 2006): 32.

50. Biblioteca Pública del Estado de Jalisco, Fondos Especiales, caja 145, exp. 6, prog. 2188, f. 5: "Hasta llegó a desabrocharse un boton de la pretina de los calsones, pero que no pecó pues la mula se resistió que hasta una patada le dio en una espinilla."

51. Brantz, *Beastly Natures*, 3.

52. It is also worthwhile noting that some of the insects and animals discussed in these chapters—namely, the locusts discussed by Few, the mosquitoes looked at by McCrea, and the seals examined by Soluri—often exhibited no visible signs of alarm or anxiety, at least none that were readable by humans, either prior to or during their killings. Both Few and Soluri register a certain degree of human surprise, as evidenced in their historical documents, at how easy it sometimes was to kill those respective creatures. Soluri, on the other hand, also mentions that, depending on the intention of the writer, the seals were sometimes described as violently resisting their human predators.

53. We thank Frederico Santos Soares de Freitas for his incisive comments on "agency" and social entities.

54. Reinaldo Funes Monzote's essay included here is a revised version of his "Facetas de la interacción con los animales en Cuba durante el siglo XIX: Los bueyes en la plantación esclavista y la Sociedad Protectora de Animales y Plantas," *Signos Históricos* 16 (2006): 80–110. Regina Horta Duarte's essay was originally published as "Pássaros e cientistas no Brasil: Em busca de proteção, 1894–1938," *Latin American Research Review* 41.1 (1996): 3–26.

55. For other important works on bullfighting in the Iberian world, see Adrian Shubert, *Death and Money in the Afternoon: A History of the Spanish Bullfight* (New York: Oxford University Press, 2001); and Elisabeth Hardouin-Fugier, *A History of Bullfighting* (Chicago: University of Chicago Press, 2010). On animals and human festivities in Spain, see also William Christian Jr., "Sobrenaturales, humanos, animales: Exploración de los límites en las fiestas españolas a través de las fotografías de Cristina García Rodero," in *La Fiesta en el mundo hispánico* (Toledo: Universidad de Castilla–La Mancha, 2004), 13–32.

56. Amy Nelson, "The Body of the Beast: Animal Protection and Anticruelty Legislation in Imperial Russia," in *Other Animals: Beyond the Human in Russian Culture and History*, ed. Jane Costlow and Amy Nelson (Pittsburgh: University of Pittsburgh Press, 2010), 95–112.

57. For a look at birds, ornithological knowledge, and avian husbandry in the Americas and early modern Europe, see Rebecca Brienen, "From Brazil to Europe: The Zoological Drawings of Albert Eckhout and Georg Marcgraf," in *Early Modern Zoology: The Construction of Animals in Science, Literature and the Visual Arts*, ed. Karl A. E. Enenkel and Paul J. Smith (Leiden, Brill: 2007), 273–317; and Norton, "Going to the Birds."

58. Animal Studies Group, *Killing Animals* (Champaign: University of Illinois Press, 2006), 4.

59. This high level of animal killing is especially evident in "postdomestic" societies, where, according to Richard W. Bulliet, "people live far away, both physically and psychologically, from the animals that produce the food, fiber, and hides they depend on, and they never witness the births, sexual congress, and slaughter of those animals. Yet they maintain very close relationships with companion animals

—pets—often relating to them as if they were human." The second characteristic of postdomesticity is the high level of consumption of animal products alongside "feelings of guilt, shame, and disgust when they think (as seldom as possible) about the industrial processes by which domestic animals are rendered into products and about how those products come to market." See Richard W. Bulliet, *Hunters, Herders, and Hamburgers: The Past and Future of Human-Animal Relationships* (New York: Columbia University Press, 2005), 3.

60. For research on insect dyes and commodification in colonial Mexico, see Raymond Lee, "Cochineal Production and Trade in New Spain to 1600," *The Americas* 4.4 (1948): 458–63; and Carlos Marichal, "Mexican Cochineal and the European Demand for American Dyes, 1550–1850," in *From Silver to Cocaine: Latin American Commodity Chains and the Building of the World Economy, 1500–2000*, ed. Steven Topik, Carlos Marichal, and Zephyr Frank (Durham: Duke University Press, 2006), 76–92. Marichal writes, "Cochineal (*grana cochinilla*) became, after silver, the most important Mexican export for over three hundred years, or down to approximately 1850" (Marichal, "Mexican Cochineal and the European Demand for American Dyes," 76).

61. Livingston and Puar, "Interspecies," 5.

62. Another take on these cases is offered by Frank T. Proctor III, "Amores Perritos: Puppies, Parties, and Popular Catholicism in Bourbon Mexico City," paper presented at the Tepaske Seminar on Colonial Latin America, Emory University, 26 May 2011.

63. For a fascinating account of animal symbolism and consumption in the Dominican Republic, see Lauren Derby, "Gringo Chickens with Worms: Food and Nationalism in the Dominican Republic," in *Close Encounters of Empire: Writing the Cultural History of U.S.-Latin American Relations*, ed. Gilbert M. Joseph, Catherine C. LeGrand, and Ricardo D. Salvatore (Durham: Duke University Press, 1998), 451–91.

64. Erica Fudge, *Perceiving Animals: Humans and Beasts in Early Modern English Culture* (Chicago: University of Illinois Press, 2002), 2.

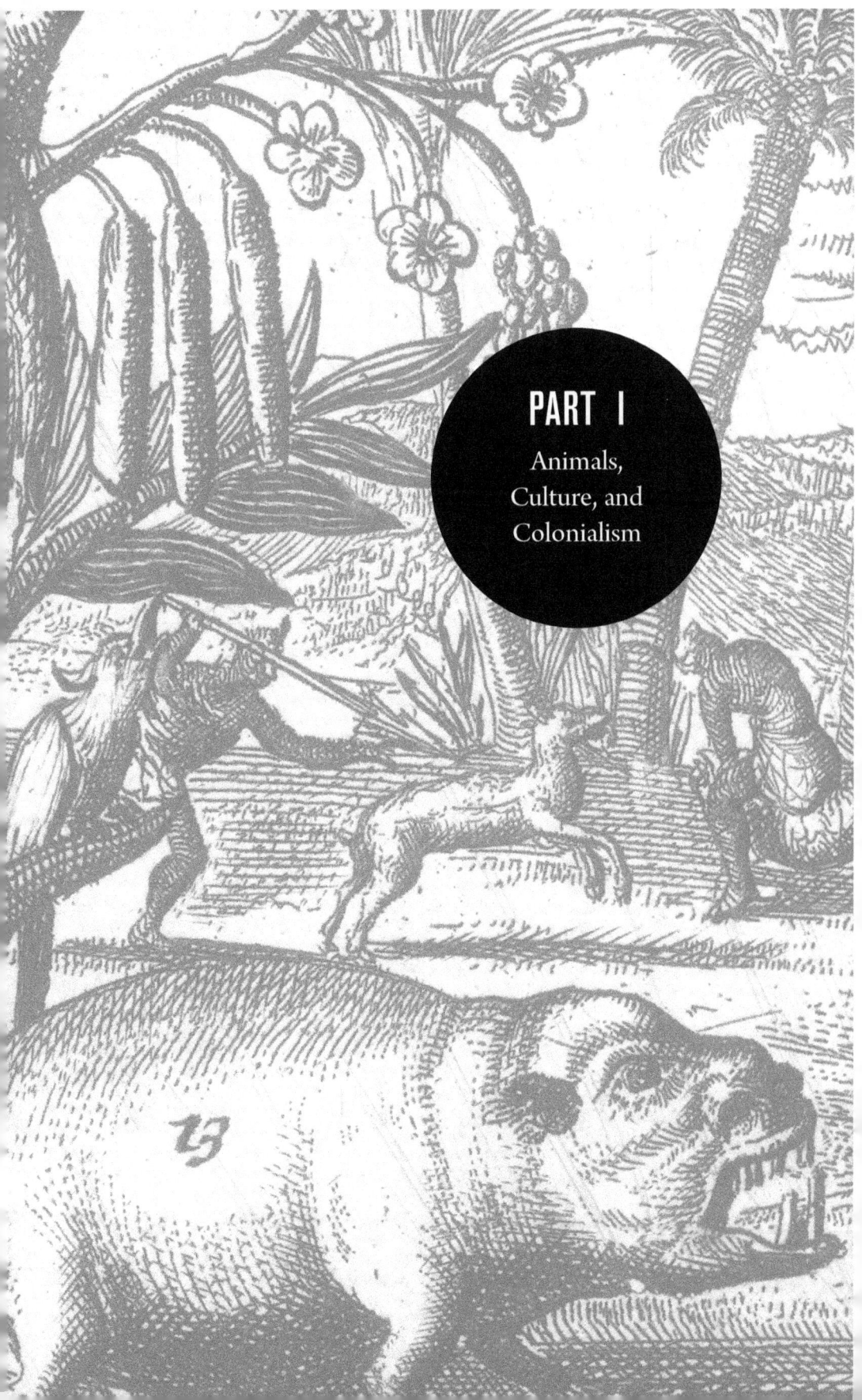

PART I

Animals,
Culture, and
Colonialism

The Year the People Turned into Cattle

The End of the World in New Spain, 1558

LEÓN GARCÍA GARAGARZA

> They were all transformed, they all turned into cattle.
> —JUAN TETON, 1558

In 1558—the year 1-Rabbit according to the traditional Mexican calendar—
the Franciscan friar Pedro Hernández apprehended a native religious leader
named Juan Teton in Xalatlauhco, a town located between Toluca and the
capital of New Spain, along with the native lords of the towns of Cohuatépec and Atlapolco.[1] The prisoners were brought before the archbishop of
Mexico to face charges of heresy and apostasy. Juan Teton was accused
of having persuaded the lords to wash away from their heads the waters of
Christian baptism, to return to the sacred tradition of their grandfathers.
Furthermore, Teton had declared to his followers that the Spaniards were
converting people into cattle, so that nobody should eat the meat of any
beast introduced to the land by the foreign invaders. Inspired by Alfred W.
Crosby's seminal work, *The Columbian Exchange*, modern scholars have
analyzed systematically the massive changes that European plants and animals inflicted in the early colonial landscape.[2] A good number of anthropologists have also touched on the symbolic representations of European domesticated animals in the Native American imaginary. By focusing on the
case of the indigenous diviner Juan Teton, I investigate one early colonial
instance of such a representation. In the sixteenth century Juan Teton was
able to articulate a coherent diagnosis of the tremendous social and ecological transformations that the irruption of the colonial biota inflicted on the

landscape of central Mexico. This diagnosis—as well as his prescription for it—was based on the traditional cosmological and eschatological notions of his people and, as such, ran directly against the interests of the Spanish invaders and the Catholic Church.

Juan Teton owes his appearance in the historical record to the survival of a late sixteenth-century Nahuatl-language manuscript known as the *Anales de Juan Bautista*.[3] The *Anales de Juan Bautista* was produced in Mexico City by indigenous scribes who, trained by Spanish friars, adapted into alphabetic writing the precontact genre of historical record-keeping: the pictographic "year count" (*xiuhpohualli*). This native tradition was easily adapted into a parallel European genre of chronological history writing, cultivated since antiquity, that of the Annals. In the sixteenth century a good number of Nahuatl-speaking city-states produced Annals in their own vernacular, usually combining alphabetic script with the pictographic conventions of the past.[4] These books recorded events that affected the community as a whole, such as the weather, omens, military conquests, religious festivals and important political matters, such as military campaigns, or the death of a ruler.

In this chapter I analyze the meaning implicit in Juan Teton's preaching, particularly his religious injunction to refrain from eating the flesh of animals introduced by the Spaniards. Because his enigmatic remarks appear only in a couple of pages written by a native employee of the Church, to make his testimony intelligible we must subject it to a critical hermeneutics, analyzing its components through comparison with the data provided by other colonial sources that deal more amply with characteristics of Mexican precontact religion, especially indigenous understanding and representations of animals. These sources are abundant, and many were written by both Spanish and indigenous writers in the decades immediately following the fall of Tenochtitlan. By using a multidisciplinary approach that takes into account the information on native religious practices provided by archaeological data and contemporary ethnography, we can begin to reconstruct the cosmovision that sustained Juan Teton's eschatological diagnosis of his time, especially as it related to questions regarding the boundaries between the human, animal, and divine. In the events of the mid-sixteenth century, Juan Teton saw the signs of the end of a cosmic era as prophesied by his ancestors, and partly represented by the colonizers' domesticated animals.[5]

Juan Teton was a diviner and ritual practitioner from the Otomi town of Michmaloyan.[6] While the *Anales de Juan Bautista* does not give us any hint of the specific term characterizing Juan Teton's religious office, we nevertheless recognize through his discourse a *tlaciuhqui* ("searcher of things" or "diviner"; pl. *tlaciuhqueh*). This Nahuatl-language term describes a person endowed with the power to prophesy the future and divine fate.[7] As both the historical and the anthropological sources indicate, divination has been articulated since ancient times within the complex of Mesoamerican ritual practices, and thus of political power. The native informants of the Franciscan friar and missionary Bernardino de Sahagún, who participated largely in the evangelization of New Spain, specifically identified the tlaciuhqui wise men as Otomi, and their description of these "sorcerers" accords very well with our description of the character of Juan Teton:

> [Among the Otomis] there were wise men, who they called, who they named, *Tlaciuhqui*. That is, he performed divination [*quitlaçihuj*]; he was equal to, he resembled [a god]; he addressed them as if [he were] a god. He addressed the gods; he informed them of that which they desired. The Otomi inquired of [the sorcerer], if it were necessary to go to war. . . . They inquired of him whether there would be rain during the year. . . . They inquired [if] famine, if perhaps sickness might come. . . . They worshipped [the tlaciuhqueh] as gods; hence were they highly esteemed.[8]

The description of the Otomi tlaciuhqueh in this passage is revealing in that it delineates the social functions of these ritual specialists: they were counselors of state, consulted by rulers and individual warriors in times of war, and indispensable advisors of the commoners, for they determined the agricultural schedule according to their powers of divination. Able to interpret the signs of time, the discourses of the tlaciuhqueh articulated standard eschatological themes in times of crisis. They were also healers. As the informants of Sahagún remarked, the tlaciuhqueh were considered human deities and instilled a fearful reverence from the collectivity. As stated in the passage of Sahagún, the tlaciuhqui was above all "the [community's] wise person." In Mesoamerica, wisdom eminently defined a person's ability to communicate—and to successfully negotiate—with the deities, practically in a condition of equality. This is because divination was only one of the attributes of the tlaciuhqueh. In more general terms, the diviners were sorcerers, persons endowed with the power to diagnose and heal illnesses, inflict harm on their victims through supernatural means, and control the weather through ritual techniques. Private persons and political rulers regularly employed these multifaceted ritualists in order to attract the favor of the all-pervading divinities, bring wealth and fertility to their polities, or, alternatively, employ their charismatic powers to prevail in battle against an enemy.

The Spanish evangelization attempted—unsuccessfully—to radically redefine the moral status of the native ritual specialists, artificially subsuming all of their social functions under the category of "sorcery," a pejorative term meant to devalue their activities, and to ultimately extirpate them from society. As charismatic prophets, the tlaciuhqueh in particular gained prominence as the early Church prepared to demolish the institutional temple framework of the precontact era. While itinerant tlaciuhqueh—that is, ritual specialists not tied to a particular local temple as priests—had played an important role in the life of indigenous rural communities since the dawn of Mesoamerican civilization, the encroachment of the Catholic Church and the consequent abolition of the institutional native priesthood made their role even more important, as the peasant communities sought them to maintain their native religious traditions, inseparable as these were from all productive activities, especially those related to the agricultural cycle. A common manifestation of the charismatic power of the tlaciuhqueh was—and still is—*nagualismo*: the ability to transform oneself into an animal at will.[9] A careful analysis of the ethnohistorical literature indicates that Juan Teton was a *nahualli*, a word that Sahagún translated as *hechicero* (sorcerer), but that in reality designates the faculty of metamorphosis (shape-shifting) possessed by powerful ritual specialists.[10] A nahualli is a person who manifests superhuman characteristics, in the sense that he or she is able to transform *at will* his or her appearance—generally into an animal, but also into a meteorological phenomenon, such as a lightning strike, a cloud, or even a ball of fire. In the Mesoamerican imaginary, the categorical borders between humans, deities, and animals are rather fluid. The Mayanist John Monaghan has perceptively remarked that in Mesoamerica the notion of personhood (Maya: *vinik*) is relational: in contrast to the European individualistic notion of an autonomous personhood, Mesoamerican ideology posits that the individual self is only a partial manifestation of humanity. The self is rather "a status that inheres in a collectivity."[11] This notion not only supports the strong emphasis on corporate rights and collective forms of worship that is distinctive to indigenous Mexican societies, but also provides a coherent taxonomy of the natural world, which affirms a fundamental continuity between all animate beings. For that reason, many Mesoamerican languages apply indistinctively the same word for person to both humans and animals.[12] This fluid individuality in turn is structured around the native notion of the soul and its destiny—in the case of the Nahuas, the soul is conceived as a tripartite entity temporarily residing in a corporeal body.[13] For my purposes, I will only remark that the *tonalli* (igneous, heavenly) soul is the operative link between the human and the animal individual. The tonalli is prone to leave the body in dreams or in moments of psychological crisis. The tonalli endows a person with his/her/its fate with an animal double, or nahualli. In the felicitous formulation of Esther Hermitte, the term *nahualli* designates the shared coessence of a human and an animal.[14] For the common human person, the coessential *nahual* is accessible only through dreams. On the other hand, a "sorcerer" can also access his or her coessential animal through ritual techniques. It is no wonder, then, that the ancient Mexicans designated the ritual specialist as a nahualli.[15]

The identity between the tlaciuhqui diviner and the nahualli shapeshifter is made clear in the fourth book of the *Florentine Codex*, in a passage that furthermore specifies that even a member of the commoners was perfectly able to become a nahualli. Thus, a person born in one of the days governed by the sign One-Rain of the Nahua divinatory calendar was fated to become a nahualli: "It was said that one who was then born became a nahualli shapeshifter, a tlaciuhqui diviner. That is to say, he was inhuman [*amo tlacatl*, meaning that his persona possessed numinous qualities]. He transformed himself [*quinaoaltia*], he turned into something else; perhaps he had a wild beast as a nahualli. And even if he were a *macehual* [commoner], his work would still be the same: to turn himself maybe into a turkey, or a weasel, or a dog. His transfiguration could be into anything, [for] he made himself into his own nahualli."[16]

As numinous personalities, the nahuallis were both feared and respected. They had the power to destroy or to heal, to guide or to delude, through magical means. I have chosen to designate Juan Teton as a tlaciuhqui, not just because this was the preferred term to designate the nahuallis in the Otomi area where he was from, but mainly because the scant paragraphs that report his activities in the *Anales de Juan Bautista* care only to describe the eschatological prescriptions that he articulated in 1558, a few months before his arrest, without mentioning any of the magic feats commonly employed by Mesoamerican ritual specialists like him. In fact the author of the *Anales de Juan Bautista* takes pains to dismiss the character of Juan Teton, describing him off-handedly as a deluded idolater of low social standing who was somehow able to convince the indigenous authorities of several towns to reject Catholicism and re-embrace the traditional religion, in the expectation of the end of time. The text states: "One little *macehual* [commoner], a man named Juan Teton, a native of Michmaloyan, corrupted with his lies the people of Cohuatepec and Atlapolco. He washed away their baptism. And this Juan was the one who washed the heads of the people, he misled them, he lied to them. He told the ones from Atlapolco who were at Cohuatepec: 'The Cohuatepecans were the first in washing [their] heads.' So when [the Atlapolcas had also] washed their heads, they sent then their warning summons [*ymamauh*] to Atlapolco, and their city became divided in two because of his lies."[17]

The mixed name Juan Teton tells us that he had already been baptized, and thus was subject to Inquisitorial prosecution.[18] Even though his surname is derived from a Nahuatl word, as a native from Michmaloyan he could well have been Otomi. The name Teton means "small stone," and reflects his low social status, as the suffix -*ton* is pejorative. The name Teton may have been calendrically associated with the day-sign *Tecpatl* ("Flint," also one of four year-bearers), but Nahua names were often irreverent and derisive.[19] Ángel María Garibay believes that the name Teton actually revealed the insurgent ideology of the owner, as one dedicated to ritually oppose the Spanish invaders.[20] In any case, charismatic sorcerers issued from all social strata, commoners included, and if Teton was actually a macehual, or commoner—as the text of the *Anales de Juan Batista* emphasizes—his arrest and prosecution for idolatry was extraordinary, as only a handful of ritual specialists of macehual origin were persecuted by the Inquisition in the sixteenth century.[21]

Teton was active in the Tepaneca area northwest of Mexico City, which still contains to this day, as in precontact times, a veritable mosaic of ethnic groups, namely the Otomi, the Mazahua, the Nahua, and the Matlatzinca.[22] Whether he was an Otomi or a Nahua, Teton would have used a Nahua name in his ministrations among Nahuatl speakers. Otomi tlaciuqueh were particularly renowned in the area since pre-Hispanic times, and Teton could well have been one of these. Whatever the case, the *Anales de Juan Bautista* portrays Teton as a recalcitrant idolater, actively opposed to the Christian conversion, and as a Mesoamerican charismatic prophet armed with a characteristic eschatological discourse that became prevalent in the decades after the Conquest. This discourse identified the Spaniards as the devouring *tzitzimimeh* deities of the end of time. The Inquisition documents of the 1530s contain two notable cases involving indigenous charismatic diviners who—like Juan Teton decades later—railed against Spanish domination and against the imposition of Catholic monotheism. These are the Inquisition cases against Martín Ocelotl and Andrés Mixcoatl, both from 1539. In the discourse of these two religious leaders the Spaniards are already identified as tzitzimimeh deities. Ocelotl and Mixcoatl were active in the area of Tetzcoco, east of Mexico City. The fact that twenty years later Juan Teton incited the people to rebel using the same mythohistorical discourse (but now in the Tepaneca area, northwest of the capital) constitutes an amazing historical phenomenon.[23] Apart from the identification of the Spaniards as tzitzimimeh deities, other stock traits of this discourse were the summons to store provisions in anticipation of an imminent drought and famine, the expectation of generalized war, and the need for a ritual purification that would counteract the insidious magic of the friars— in other words, for an active apostasy of Catholicism and a reversion to the traditional religiosity of the ancestors. Otherwise, the recorded discourse of Juan Teton has its own unique specificity: unlike his predecessors Martín Ocelotl and Andrés Mixcoatl, Teton actually articulated his eschatology around the imminent end of the fifty-two-year cycle that traditionally raised the specter of the world's end.

The year in which Teton was arrested, 1558, represented an exceedingly important calendar date for the Indians of central Mexico, who counted it as the year 1-Rabbit, the traditional date of the "Tying of the Years" (*xiuhmolpilli*). This festival marked the beginning of a new fifty-two-year cosmic cycle, but also raised the ominous possibility that the world would end. The native chronicles recorded that in the year 1-Rabbit of 1454, the gods punished the people with a terrible drought. A famine ensued, killing thousands. That year, in order to survive, many desperate inhabitants of Mexico's central plateau sold themselves and their own children as slaves to merchants from the lowlands.[24] While at least one early colonial source—the *Codex Telleriano-Remensis*—states that Moctezuma II ordered that the Tying of the Years ceremony be moved one year ahead, from 1-Rabbit (1506) to 2-Reed (1507), this does not mean that other cities agreed with this change.[25] Due to their common historical experience, all Nahuatl speakers from the central plateau dreaded the arrival of the year 1-Rabbit as a recurring harbinger of calamity. The fact that Juan Teton prophesied doom for the year 1-Rabbit of 1558 is an indication that the calendrical conventions still existed among the people of his area, including the belief in the special danger of the period starting in the year 1-Rabbit and culminating in the following year, 2-Reed. In any case, Teton diagnosed the collective metamorphoses of the people into European animals in terms of the traditional eschatological prescriptions used by the Mexica themselves—and most likely by other cities as well—before the calendrical reform of Moctezuma II.[26]

The liminal period marked by the transition from the year 1-Rabbit to the year 2-Reed was a time of danger and catastrophes. The festival of the Tying of the Years enacted the very real possibility that the world would end. That night, all fires were extinguished. In Mexico City a procession of priests with deliberate slowness marched south, from the temple of Huitzilopochtli to the Huixachtepetl (Hill of the Huizache Tree), on the peak of which they waited for the Pleiades to climb to the zenith as the sign that the sun would continue to shine for another fifty-two years. Then they sacrificed a human victim and lit the New Fire on his chest, obliterating with this vital spark the absolute darkness that until that moment had enveloped this dreadful night. The fire reborn was then transported to every house and temple, consecrating the rebirth of the sun—the renewal of the world itself. The date of the Tying of the Years was thus one of acute collective anxiety, and was marked by a number of taboos, the violation of which could result in being transformed into animals or monsters. For instance, during the night vigil, pregnant women were kept inside clay bins normally used for storing maize. An armed man kept watch outside the bins, for that night the pregnant woman inside could well be transformed into a *tzitzimitl* (pl. *tzitzimimeh*), one of thousands of ravenous skeletal deities who were ready to fall headfirst from the stars, like a rain of arrows, if the New Fire failed to materialize. That night also, children were prevented from falling asleep, for if they did they would be transformed into mice.[27] The cosmic meaning of the night of the Tying of the Years was centered around the question of whether the New Fire would materialize, or whether the god Tezcatlipoca would devour the sun while the tzitzimimeh deities rained down from the darkness to devour the people, some of whom would have already been transformed into animals.[28]

The Mexica under Moctezuma II successfully celebrated the last Tying of the Years in the year 1507, only twelve years before the arrival of Spanish conquistador Hernán Cortés. That year would witness the last public lighting of the New Fire in Mexico, for the following 52-year cycle brought a major catastrophe to the people: from 1519 onwards, the Spanish Conquest overturned the cosmic order, destroying the native temples, so—in the eyes of the indigenous *tlaciuhqueh* and their followers—the traditional rituals that had sustained and nourished the gods were fatally neglected. In 1558, when Juan Teton was arrested, he could look back on a period of devastation dating from 1524, when the Franciscans launched a systematic destruction of native temples. For him, as well as other native healers, wise men, ritual specialists, and surely a large number of their followers, the new

Spanish rulers seemed like the prophesied *tzitzimimeh* deities, charged with devouring the people at the end of time.

In a way, this native perception was accurate: the people were indeed being devoured by the combined onslaughts of war, forced labor, displacement, and especially by the introduction of new and terrifying diseases. A large segment of the Indian population—close to 90 percent according to modern estimates—perished in the decades following the Conquest of Mexico-Tenochtitlan.[29] For the people of the Tepaneca region, to whom Teton addressed his speeches, it might well have seemed plausible that the return of the year 1-Rabbit in 1558 would fulfill the old prophecies of the world's end, ushering in the final coda to the protracted process of ruin and extinction that had befallen the land since the arrival of Cortés. Juan Teton's mission was not simply to bear bad tidings; he was evidently attempting to ensure that his people would survive the transit into the next cyclical era. If they followed his ritual prescriptions, he assured them, they would be spared the fate of the peoples of the previous four creations—that is, they would survive the passage into the next cosmic cycle as human beings instead of being transformed into the strange animals brought by the Spanish tzitzimimeh. Since for the Nahuas—and for the Otomis as well—witchcraft and religion were not distinguishable categories, but rather were conceived as a single whole, we can surmise that Teton and his indigenous audience necessarily perceived the Spanish friars as creatures endowed with terrifying spiritual powers, able to "bewitch" (i.e., harm through magical means) their enemies as much as the charismatic traditional priests of the native temples had always been able to do. Moreover, the power of the friars revealed, in Teton's eyes, the disguised presence of terrifying deities: the tzitzimimeh of the end of time.

The following are the words of Juan Teton as reported by Juan Bautista.

> Listen, see what you think: you know what our grandfathers said, that when our years be tied, darkness will prevail and the tzitzimimeh will descend to eat us and then there will be a transformation of the people. Those who were baptized and believed in the Christian God will be transformed.[30] Those who eat the flesh of cows [*huacaxnacatl*] will be transformed into that.[31] Those who eat the flesh of pork [*pitzonacatl*] will be transformed into that. Those who eat the flesh of lambs [*ychcanacatl*] will be transformed into that.[32] The same will happen to those who dress in shirts made of wool. Those who eat Castilian chickens [*quanaca*] will be transformed into that.[33] All that which is the food of those

who live around here [the Spaniards], if eaten, will transform everyone, they will be destroyed, nobody will exist anymore, [for] the end of their lives and their reckoning is up; those who stood up and went to the woods, to the prairies, they will be precipitated deep into the ravines, they will go about lamenting. Look at the people from Xallatlauhco, who converted first into Christianity, the sons of Don Alonso were transformed into his cape and his hat. As for those who led the people, they were all transformed, they all turned into ruminants [*quaquaque*]. The town and its people [the *altepetl*] is no more. Those who remain are just [ambling about] in the valley and in the forests, everywhere there are only cows [*vacastin*].[34]

While the acculturated scribe of the *Anales de Juan Bautista* disparaged Juan Teton as a deluded commoner who communicated the lies inspired in him by the ancestral gods—now refashioned as "devils"—the *tlaciuhqui* himself was actually complaining that several native leaders had been fatally transformed into "ruminants" (quaquaque) because they had abandoned the traditional religion to adopt instead the customs of the Spaniards.[35] Knowing the signs of the time, Teton decided instead to lead his people to safety just as the world was about to collapse.

The *Anales de Juan Batista* reports that Teton persuaded the native authorities of Cohuatépec and Atlapolco to apostatize, submitting to a ritual washing of their heads, that is, to a lustral ceremony meant to annul the magic efficacy of the Christian baptism they had previously endured. The power of water is washed away by water: with this marvelous economy, perfectly coherent within the structure of traditional native ritual practices, Teton cleansed his followers, making them fit to survive the imminent trials of the world's end: first, he ritually cleansed the people from Cohuatépec. Then he sent a message to the authorities of Atlapolco, asking them to submit to his counterbaptism. On the receipt of this message, the authorities of this town became divided between those who submitted to Teton, like the ruler of Atlapolco don Pedro Xico and the fiscal officer Juan Tecol, and an opposing faction, who raised the alarm to the Catholic authorities, resulting in the arrest of all the apostates.[36]

But before Juan Teton fell into the hands of the Church authorities, he went about demanding that his followers submit to an ascesis based on dietary restrictions. He pointed out that the town of Xalatlauhco—in his words, the first to embrace Christianity in the region—was now a depopulated wasteland. Where there had been many people, now there were only brutish cows and sheep. There, the local leaders had been magically transformed into "three capes and three hats." The lords of Xalatlauhco had dissolved, leaving only the emblematic garments of the foreign rule that they had so shamefully accepted—to be then reborn as cows. Teton saw in this awful transformation the agency of the traditional cosmic economy: the Spaniards were disguised tzitzimimeh deities using sorcery to effectively devour the people. This magic proceeded in two stages: first, Christian baptism rendered the recipient vulnerable to dissolution as a human being; then the consumption of Spanish domestic animals triggered a fateful metamorphosis, one operating with a strict and precise reciprocity—a person would be transformed in the animal that he or she ate. Teton declared that even the mere contact with the shirt made of the wool of a sheep would turn the wearer into this animal. What was the basis of these beliefs? What did Teton's audience, which included the leadership of Atlapolco, make of Teton's claims?

Animal Conquistadors

The Spanish siege of Tenochtitlan inaugurated a series of devastating epidemics that depopulated large swaths of New Spain. In the space of a couple of decades—from 1520 to 1542—the landscape of Mexico was radically transformed. Following the fall of Tenochtitlan in 1521, the Spanish introduced cattle and sheep ranching; thus, as the human population crashed, the livestock population soared. By 1558 Juan Teton and his followers could see that areas that had been extensively populated and intensively cultivated twenty years before were now pastures upon which herds of strange, foreign animals roamed freely, stripping the land of vegetation.

The Spanish introduced many domestic animals into the New World, among them cattle, horses, donkeys, sheep, goats, pigs, and chickens. Cattle and sheep husbandry had a crucial economic and dietary role in the Old World—these animals had served there as a measure of wealth, prestige, and nutrition since early Neolithic times. Cattle husbandry in the New World was associated ab initio with land colonization. Moreover, the astonishing reproduction rates of cattle in the Americas ran parallel with the spread of human pathogen agents. The increase of cattle and sheep ranching was also accompanied by a reduction of the indigenous labor force. Even as early as 1527, cattle herds were reproducing at an incredible rate across Mexico, devastating the corn fields that had sustained human civilization in Mesoamerica for centuries. According to the environmental historian Elinor G. K. Melville, the years 1530–65 mark a period of uncontrollable expansion of cattle and sheep herds in New Spain. Melville focused her analysis on the Valley of Mezquital, the area where Michmaloyan—Juan Teton's birthplace—is located. Melville calculated that by the 1550s the population in the Tula River watershed had declined by two-thirds from the level of the precontact era.[37]

The year 1542 was a legal landmark in the history of the Spanish colonies. Adding to the comprehensive corpus of new laws that changed the character of land tenure, the Crown began offering royal grants—*mercedes reales*—to landless conquistadors who had distinguished themselves in battle. These vast land grants were given specifically for raising livestock, and became the kernel for the future development of private haciendas. In the Tula River watershed area, thirteen mercedes reales were granted in the 1550s, with 108 estancias for raising animals.[38] In the beginning, the introduction of small herds was scarcely a matter of concern for the natives. As time went by, however, the herds started to venture beyond their foraging areas, which they depleted, to invade the cornfields (*milpas*). The archives show an increase in the number of complaints from Indian towns about European animals, particularly sheep, destroying their cornfields.[39] For instance, in 1551 the Indians of Xilotepec, a town between Michmaloyan and Xalatlauhco, petitioned the Crown for the removal of a great number of animals, arguing that the people were abandoning their lands to escape destruction.[40] Colonial documents abound with examples of this socioecological catastrophe, which reached its nadir at the end of the sixteenth century. In 1589, for example, Don Juan Suárez de Peralta, nephew of the second viceroy of New Spain, praised the beneficial social effect that the introduction of pack animals had in New Spain as they took upon their backs the heavy loads traditionally carried by native human porters (*tamemes*). Suárez de Peralta, however, could not but complain about the astonishing reproduction of wild livestock and horses, which ambled about, aimlessly destroying the cultivated fields of the natives. The animals were so many, complained the aristocrat, that the countryside was full of their stinking corpses—an endless feast for the wild dogs and for the "hideous" vultures of the land, which at least prevented the spread of disease caused by the bad odor of the corpses.[41] This aristocratic lens, of course, was unable to perceive the connection between the parallel fall of the human native population and the corresponding increase of European rogue livestock. The native accounts only confirmed the phenomenon of the uncontrollable reproduction of large European animals. In 1609 the indigenous rulers of Otumba redacted a document testifying about the accuracy of the historical account that their illustrious peer Don Fernando de Alva Ixtlixóchitl had written about their town, exhorting him to tell the king of Spain that the grievances of obligatory personal service to the Spaniards were matched by the destruction that the European animals inflicted on their fields.[42]

This destruction was already more than evident in the 1550s. In 1555 the Franciscan friar known as Motolinia stated in a letter to the emperor that the authorities of New Spain were finally getting under control the relentless expansion of the *estancias de ganado* (cattle ranches granted to Spaniards), thus slowing the consequent destruction that the rogue livestock had been inflicting on indigenous lands. The Franciscan otherwise enthusiastically praised the growth of the sheep estancias all across central Mexico, from Toluca to Tlaxcala, noting that the country had many unused lands (*valdíos*) where the small cattle would thrive. Motolinia then asked the king to make illegal the indigenous use of horses, since the minoritarian Spanish population was at risk of being attacked by bands of indigenous horse riders.[43] While the Indians in New Spain were allowed to raise small domestic animals, large cattle ranching was reserved for Spaniards. Indians were expressly prohibited from riding horses or handling cattle, as these activities were restricted to Spaniards and their African slaves.[44] The traditional Spanish-guild-like associations (the *Mestas*) that regulated the ownership, transhumance, and marketing of sheep and cattle were established in New Spain in 1529, only seven years after the Conquest.

The expansion of ranching and the population explosion of free-range grazing animals created considerable social tension. The Indians always got the worst of it, as sheep and cattle ranged through the countryside and trampled their fields. In response to numerous complaints about the destruction of native cornfields, in 1536 the Viceroy Antonio de Mendoza issued the first decrees concerning stables and the spaces designated for cattle breeding, granting more land to the Indians. By the mid-sixteenth century, some Spanish ranchers individually owned more than 100,000 heads of cattle. Scholars estimate that central New Spain during this period was populated with some 1,300,000 cows, about three times the number of cows found in the tropical lowlands. Time and again, Teton was, at least metaphorically, correct: the expansion of cattle and sheep is a mirror image of the contraction of the human indigenous population.[45]

The acute population crisis to which Juan Teton indirectly refers in his 1558 speech was thus a direct result of the uncontrolled reproduction of European herds. The Spanish ranchers took advantage of a new system of labor relations that since the early 1550s had begun to replace the old encomiendas (which were essentially grants of native labor and tribute given to a Spaniard): under the new system of the *repartimiento*, Spaniards were allowed to raise livestock without obeying the older covenants concerning land usage or the need to solicit native labor that had been part of the encomienda system. The repartimiento facilitated the change from intensive cultivation to an extensive grazing agricultural economy based on animal husbandry. As a result of the overabundance of cattle and sheep, the land was not only stripped of centuries worth of cultivated plots, which had become vital to maintaining the soil, but it was overgrazed as well, which resulted in large-scale erosion.[46]

What drove the livestock economy, in 1558, was not the use of the animals for New Spain's food stock, but the export in cowhides and wool, which had started to develop to satisfy the demands of the Spanish armies in Europe. An informal economy arose consisting of smugglers and thieves, who took advantage of the abundance of feral cattle in Mexico to cull the often unsupervised herds. Given the abundance of meat, New Spain's internal meat market started to develop as well: Indians began to consume bovine meat, and butcher shops started to appear in Indian towns. Due to the increase in demand, meat prices in New Spain rose at the end of the sixteenth century. Meanwhile, the rate of increase in the number of cattle began to peak in the last third of the sixteenth century.[47] The words of Juan Teton indicate that by 1558 the indigenous population of central Mexico had already incorporated pork, sheep, and chicken into their diet. However, it appears that beef was a product reserved mostly for the Indian nobility, if only because its members occasionally partook of the Spaniard's table.

A Diet for the End of Time: Native Perspectives

In Mesoamerica, civilization thrived on an economy based on the cultivation of maize. The absence of domestic ungulates in the pre-Columbian world resulted in a civilization unique in its restriction of animal husbandry to a few endogenous small animals, the most important of which were the turkey and the dog. The Mesoamerican peoples, however, had abundant sources of protein, as demonstrated by Bernard R. Ortiz de Montellano: water fowl, small mammals, snakes, iguanas, fish, frogs, insects, and larvae—the ecosystems of Mesoamerica provided a varied and plentiful source of protein to the indigenous diet.[48]

Animals had a preeminent place in the cosmovision of Mesoamerica as well. There, animated beings were seen as inherently protean, constantly transforming from one shape to another. Accordingly, animals, gods, and men shared characteristics and were prone to transform into one another.

The Nahuas used the term *in yolque, in manenemi* (those who have a heart, those who walk) to refer to living, terrestrial creatures.[49] This was due to the fluid nature of the soul, particularly the life force, or tonalli, one of three soul entities that was found in—and shared by—humans, animals, plants, and "things" such as celestial bodies, boulders, or mountains.[50] In that sense, the Mesoamerican cosmos was—and still is—truly a living one. In 1558, when Juan Teton proclaimed that the people were being converted into sheep and cows, he and his audience may well have seen the Spanish occupation and the massive transformation of the landscape around them in terms of this well-known "natural" mechanism, which found its articulation in the traditional cosmogony.

At the time of the Spanish Conquest, all the cities in Mesoamerica shared a cosmogony that told of five consecutive creations. Only the last one—ours—resulted in the creation of fully endowed human beings—that is, of intelligent creatures able to reciprocate the divine sacrifice that had created the world and given them life.[51] The ability to intelligibly praise their creators was in fact the main quality that differentiated human beings from other creatures of flesh and blood. According to the myth, at the beginning of each creation cycle, humans and animals shared similar traits. One version of the myth, found in the sixteenth-century manuscript known as the *Historia de los mexicanos por sus pinturas*, relates that in each successive creation the gods created macehuales, that is, people (commoners) who were supposed to worship their creators.[52] However, the macehuales of each of the four previous eras showed a fatal deficiency that convinced the gods to metamorphose them into animals. According to a Maya tradition, this deficiency was preeminently semantic: since the privileged mechanism of creation was language, those creatures that could not produce this attribute were doomed to serve as mere foodstuffs for other creatures.[53] The Popol Vuh of the K'iche' Maya states that the first creatures became animals because their mindless chatters and howls were inarticulate and did not praise the work of their creators. The creatures of the fifth era finally achieved the ability to reciprocate with speech, the divine gift of life. It is worth noting, however, that both animals and humans had the common duty to nourish their creators with their blood, because the gods had given their own blood to animate their creation.

As Alfredo López Austin observes, the peculiar characteristics of each animal species were imprinted at the cataclysmic moment of each world's destruction, the moment in which people were transformed into the animals that we see today.[54] In the Nahua versions of the myth, the deciding element in the fate of the created people (macehuales) was the particular foodstuff it preferred in each successive creation, so that plant consumption was conjoined with verbal skill as the determining trait that situated the position of each creature in the cosmos. This notion seems to reflect the ethos of a fully agricultural society, which looked on nomadic societies (generically known as "Chichimecas") as inferior because they did not base their way of life on the production of corn. To understand the way in which Teton's words play on the close cultural associations that bound together the proper foods, the definition of the human, and the sacrifices demanded by the gods, one can compare the metamorphic dietary features contained in three sixteenth-century Nahua versions of the cosmogonic myth of the Five Suns: the *Leyenda de los soles* (found in the *Codice Chimalpopoca* and written around 1570), the *Hystoire du Mechique* (probably written by Fray Andrés de Olmos, 1485–1571), and the *Historia de los mexicanos por sus pinturas* (most likely written in the 1530s). The dates are close enough to Juan Teton's time to give us a representation of his background assumptions. These texts thus illuminate the somewhat cryptic pronouncements of Teton as recorded by Juan Bautista.

The *Leyenda de los soles* indicates that in each creation the Sun god gave form to creatures, "and gave food to each one of them."[55] This source gives only the esoteric, calendrical name of the plants consumed by the people of each successive creation. This particular taxonomy reflects the ritual lore associated with planting in Mesoamerica and is sometimes difficult to interpret.[56] According to this version, the people of the first creation used to eat *chicome malinalli* (seven-coiled herb), a product that Primo Feliciano Velázquez identifies with pine nuts.[57] Both the *Leyenda de los soles* and the *Historia de los mexicanos por sus pinturas* tell us that "tigers" devoured the people of the First Sun.[58] The latter version provides details absent in the *Leyenda de los soles*, specifying that the world was then peopled with giants and that Tezcatlipoca, the First Sun, became a tiger after being pummeled by his rival Quetzalcoatl. As he fell, Tezcatlipoca multiplied into many tigers and then ate his children, the earthly giants.[59] The lore of the jaguar in Mesoamerica is extensive. The jaguar incarnated the shape-shifting powers of shamans, many of whom claimed to have this animal as their double *nagual*. It was, on the divine level, the premier manifestation of the Divine Sorcerer Tezcatlipoca. Accomplished warriors could also partake of the powers of the jaguar, earning the right to dress in the attire of this animal.

An alternate version of the myth, the *Hystoire du Mechique*, gives us another version of the people of the First Sun, who used to eat *acicintli*, a "river herb" (literally, "water corn"). They were then annihilated in a flood and were turned into fish.[60] Both fresh and saltwater fish were abundant, of course, in Mexico, and played a considerable role in the diet of the people. Then, the *Hystoire du Mechique* continues, the people of the Second Sun ate the herb called *cencoccopi*, described by Fray Andrés de Molina, in his *Vocabulario* (1555), as a noxious weed that looks deceptively like corn, and that probably designates a variety of primitive corn, or *teocintle*. The people of this era were burned by a rain of fire and became turkeys, butterflies, and dogs.[61] For its part, the *Leyenda de los soles* version places the rain of fire at the end of the Third Sun, specifying that the people had been eating then *chicome tecpatl* (7-Flint) until they were turned into turkeys (*totolme*), and then annihilated as children (chicks), so turkeys were called "baby children," or *coconepipilpipil*.[62] The Nahuatl word for "turkeys" is *pipillo*.[63] Nahuatl language homologized the reduplicated term *pipilpipil* to signify both "turkey chicks" and "young lads." And the word *pipiltin* means both "children" and "nobles."[64] The author of the *Leyenda de los soles* takes advantage of the triple synonymy of the Nahuatl for "children," "nobles," and "chicks" to establish a link between turkeys and nobles, which may have the genealogical implication that the people of the third era—today's turkeys—were the ancestors of the present-day nobility. Supporting this notion, the scholar Guilhem Olivier relates that Tezcatlipoca was frequently represented as a turkey, adding that the patron of the nobility, *piltzintecuhtli*, received turkey-feather offerings in substitution of human sacrificial victims, and that some prisoners of war destined for sacrifice mimicked the turkey, signaling by their performance that they were ready to be ritually immolated.[65] This suggests that the turkey was understood to be a theriomorphic guise of Tezcatlipoca in his role as the most prized sacrificial victim, the prisoner of war. In my research, I have found evidence that the metonymic association between turkeys and nobles was reinforced through sumptuary laws in some communities. For instance, the Nahuas of Cuetzalan did not allow commoners to eat turkey—only the members of the nobility were allowed to eat this delicious meat.[66] The same restriction prevailed among the Zapotecs and Nahuas who lived in the towns subjected to Tecuicuilco in the Valley of Oaxaca.[67] However, while this dietary taboo fits in the general logic of pre-Hispanic sumptuary legislation in Mexico, there is evidence that it was not universally observed. Commoners in some places seem to have raised turkeys not only for the noble table, but for their own consumption, unconcerned with sumptuary restrictions. The subject raises crucial issues of class and hierarchy.[68] The point is that the native imaginary established an intimate link, reinforced through cosmogonic argument and sumptuary legislation, between the turkeys and the nobles. The evidence for this link still prevails among some ethnic groups such as the Teneek (Huastecs) of eastern Mexico, who today use the same term to designate the turkey chicks and the "Spaniards" (the mestizos): *eejek*. Furthermore, the anthropologist Ariel de Vidas remarks that the Teneek detect the aristocratic (read: bellicose) nature of the eejek by observing the common eating habits of the turkeys and the Spaniards: both are hand-fed when young, and eat the choicest foods (for the turkeys, that would be the corn mash). Both creatures are born to be served and pampered, and grow up to become violent creatures. In a fascinating reversal of historical ecology, the Teneek identify themselves with Old World chickens, resilient creatures that thrived on leftovers and had to stand their ground before the aggressive native turkeys.[69] It is highly probable that the precontact Nahuas followed a similar logic to determine that the nobles and the turkeys shared a common nature.

Returning to the cosmogony of the Five Suns, the *Hystoire du Mechique* has it that, during the fourth creation of the *Ehecatonatiuh* (Wind-Sun), the people consumed the fruit of the *mizquitl* tree—or mesquite.[70] At the end of the cycle they were swept away by the wind and turned into monkeys, clownish animals related to musicians and jesters.[71] For its part, the *Leyenda de los soles* claims that the world ruled by the Wind (4-Wind) was the second creation, agreeing otherwise with the *Hystoire du Mechique* that the people were turned into monkeys, that is, into a parody of mankind—"mere manikins, mere woodcarvings," as the Popol Vuh puts it.[72] In spite of this disparaging gloss, monkeys were revered for their particular sacred endowments: Ozomatli, the monkey god, ruled the central segment of the divinatory calendar and was revered as the patron of music, dance, and lust, performing thus a crucial role in the cosmos.

The fourth creation, according to the *Leyenda de los soles*, yielded the transformation of the ancestors of mankind into dogs, a metamorphosis that is narrated in much more detail than were the previous ones.[73] This narrative interest most likely reflects the intimate association between humans and their companion dogs. The myth specifies that the first father and the first mother—Tata and Nene—transgressed a dietary rule when they decided to cook a fish inside the tree of Tamoanchan while they waited for the waters of the flood to subside, and they were then transformed into dogs. The affiliation between present-day humans and their dog-ancestors is reflected in the fact that a number of ethnic groups in Mesoamerica still consider themselves descendants of the dogs; in fact, the term *Chichimeca* may mean "People of the Dog-Lineage."[74] The dog is found in several places in the Mesoamerican semiotic space. It was also one of the calendrical signs that determined the fate of the people. Several varieties of domestic dogs, some of which were hairless—the *xoloitzcuintli* and the *teuih*—were originally bred in Mesoamerica.[75] Dogs were part of the Mesoamerican diet and were sold in the markets as food.[76] As in other cultures, dogs were prized companions as well, playing a large role in the mythology of the afterlife. A dog was often sacrificed during a funeral ceremony in order to help the soul of the deceased through its journey.

The other animal metamorphosis mentioned in the myth was the transformation of humans into butterflies. As in many other regions of the world, in Mexico the butterfly was revered as a harbinger of rebirth. One myth told that dead warriors traveled to the sun and returned to earth after four years in the shape of butterflies and hummingbirds. "Night butterflies" (our moths) on the other hand belonged to the shadowy world of the tzitzimimeh, and were feared as harbingers of death and disease, especially under the dreadful aspect of the goddess Itzpapalotl, "Obsidian Butterfly," who was the choleric grandmother of the gods.[77] Finally, the Fifth Sun—our own world—was marked by the creation of a species which obtained a superior means of nourishment—maize—from the creator, thus attaining full humanity.

The perceptive reader will have noticed that the transformations of the previous eras all involved consuming plant species that proved insufficient to the task of nourishing creatures with sufficient intelligence to fulfill their ritual responsibilities. Full-fledged human beings were formed only in the last creation, when Quetzalcoatl mixed his blood with the ground bones of the dead ancestors and then stole the corn to nourish his creatures. All of the foodstuffs of the previous creations were edible wild plants suitable only for marginal consumption.[78] Reliance on such a primitive diet inevitably resulted in the transformation of the ancestors into different varieties of animals—fish, turkeys, monkeys, butterflies, and dogs—at the end of each era. Our own era, called 4-Movement, is destined to conclude with violent tremors: the jaguar Tezcatlipoca will devour the sun, and the skeletal tzitzimimeh deities will invade the earth to devour the people. According to this cosmology, humans are obliged to punctually fulfill their ritual obligations vis-à-vis the gods through ritual offerings of blood in order to delay as long as possible the inevitable fulfillment of this fate.

This is the background against which Juan Teton appears in the *Anales de Juan Batista*. Given this grand narrative of world creation and destruction, he could easily conclude that the recently introduced foreign animals were a distinctive sign of the end of time. And as in those stories, those who ate incorrectly—that is, those who consumed the animals of the recently arrived Spanish colonizers—were doomed to become animals. He conjectured that the meals made from the meat of animals that were visibly wreaking destruction on the central food source, maize, must be relinquished in order to save the people of the area. The idea that the contents of someone's diet procured specific transformative capacities is an intriguing feature of the cosmovision of Mesoamerica. If Juan Teton's particular prescriptions were a response to an eschatological situation precipitated by the realities of colonial imposition, the truth is that examples of human-animal transformation—nagualismo—were normative in the Nahua worldview, and nagualismo naturally included diet, the absorption of the soul of the eaten creature, as an integral component of its operative mechanism. Juan Teton's interpretation of the transformation of the local environment was well articulated within the cosmological narrative of his ancestors.

The fact is that in the native cosmovision, the boundary between an animal and a human existence was always porous: human-animal metamorphosis could occur at any time, but especially after death. Sahagún's informants reported that the people of the Teotlixco believed that when they died, their heart-souls (*teyolia*) were reincarnated in the form of a pleasant singing bird, the "heart-bird" (*yollototl*).[79] The Teotlixcans, however, expressed their reverence for their beloved totemic bird in a rather more surprising way: the informants said that the heart-bird was an edible bird, suggesting that a form of sublimated cannibalism was an acceptable component of the cosmic economy. In contrast, Juan Teton saw the Spaniards as the terrible tzitzimimeh deities with bewitched animals that ate up the land and transformed those with whom they came into contact, dehumanized through a circular process of consumption. To eat a Spanish animal in that sense would be for him an *unacceptable* form of metamorphic cannibalism, because the metamorphosis here signaled the destructive sorcery of the tzitzimimeh, the enemies of mankind, who had to be counterattacked with the appropriate ritual means.

Juan Teton observed that the people of Xalatlauhco had disappeared, that foreign animals now occupied the landscape instead, and that the year 1-Rabbit was around the corner. All of these things could be understood through a long-established Mesoamerican hermeneutic. The conclusion was obvious: the Spaniards—destructive tzitzimimeh deities in human guise—were using powerful magic to destroy the people. Juan Teton warned the nobles of several neighboring towns that those who persisted in eating European meats and did not wash off the Christian baptism would inevitably be transformed into those strange animals, to be devoured at the end of time, which was coming near. With this declaration, Teton embedded the facts of the ecological and demographic catastrophe precipitated by the Spaniards into the traditional elements of native eschatology with surprising ease.

In 1558, to prepare for the imminent drought at the world's end, Teton asked his listeners to store the traditional foodstuffs: the tender maize, with all its different parts (hair, leaves, ears, etc.), the tomatoes, the squash, and an as yet unidentified preparation of pulque (the traditional alcoholic beverage of Mexico, the product of fermented maguey juice).[80] The tlaciuhqui also counseled his listeners to store the traditional paper, which is widely used in ritual even today. He urged the people to collect one edible wild product, the "forest mushroom."[81] Revealingly, Teton included only one animal in his list of provisions for the end time: the autochthonous turkey.[82] Nutritional and medicinal, edible or psychoactive, these products were all indigenous products that had been used for millennia in Mesoamerica as food staples and ritual offerings. They were the very basis of the world of the Fifth Sun, sanctioned by the gods, and proven to sustain one's own humanity.

For that reason, Juan Teton urged his followers to eat only the traditional fruits and animals in order to survive the coming ordeal of the end of the era of the Fifth Sun and the beginning of the next cosmic cycle. More than objects of diet, the items on Teton's list were living spirits ready to be activated with the means sanctioned by ancestral religious practice to express their apotropaic properties. In contrast, the products introduced by the Spaniards lacked a place in the ritual and cosmogonic tradition of the natives; since they were radically strange, as strange as the time in which he was living, Teton diagnosed them as inherently lethal, falling back on a tradition in which the people in various other eras had been led astray by eating the wrong foods. In this way, the Spanish changes to the nutritional culture of Mexico could be inserted into the logic of traditional cosmology.

Even as Teton was warning the people of Xalatlauhco, the priests and missionaries of the church were busy inculcating the Indians with a new cosmovision based on the Bible. In this cosmovision, major symbolic roles were played by the European animals. Although originally Jesus Christ, who was continually compared to a lamb and who used herding metaphors, would have been incomprehensible within a strictly Mesoamerican framework, eventually the natives of Mexico would learn to understand the place of the European animals in the world according to biblical lore, with its seven days of Creation and its Ark of Noah. It is clear, however, that in 1558, in the area of Xalatlauhco, the radical shift from the traditional cosmovision to Catholic natural theology still had little grip on the people—at least Teton and his followers seem completely oblivious to it. On the other hand, even supposing that the friars had been to able to explain Genesis to the area's natives, our glimpse of Teton shows that he would most likely have found it utterly useless, predicated as it was on the exclusion and execration of his own ancestral gods. In any case, Teton's discourse incorporated the European animals squarely into the Mesoamerican cosmic economy: they were agents of transformation at the end of the Fifth Sun.

For Teton, the continued existence of humankind depended on the proper ritual behavior of his followers. The native prophet envisioned a last pilgrimage with his followers through a landscape soon to be devastated by drought and war: according to his prophecy, in the year 1-Rabbit the Earth deity known as Tlaltipaque (Owner of the Earth) would dry up all means of sustenance, triggering a massive migration of peoples and general war. Teton and his followers, the untransformed human survivors, would then have to face a last ordeal: the encounter with Tlantepozilama (Old Hag with Copper Teeth), the telluric tzitzimitl goddess manifest in her most frightful aspect, as the ravenous devourer of the world.[83] To avoid her terrible jaws, Teton prescribed a curious performance: he asked the people to parade before the old goddess with a jar of dry pulque placed over their bellies. The goddess would then allow them to pass unharmed.[84] Finally, the people would arrive to the safety of Chapultepec (Grasshopper Hill), where apparently they would survive the transit into the next cosmic cycle.

To summarize: the words of Juan Teton, which swayed the rulers of some of the towns in the valley of Toluca, provide a glimpse into the balance of forces in the rapidly changing physical environment and cultural universe of Mexico in the mid-sixteenth century. As part of their ongoing acculturation, the Indians of central Mexico were already consuming European meats— sheep, beef, chicken, and pork—as those animals appeared on the land- scape and the old agricultural traditions, centered on the cultivation of maize, were being literally uprooted. Juan Teton claimed that the depopula- tion of the region was caused by the consumption of European animals and by the acceptance of Christian baptism, which turned humans into beasts. He then articulated a coherent program of resistance involving relinquish-

ing European animals, ritually deactivating Christian baptism, and returning to the proper symbols and rituals that would allow the people to survive
the end of the Fifth Sun era.

Historians have taught us that the colonial enterprise, with its new forms
of economic exploitation, cultural appropriation, and epidemiological exchange, caused the massive demographic collapse of the sixteenth century.
Juan Teton, on the other hand, *knew* that the disappearance of the people
was the result of the insidious magic of the Spaniards, which he unmasked
as the terrible tzitzimimeh deities, that is, as fully recognizable—and hence
intelligible—actors in the traditional indigenous cosmology. These terrible
supernatural beings were transforming the people into herds of strange animals. The traditional eschatology—sacred stories of the cyclical creations
of the world—provided Juan Teton with a hermeneutics through which this
catastrophe of human to animal transformation could be read for what it
was: the signal of the end of the Fifth Sun era. With this knowledge, he
urged his people to return, before it was too late, to the proper human diet
and the proper rituals in order to ensure their survival as the world ended.
He and his people would then be able to safely pass into a world renewed.

Notes

1. At the time of the Conquest, when Hernán Cortés conquered and appropriated
   the area, the towns of this region paid tribute to the Mexica. From 1528 on,
   Atlapolco and Xalatlauhco belonged to a private *encomienda*. In 1557 the area
   became a Visita of the Franciscan *doctrina* of San Pedro and San Pablo Calimaya.
   Both Atlapolco and Xalatlauhco became *cabeceras* of their respective jurisdictions
   and by 1569 had resident secular priests. See Peter Gerhard, *Geografía histórica de
   la Nueva España, 1519–1821* (Mexico City: Universidad Nacional Autónoma de
   México, 1986), 278–28. In the account of Juan Teton in the *Anales de Juan Bautista*,
   the list of the apostates arrested in 1558 included the *fiscales* of Atlapolco and
   Cohuatépec, a sign that in these towns the church had already established local
   native officials. See Luis Reyes García, *Anales de Juan Bautista* (Mexico City:
   CIESAS, 2001), 158.
2. Alfred W. Crosby, *The Columbian Exchange: Biological and Cultural Consequences
   of 1492* (Westport: Greenwood, 1972).
3. I will refer throughout to the manuscript that was transcribed and translated into
   Spanish by Luis Reyes García under the title *Anales de Juan Bautista*. Reyes García
   included a photographic reproduction of the manuscript. No other traces of Juan
   Teton's activities, or of his trial, have been found in the historical record.
4. The *Anales de Juan Bautista*, however, lacks any pictographs, as it was produced
   late in the sixteenth century, when the tradition of pictographic writing was

becoming obsolete in the capital of New Spain. Among other important colonial indigenous Nahuatl annals we find the *Anales de Cuautitlan*, the *Anales de Tlatelolco*, and the *Codex Aubin*, which contains records from Mexico City.

5. Alfredo López Austin defines a cosmovision as "a structured group of ideological systems that evolve from different areas of social action and which returns to them, providing a reason for their principles, techniques, and values. Its rationality is enhanced by its operation in different areas of social activity. Since a cosmovision is built on all of the daily practices, the logic of those practices is transferred to cosmovision and impregnates it" (*Los mitos del tlacuache* [Mexico City: Universidad Nacional Autónoma de México, 1996], 11). For the multidisciplinary approach methodology applied to the study of Mesoamerican religion, see Davíd Carrasco, *Religions of Mesoamerica* (Harper: San Francisco, 1990), 11–19.

6. Reyes García, *Anales de Juan Bautista*, 156–57. Gerhard identifies Michmaloyan as an Otomi town near Tula, to the north of Xalatlauhco. Otomi is a Nahuatl word that designates an ethnic linguistic group, that of the Ñahñu (*Geografía Histórica de la Nueva España*, 341).

7. Alfredo López Austin, "Cuarenta clases de magos del mundo náhuatl," *Estudios de Cultura Náhuatl* 7 (1967): 87–117.

8. Bernardino de Sahagún, *Florentine Codex: General History of the Things of New Spain*, trans. Arthur J. O. Anderson and Charles E. Dibble (Santa Fe: School of American Research; Salt Lake City: University of Utah Press, 1950–82), book 10, 177. Anderson and Dibble translate *quitlaçihuj* as "he [the *tlaciuhqui*] performed sorcery for [the god]." I prefer the original sense of the verb as divining the will of the deities.

9. Nagualismo is one of the distinctive characteristics of Mesoamerican cosmovision, and the literature about the phenomenon is quite extensive. In Teton's area of Xalatlauhco, all the way to the Popocatepetl to the southeast, the modern tlaciuhqueh are still active and are known by the Spanish term *graniceros* (hailmakers).

10. Sahagún, *Florentine Codex*, book 10, 31. See also López Austin, "Cuarenta clases de magos en el mundo náhuatl."

11. John Monaghan, "The Person, Destiny, and the Construction of Difference in Mesoamerica," RES: *Anthropology and Aesthetics* 33 (spring 1998): 137–46.

12. Monaghan, "The Person, Destiny, and the Construction of Difference in Mesoamerica," 140. The Maya word for person is *vinik*. In Nahuatl the word *macehualli* denotes the "human" person, while the word *yolqueh* (owner of a heart) is used to designate animals. However, all animals were created ab initio as *macehuales*. The human animal, of course, also has a heart.

13. These are the tonalli, a soul entity residing in the head: the *ihiyotl*, whose center is in the liver; and the *teyolia*, residing in the heart. For a comprehensive discussion of the Nahua concepts of the soul, see Alfredo López Austin, *Cuerpo humano e ideología: Las concepciones de los antiguos nahuas*, 2 vols. (Mexico City: Universidad Nacional Autónoma de México, 1980); and Jill Leslie McKeever Furst, *The*

Natural History of the Soul in Ancient Mexico (New Haven: Yale University Press, 1997).

14. Esther Hermitte, "El concepto de nahual de Pinola," in *Ensayos antropológicos en los Altos de Chiapas*, ed. Norman McQuown and Julian Pitt-Rivers (Mexico City: Instituto Indigenista Interamericano, 1989), 371–90.

15. As Julian Pitt-Rivers observes, in spite of our Christian-based taxonomy of magic and religion, it is fundamentally wrong to equate a Mesoamerican ritual specialist (a nahualli) with a sorcerer, for the indigenous nahualli does not obtain his or her power through a pact with the devil, but rather through his or her own fate (tonalli). Being fundamentally solar, the power of the tonalli only increases with age. See Julian Pitt-Rivers, "Spiritual Power in Central America: The Naguals of Chiapas," in *Witchcraft Confessions and Accusations*, ed. Mary Douglas (London: Routledge, 2004), 191.

16. Sahagún, *Florentine Codex*, book 4, 42.

17. Reyes García, *Anales de Juan Bautista*, 156.

18. The early Mexican Inquisition had jurisdiction only over baptized persons. In the 1530s, the Apostolic Inquisition of New Spain under Zumárraga launched an intense campaign against "idolatrous Indians," which culminated with the execution of the noble Tetzcocan idolater Don Carlos Ometochtzin. The execution provoked a scandal that resulted in Zumárraga's downfall, and in the relaxation of the campaign against the natives, at least in Central Mexico. We don't know if Juan Teton was formally prosecuted after his arrest in 1558, since no file bearing his name has survived in the records of the Inquisition.

19. See James Lockhart, *The Nahuas after the Conquest: A Social and Cultural History of the Indians of Central Mexico, Sixteenth through Eighteenth Centuries* (Stanford: Stanford University Press, 1992), 118–22.

20. Names associated with stones and flints had a martial character. There is a slight possibility that Teton may have actually been called *Tetonal* ("*tonalli* of the people": a compound metaphor designating the divine origin of noble attributes), but that the author of the *Anales de Juan Bautista* preferred to emphasize the insignificant character of the diviner by calling him Teton. This, however, is purely speculative. See Ángel María Garibay, "Temas Guadalupanos 2: El diario de Juan Bautista," *Ábside: Revista de Cultura Mexicana* 2 (1945): 155–69.

21. I want to thank David Tavárez for pointing this out to me. The most notable macehual idolater tried by the Inquisition was Andres Mixcoatl, in 1539. See Luis González Obregón, *Procesos de indios idólatras y hechiceros* (Mexico City: Archivo general de la Nación, 1912). In the 1530s the charismatic diviners Martín Ocelotl and his disciple Andrés Mixcoatl articulated an anti-Christian message with the same eschatological features as the one pronounced twenty years later by Juan Teton.

22. The terms *Otomi*, *Nahua*, and *Matlatzinca* designate linguistic groups. However, in precontact Mexico, ethnic identity was not language-based, but predicated on the allegiance to one's own city-state, or *altepetl*, and, more broadly, to one's lineage. For instance, the term *Tepaneca* refers to a Nahuatl-speaking lineage

whose members lived in independent cities situated mostly on the Western shores of Lake Texcoco. The Tepaneca city of Azcapotzalco had been the dominant city-state in the Basin of Mexico until the Mexica from Tenochtitlan defeated it in the fifteenth century. The Aztec Triple Alliance included the Tepaneca city of Tlacopan as a junior partner.

23. The Inquisition cases against Ocelotl and Mixcoatl were published in 1912 as Luis González Obregón, *Procesos de indios idólatras y hechiceros*, vol. 3 (Mexico City: Archivo General de la Nación, 1912). For a more recent analysis of these cases, see Serge Gruzinski, *El poder sin límites, cuatro respuestas indígenas a la dominación española* (Mexico City: Instituto Nacional de Antropología e Historia, 1988).

24. Sahagún, *Florentine Codex*, book 7, 21–24. See also Diego Durán, *Historia de las Indias de la Nueva España* (Mexico City: Porrúa, 1967), 241–44; and Eloïse Quiñones-Keber, ed., *Codex Telleriano-Remensis* (Austin: University of Texas Press, 1995), 67 and 217. The year 1-Rabbit of 1454 started a drought that afflicted Mexico for three years. See John Bierhorst, *Codex Chimalpopoca: The Text on Nahuatl with a Glossary and Grammatical Notes* (Tucson: University of Arizona Press, 1992), 61–64.

25. Quiñones-Keber, *Codex Telleriano-Remensis*, fol. 42r. For a discussion of the political strategies involved in changing the date of the New Fire festival, see Ross Hassig, *Time, History, and Belief in Aztec and Colonial Mexico* (Austin: University of Texas Press, 2001), 70–109. On the Tying of the Years ceremony, see Alfredo López Austin, *Hombre-Dios: Religión y política en el mundo Náhuatl* (Mexico City: Universidad Nacional Autónoma de México, 1989), 99.

26. The *Anales de Cuautitlan* specifies that the gods established the sky in a year 1-Rabbit, when the fire-drill of Tezcatlipoca first appeared, during the flood that destroyed the era preceding ours. The sky was smoked that very same year when the ancestors of mankind lit a fire in secret to cook a fish. Angered by their transgression, Tezcatlipoca cut their heads off and sowed them in their rumps, giving origin to dogs. Darkness ensued for twenty-five years, until Tezcatlipoca drilled a New Fire in a year 2-Reed, giving birth to the sun. See John Bierhorst, *History and Mythology of the Aztecs* (Tucson: University of Arizona Press, 1992), 144–45. The period between 1-Rabbit and 2-Reed was thus dreaded as a liminal onset of darkness, one that could well be definitive.

27. A native pictorial depiction of the *xiuhmolpilli* ritual can be found on page 34 of the *Codex Borbonicus*, an early colonial pictorial codex. Chapters 9 and 10 of book 7 of the *Florentine Codex* have the most elaborate description of this festival. In a 1999 essay Miguel León-Portilla paired the description of the festival provided by these two sources. See Miguel León-Portilla, *Fray Bernardino de Sahagún en Tlatelolco* (Mexico City: Secretaría de Relaciones Exteriores, 1999), 69–80.

28. Sahagún, *Florentine Codex*, book 7, chap. 10, 27.

29. For a summary of the research on the demographic collapse of the sixteenth century, see Elinor G. K. Melville, *Plaga de ovejas: Consecuencias ambientales de la Conquista de México* (Mexico City: Fondo de Cultura Económica, 1999), 58.

30. *Mocuepazque*: "They will be transformed." Fray Andrés de Molina, in his *Vocabulario en la lengua castellana y mexicana y mexicana y castellana* (Mexico: Porrúa, 1977 [1555]), translates the verb *cuepa* as "to turn." As in English and Spanish, the Nahuatl verb *cuepa* can be used in the sense of "to turn" (*volverse*) or "to be transformed."

31. *Huacax + nacatl* = "cowflesh." The noun borrows from the Spanish for cows: *vacas*. Soon after this, Teton uses another variant of this borrowing, with the plural suffix: *vacastin* (cows).

32. Literally translated, *ychcanacatl* means "cottonflesh." Enchanted by the sheep's white fleece, the Nahuas associated these European domestic animals with the white flower of the cotton plant.

33. Molina translates *quanaca* as "gallo o, gallina de Castilla." A pig was called *pitzotl*, a natural extension of the word used to designate the autochthonous wild peccary. There were no domestic pigs in Mexico before the Conquest.

34. Reyes García, *Anales de Juan Bautista*, 156–59.

35. The paradox of the new moral Catholic economy regarding the Mexican gods is that these gods were not denounced as false entities, but rather viewed as real spiritual agents, charged with bringing evil into the world. In short, they were devils. This taxonomy perversely reinforced the belief in the veracity of the gods' existence.

36. Reyes García, *Anales de Juan Bautista*, 157–59. Ritual lustral baths were common in precontact Mesoamerica, and were practiced in all sorts of initiatory rituals, from birth to death. For instance, four days after an infant's birth, the Aztec midwife washed the baby, ridding the newborn of the earthly *tlazolli*, or "filth," which he or she had acquired during the transit from the highest heaven to our world. See *Florentine Codex*, book 6, chap. 32. The bath protected the recipient from the ravenous appetite of the tzitzimimeh. The Nahuas of Chicontepec still practice today traditional lustral ceremonies on many occasions. For lustral baths during marriage ceremonies in pre-Hispanic times, see Fray Gerónimo de Mendieta, *Historia eclesiástica Indiana* (Mexico City: Porrúa, 1980), 128.

37. Melville, *Plaga de ovejas*, 161.

38. Melville, *Plaga de ovejas*, 161. Michmaloyan had been granted as an encomienda to the conquistador Juan Zamudio. In 1529 the Franciscans had already built a convent in Tula, a few miles to the south. See Gerhard, *Geografía histórica de la Nueva España*, 342.

39. Melville, *Plaga de ovejas*, 161. Melville speculates that by then the Indians were already taking care of their own small herds, although they did not receive concessions of land for this purpose, and they do not appear in the documents as owners of estancias. It is only in the decade of the 1560s when the Indians first appear officially in the archives as owners of sheep. Melville points out that the beneficiaries of these concessions were the caciques, not the macehuales.

40. Melville, *Plaga de ovejas*, 160. Eight Spanish *encomenderos* were responsible for this devastation. In contrast, in the English colonies during the seventeenth century

cattle were often attacked and hunted by the native peoples, prompting a swift retaliation. See Virginia DeJohn Anderson, *Creatures of Empire: How Domestic Animals Transformed Early America* (New York: Oxford University Press, 2004), 228–29.

41. Juan Suárez de Peralta, *Noticias históricas de la Nueva España* (Madrid: Imprenta de M. G. Hernandez, 1878 [1589]), 3–4.

42. Fernando de Alva Ixtlixóchitl, *Obras Históricas* (Mexico City: Oficina tip. de la Secretaria de fomento, 1891), 466.

43. Toribio de Motolinia, "Carta de Fray Toribio de Motolinia al emperador Carlos V, 2 de Enero de 1555," in José Fernando Ramírez, ed., *Colección de documentos para la historia de México*, vol. 1 (Mexico City: Porrúa, 1971), 177.

44. Renée González-Montagut, "Factors That Contributed to the Expansion of Cattle Ranching in Veracruz, Mexico," *Mexican Studies / Estudios Mexicanos* 15.1 (1999): 101–30. In contrast, the English colonists of seventeenth-century North America attempted to promote cattle husbandry among the natives in order to instill "civilized" habits in them. See DeJohn Anderson, *Creatures of Empire*, 213.

45. González-Montagut observes that, in comparison to a regime of work based on intensive agriculture, cattle husbandry requires relatively little labor force: "The extensive use of land and the epidemics that reduced the population allowed for the recovery of areas that had been intensively used during pre-Hispanic times" ("Factors That Contributed to the Expansion of Cattle Ranching in Veracruz," 105). In the 1550s the vice-royal government tried to protect the Indian land rights by forbidding cattle and horses in the most populated, central areas. However, small animals, such as sheep, goats, and pigs, were not expelled. For her part, Melville observes that when the large animals were removed from the Valley of Mezquital, the population of sheep exploded, quadrupling in number from 1550 to 1560 (*Plaga de ovejas*, 65). The consequences for indigenous traditional agriculture were devastating.

46. Melville, *Plaga de ovejas*, 110, 145.

47. González-Montagut, "Factors That Contributed to the Expansion of Cattle Ranching in Veracruz," 104–7.

48. See Bernard R. Ortiz de Montellano, *Aztec Medicine, Health, and Nutrition* (New Brunswick: Rutgers University Press, 1990), 115–19.

49. Sahagún *Florentine Codex*, book 11, chap. 1. The category can, of course, include people as well. Nahuatl has a distinct category for creatures that harm man: *tecuani*, or "people-eater." DeJohn Anderson observes that the Indians of North America, like the Mesoamericans, did not place nonhuman creatures into a distinct category: "Indians . . . did not relegate animals to a purely physical existence, as colonists did, but also accorded them spiritual significance" (*Creatures of Empire*, 18).

50. The tonalli is the hot and luminous "soul" bestowed by the gods of the highest heaven to all the creatures. It has the property of traveling away from the body at certain times, and usually appears in a double form, so that each human being has

an animal companion, or *nagual*. The other two soul entities are the heart (*teyo-lia*) and the *ihiyotl*, which is a gaseous entity that resides in the liver. For a comprehensive discussion of the Nahua soul entities, see López Austin, *Cuerpo humano e ideología*, chaps. 5 and 6.

51. This notion is clearly articulated in the K'iche' Maya creation myth consigned in the Popol Vuh. See *Popol Vuh: The Definitive Edition of the Mayan Book of the Dawn of Life and the Glories of Gods and Kings*, trans. Dennis Tedlock (New York: Touchstone, 1996), 66–73.

52. Angel Maria Garibay, *Teogonía e historia de los mexicanos: Tres opúsculos del siglo XVI* (Mexico City: Porrúa, 1996), 27. The Nahuatl word *macehual* (commoner) means "(one who) is deserving."

53. *Popol Vuh*, 67–68.

54. López Austin, *Los mitos del tlacuache*, 57.

55. Primo Feliciano Velázquez, trans., *Códice Chimalpopoca* (Mexico City: Universidad Nacional Autónoma de México, 1992), 119 and 129. The text says, "Çeçentetl in tlamamaca" (He [the Sun] provided food to each one [of his creatures]).

56. Ruiz de Alarcón records the calendrical names of some cultivated plants, in *Treatise on the Heathen Superstitions That Today Live among the Indians Native to This New Spain, 1629*, trans. and ed. J. Richard Andrews and Ross Hassig (Norman: University of Oklahoma Press, 1984). Bierhorst was unable to identify the foodstuffs mentioned in the *Leyenda de los soles*, but he conjectures that they represent "early staples . . . wild seeds or primitive grain, becoming progressively more like corn" (*History and Mythology of the Aztecs*, 142).

57. Feliciano Velázquez, *Códice Chimalpopoca*, 130.

58. The Spanish called "tigers" the big cats we know today as "jaguars" (*Panthera onca*). The Nahuatl name is *ocelotl*. The other big cat of Mesoamerica—the puma (*Puma concolor*)—was called "lion" by the Spaniards. Its Nahuatl name is *miztli*. Generically, all big cats are called *tecuanimeh* (people-eaters) in Nahuatl.

59. Garibay, *Teogonía e historia de los mexicanos*, 30. In the 1530s, the rebellious tlaciuhqui Andrés Mixcoatl invoked this cosmogonic episode to explicitly deny the truth of the version of Genesis taught by the friars. See González Obregón, *Procesos de indios idólatras y hechiceros*, 62. Some Nahuas believe that a jaguar devours the sun during solar eclipses. See Alfredo López Austin, "La magia y la adivinación en la tradición mesoamericana," *Arqueología Mexicana* 7.69 (2004): 23.

60. Garibay, *Teogonía e historia de los mexicanos*, 103.

61. Garibay, *Teogonía e historia de los mexicanos*, 103.

62. "Inic poiluhque pipiltin catca yei ca axcan in monotza coconepipilpipil" (Bierhorst, *Codex Chimalpopoca*, 87). Bierhorst's own translation of this passage fails to adequately identify the transformed turkeys (*totolme*) with the baby chicks (*coconepipilpipil*) (*History and Mythology*, 143).

63. Another term for "turkey" is *huey xolotl* (the old clown-twin), a term fraught with mythological allusions that in time was taken up into Mexican Spanish as *guajolote*. Spaniards used the more generic noun *pavo* (*de Indias*).

64. Which makes the Nahuatl term analogous to the Spanish *infante*.

65. Guilhem Olivier, *Tezcatlipoca, burlas y metamorfosis de un dios azteca* (Mexico City: Fondo de Cultura Económica, 2004), 71–72.

66. "Relación de Cuezala," in *Relaciones geográficas del siglo XVI: México*, ed. René Acuña (Mexico City: Universidad Nacional Autónoma de México, 1985), 1:317. The passage is illustrative of the nostalgia for the precontact past felt by the natives in the 1580s. It reads: "Vivían [los de Cuezala en] tiempo pasado, más sanos y más, y dicen ser la causa, a lo que entienden, que ahora comen mucho y trabajan poco, y que, en su gentilidad, ningún indio común podia comer sino tamales y un poco de atole, y no gallinas, y trabajaban mucho más que ahora."

67. On the sumptuary legislation, see "Relación de los pueblos de Tecuicuilco, Atepeque: Zoquiapa y Xaltianguiz," in *Relaciones geográficas del siglo XVI: Antequera*, ed. René Acuña (Mexico City: Universidad Nacional Autónoma de México, 1984), vol. 45, 97. This 1580 source adds that now the commoners eat not just turkeys, but also sheep and goats, "when they can afford it": "Antiguamente . . . los macehuales no podían, comer gallinas, sino sólo los principales: hoy son comunes a todos, y también comen ovejas y carneros, cuando alcanza."

68. Bernardino de Sahagún lists the culinary habits of the Aztec nobility. While inconclusive, the language seems to affirm the sumptuary restrictions that limited the consumption of turkey dishes to the nobility, as in Cuetzala. Attending commoners, however, were allowed to eat the uneaten remains of food served in noble banquets. See Sahagún, *Historia general de las cosas de la Nueva España* (Mexico City: Porrúa, 1989), book 8, chap. 13, 463–64.

69. Anath Ariel de Vidas, "A Dog's Life among the Teenek Indians (Mexico): Animals' Participation in the Classification of Self and Other," *Journal of the Royal Anthropological Institute* 8.3 (September 2002): 531–50.

70. *The Leyenda de los soles* gives *matlactlomome cohuatl*, or 12-Snake, as the calendrical name of the plant food eaten by the people of this era. It is likely the mizquitl tree. See Bierhorst, *Codex Chimalpopoca*, 87.

71. Garibay, *Teogonía e historia de los mexicanos*, 104.

72. *Popol Vuh*, 73. The book says that the gods failed in their attempt to make humans during this era because they carved their creatures from wood.

73. However, the *Hystoire du Mechique* version lumps together the transformation of people into turkeys, dogs, and butterflies during the second era.

74. From the Nahuatl *chichi* (dog) and *mecatl* (rope, lineage). It must be said that not everyone accepts this etymology. However, the fact is that some groups that in the past were cataloged as Chichimeca, such as the Huicholes from western Mexico, consider themselves the descendants of sacred dog-ancestors.

75. Sahagún, *Florentine Codex*, book 11, chap. 1, 17–18.

76. Francisco Hernández, *Antigüedades de la Nueva España* (Madrid: Dastin Editores, 2000), 114–15.

77. For a comprehensive treatment of the role of the butterfly in ancient Mexico, see

Carlos R. Beutelspacher, *Las mariposas entre los antiguos mexicanos* (Mexico City: Fondo de Cultura Económica, 1989).

78. Except for the "myrrh" mentioned in the *Hystoire du Mechique*, perhaps a misinterpretation of the myth, unless the original claimed that *copalli* was nourished to the *macehualtin* commoners. Copalli smoke, on the other hand, is today considered a subtle nourishment of the gods.

79. Sahagún, *Florentine Codex*, book 11, chap. 2, 25.

80. Teton urged his followers to "go dry the hail-pulque" (*xichuatzacan in teçiuhoctli*). See Reyes García, *Anales de Juan Bautista*, 158.

81. Reyes García, *Anales de Juan Bautista*, 158. The *cuauhnanacatl* mushroom (woodflesh) is defined by Sahagún: "Hay otras getas que se llaman cuauhnanacatl, porque se nacen en los árboles. Son buenas de comer, asadas y cocidas" (Sahagún, *Historia general de las cosas de la Nueva España*, book 11, chap. 7), 132.

82. Reyes García, *Anales de Juan Bautista*, 158.

83. Reyes García, *Anales de Juan Bautista*, 158. An extensive analysis of this goddess can be found in Guilhem Olivier, "Tlantepuzilama: Las peligrosas andanzas de una deidad con dientes de cobre en Mesoamérica," *Estudios de Cultura Náhuatl* 36 (2005): 245–71.

84. Reyes García, *Anales de Juan Bautista*, 158–59.

2

Killing Locusts in Colonial Guatemala

MARTHA FEW

Aside from a few insects that we see as beneficial to humans—such as lady-bugs who kill other insects destructive to flowers and crops, or bees that pro-duce honey—when we think of insects, we often want to kill them. We kill lice that infest our children's hair, exterminate bedbugs that colonize the mattresses we sleep on, and "dip" our pets to eradicate fleas and ticks that hide in their fur. We buy and use products, such as Terro Ant Killer II and Combat Roach Killing Gel, whose names underscore the uneasy and violence-tinged relationship humans have with insects. Historically, govern-ment agencies have conducted numerous insect-extermination campaigns in the United States and internationally, such as the gypsy moth eradication campaign of the 1890s in Massachusetts, Mexican programs that utilized DDT against mosquitoes in the 1950s, and the California state government's use of malathion aerial-spraying missions to kill the medfly that threatened the state's multibillion dollar agribusiness industry in the 1980s and again in the 1990s.[1] From society's perspective, insect-extermination campaigns have had high stakes, from protecting public health to ensuring stable food sup-plies and the well-being of local and regional economies.

Coordinated human efforts to exterminate specific insects prior to the late nineteenth century, before the development of chemical insecticides and the professionalization of entomology as a modern field of scientific endeavor, have an important but little studied history.[2] This is the case for state-directed locust-extermination campaigns in colonial Guatemala, a geographic area that roughly comprises what are today southern Mexico and the nation-states of Central America, and that was the site of pre-Columbian Mesoamerican civilizations that included the Maya.[3] During three-plus centuries of Spanish colonial rule, from the early sixteenth century to the early nineteenth, devouring locust "clouds," "plagues," and "swarms" consumed and transformed Guatemala's landscape, destroying food crops and stripping the countryside of living vegetation. Guatemalans considered these insects as invaders who threatened their livelihoods and their very lives.

Yet for the most part historians have focused on insects such as locusts only in supporting roles in a greater narrative describing the history of agriculture and public health.[4] Even those working in the growing field of human-animal studies have rarely paid much attention to insects or placed them at the center of historical analyses. A welcome exception is Eric C. Brown's edited volume *Insect Poetics*, in which contributors explore modern insect representations and symbolism, read through literature, visual media, and popular culture.[5] In addition, the journal *Antennae* that focuses on visual cultures related to nature has three exclusively insect-themed issues.[6] However, the research in these publications tends to emphasize insects as displayed or represented in art, museums, film, and literature, and not the histories of human relationships with the actual insects themselves.

Under the framework of recuperating the presence of insects in history, focusing on the locust provides a useful starting point. Locusts have long figured in the imagination of various cultures as an important and potentially ominous species. Ancient Assyrians prayed to the god Ashur, whose iconography includes a locust familiar.[7] Locust plagues similarly afflicted ancient Greece, where Athenians erected a statue of Apollo Parnopios in recognition of his skill at repelling the insects.[8] Islamic cultures attribute important roles to insects, including locusts, as part of their sacred texts.[9] Judeo-Christian cultures also give locusts prominent roles in their sacred history from ancient times up to the present day. The Old Testament is filled with references to locust plagues, such as the plague visited upon Egypt when the Pharaoh refused to free the Israelites referred to in Exodus.[10] Similarly, the prophet Joel describes a visitation of locusts as an apocalyptic symbol associated with the invasions of Babylon and Assyria. The New Testament also contains references to locusts as literal harbingers of the apocalypse, as in Revelations 9:3: "Then over the earth, out of the smoke, came locusts, and they were given the powers that earthly scorpions have."[11] Into modern times, Christian cultures in Europe and the Americas have read locust plagues as evidence of divine wrath and as apocalyptic portents, and used a variety of strategies including locust excommunication in efforts to halt the insect swarm's spread.[12]

Writing insects into history is methodologically and theoretically challenging in part because insects do not fit into historically and culturally constructed binaries that historians of human-animal relationships wish to critique, such as human-animal, wild-domestic, and historical categories of animal predators and pets.[13] Moreover, insects have for the most part been left out of the politics of writing histories of human-animal relationships in works that promote species protection, the ethical treatment of animals (in zoos and shelters, as pets, as parts of our food supply), and as part of a broader vision of environmental conservation.[14] While many scientists and academics have critiqued the use of pesticides and other chemical and biological tools used against insects for their detrimental effects on the environment and on other species, they have not necessarily questioned insect extermination as the desired end result.[15] Nor have they reflected on the ethics of the mass killing of insects in contrast to other, perhaps more appealing, animals.

Writing insect histories is also challenging because of the kinds of sources available to historians for analyzing locusts in colonial Latin America. In this chapter I use accounts written by political officials, Indian elites, farmers, priests, and European travelers, who frequently described the locusts in epic, even apocalyptic terms. These sources tend to be elite or government-generated, written as part of state and community responses to the infestations. Other sources, written by European travelers, include descriptions of locust plagues in ways that work to further exoticize the New World and its animal species for its European audiences.[16] And writers have primarily described locusts only as the insect swarms threatened communities or in the midst of an infestation, and so have used ramped-up language to describe the plagues and their devastation. Written sources also tend to focus on how locust plagues have affected people, their domestic animals, and community responses designed to lessen the potentially catastrophic effects, rather than on the insects themselves. Nevertheless, it is possible to show that locusts, by periodically joining to creating mass streamways and traveling hundreds of miles, have played a significant role in the history of colonial Guatemala. And Guatemalans have considered locusts to be significantly embedded in a wide range of colonial economic, political, and religious processes, processes that historians have deemed central to researching the history of colonialism in Latin America.

Colonialism was decisively shaped by locusts and by campaigns against the locust economically, politically, and socially. Aspects of Guatemalan culture that were touched by the locust include the production of agricultural goods for regional markets and for export to Europe. Swarms of this insect also forced periodic diversions of Indian tribute labor and other colonial wealth away from agricultural production to battle locusts and address the destructive aftermath of food shortages and famine. Locusts were linked to the emergence of epidemic diseases and periods of hunger or famine that contributed to massive Indian population decreases during the Conquest period, and that continued to afflict Spain's American colonies over the course of the colonial period. Finally, the arrival and disappearance of cycles of locust plagues affected the construction and maintenance of the colonial symbolic order organized around Christianity and the authority of colonial rule that pervaded the everyday life of local populations. And so locusts as agents not only were shaped by these economic, medical, labor, and religious processes, but also helped to shape them.

In this chapter I first analyze the role of locusts in the lived experiences of Guatemalans in the colonial period as informed by European and Mesoamerican exposure to the insect. I then examine the two prongs of colonial-era locust-killing efforts: Christian evangelical strategies and local, community-organized locust-killing campaigns. Finally, I consider the rise of increasingly uniform, state-directed campaigns of extermination over the course of the long eighteenth century. A new kind of professional exterminator led these campaigns—the Spanish man, usually a political officeholder in an agriculturally productive area, seen as having a particular experience and skill at killing locusts. The goal of these new state-directed, regional campaigns led by specialists was to exterminate locusts and clear them from Guatemala to ensure the continued success of European colonization there, especially the colony's model of economic development, and to simultaneously ensure the survival of colonial peoples, especially tributary Indians, the mainstay of the colonial labor force.

New World Locusts

When Guatemalans used the term *locust* to describe the swarming, devouring insects of colonial Central America, what exactly did they mean? Documents from the colonial period show two terms primarily used for locust in colonial Guatemala, *langosta* and *chapulín*, reflecting both Old World continuities and New World experiences with this insect. *Langosta* comes from the Latin *locusta* and was one of the sixteenth-century Spanish terms for the insect, along with *cigarra* in Spain due to the "cigarlike" shape of the insect's body.[17] Inhabitants on colonial Guatemala used the word *chapulín* (pl. *chapulines*) in written sources to describe both nonswarming grasshoppers and grasshoppers who, under certain conditions, swarmed to become locust plagues. *Chapulín* comes from the Nahuatl word *chapolin*, literally meaning "destroyer by the mouth" (*destructor por la boca*), or devourer.[18] Also in circulation in colonial Central America was the K'iche' Maya word *gag* to describe a locust, but this term was not as broadly utilized in archival documents as *langosta* and *chapulín*, the primary words in use at that time for naming the insect.[19]

Mesoamerican culture, like European culture, had its own memory of locust plagues and experiences with monstrous locust swarms even before the arrival of Europeans. In 1454 the Valley of Mexico, under the control of the Mexica empire, experienced a devastating succession of locust visitations that contributed to widespread famine, what Ross Hassig has called "the worst on record" for the Mexica.[20] Place names incorporated insects important to Mesoamerican cultures, such as the Nahuatl place name Chapultepec, which translates into English as "locust hill."[21] Mesoamericans have eaten chapulines from the ancient past to the present day. A common *chapulín* preparation begins with first boiling the insects, then frying them or toasting them on a *comal* (a cooking stone, also used to toast tortillas). Then they are eaten with lime and salt in tacos.[22] In case readers think that eating locusts is an exotic Mesoamerican tradition, the Old Testament also contains references to locusts eaten as food, as in Leviticus 11:22: "Of these you may eat every kind of great locust, every kind of longheaded locust, every kind of green locust, and every kind of desert locust."[23] In the New Testament, John the Baptist is described as eating locusts as evidence of his piety in Mark 1:6: "John was dressed in a rough coat of camel's hair, with a leather belt round his waist, and his food was locusts and wild honey."[24]

The Mesoamerican *chapulín* and the Spanish *langosta* were used either interchangeably or together (such as *langosta de chapulín* and *langosta que llaman chapulín*) to signify locusts throughout the colonial period and even into the twentieth century.[25] The Dominican friar Francisco Xímenez, in his monumental *Natural History of the Kingdom of Guatemala*, helpfully explains to his readers that "here they call *chapulín* what in Spain is called *langosta* and *cigarra*. The [insect] called *langosta* is the same one that causes harm to the countryside, [but] it is shorter [in length], and when this plague comes, everything is destroyed."[26] Most writers in colonial Central America and southern Mexico referred to locusts interchangeably as *chap-*

ulín and *langosta*, even as they acknowledged that the insect differed in size and other aspects from the European langosta.

Ximénez distinguished between the European and New World locust not only in terms of relative size, but also in terms of behavior. He noted that some locusts swarmed and caused widespread destruction to agricultural fields, and other locusts did not: "There are others who are small that do not cause damage, and others that are extremely small, look like mosquitoes, and jump around a lot."[27] In contrast, Thomas Gage, an English priest and traveler writing for a European audience in the seventeenth century about his experiences in Central America, collapsed New World and Old World locusts together as the universal, biblical locust of the Old Testament, whose presence indicated divine wrath: "The first year of abiding there [in Guatemala] it pleased God to send one of the plagues of Egypt to that country, which was of locusts, which I had never seen until then. They were after the manner of our grasshoppers, but somewhat bigger."[28]

But aside from these general descriptions, eyewitness reports of locust plagues contained curiously few details about the locust's appearance, its exact size, color, or the shape of its head, eyes, wings, and legs.[29] It seems that writers who warned of impending insect invasions or described anti-locust measures felt that detailed descriptions of the insect were unnecessary since the locust's presence was pervasive enough that both people born in the Americas and those born in Europe knew a locust when they saw one. This suggests a broad familiarity with locusts across colonial society, one that bridged the Old and New World peoples and cultures.

The locust was thus symbolically assimilated to the European outlook in a way that contrasts with other American species for which the Spanish had no European or classical cognates or equivalents. At the same time, locusts were not completely alien to Amerindian peoples as were some imported European species such as cows and horses. Under the sign of this apparent cross-cultural understanding of the insect, colonial accounts could focus instead on describing locust movement and locomotive behavior in detailed ways that showed an extensive experience with the insects during periods of infestation. For it was when locusts swarmed that they intruded themselves most visibly into the colonial worldview.

The 1734 *Real Academia Española* dictionary definition for *langosta* focuses on locomotion, describing the body parts that gave the locust its impressive ability to both jump and fly: "It has four wings, some above the other, and six legs, the four front legs small, and the two very long rear [legs], used for jumping. The tail is in the form of a stinger."[30] A young locust without developed wings was known in the colonial period as a *saltón*. This phase of the locust's lifecycle was distinguished mainly by the fact that it could move only by jumping. Yet even without wings, these young locusts by virtue of their sheer numbers could still cause extensive damage to plants and fields, as Gage recounted: "[The langosta] hung so thick upon the branches, that with their weight, they tore them from the body."[31]

Swarming adult locusts with mature wings, however, posed the real menace. These locust swarms traversed considerable distances and devastated farmlands. Modern entomological studies show that locust swarms can reach sizes of up to four hundred square miles and travel as far as three thousand miles from their point of origin.[32] In 1706 a colonial official warned his superiors in the capital city of Santiago de Guatemala that "growing numbers of flocks of locusts that they call chapulín" had arrived in the towns surrounding the city.[33] Eyewitnesses described locust swarms as seeming to follow royal roads and other roadways, which provided access for the transport of farm goods through the most fertile areas of agricultural production: "The langosta, commonly called chapulín, has been found in abundance very close to this city [Santiago de Guatemala], and already arrived in the lands of the towns of San Juan Amatitán and San Cristobal Amatitán, from where they have been killed and expelled and are now heading toward this city."[34] In August 1804 the colonial-era newspaper *Gazeta de Guatemala* warned of "profuse clouds of flying chapulín" working their way across the province of Suchitepéquez toward the capital.[35]

When locust swarms passed through an area, they flew into the eyes and covered the bodies of humans and animals, causing unsettling tactile experiences for people caught in their path. The English priest Gage wrote: "The highways were so covered with [locusts] that they startled the travelling mules with their fluttering about their head and feet. My eyes were often struck with their wings as I rode along. And much ado that I had to see my way, what with a Montero [a type of hunting cap] wherewith I was fain to cover my face, what with the flight of them which were still before my eyes."[36] Teeming clouds of locusts covered the fields, air, sun, ocean in an almost otherworldly way, according to Gage: "They [the locusts] did fly about in number so thick and infinite that they did truly cover the face of the sun, and hinder the shining forth of the beams of that bright planet."[37] Eyewitnesses testified that locusts' endless eating aggravated the horror caused by enormity of the swarms. Writing in the 1660s, the Dominican friar Antonio Molina observed that the locust swarm "satiated itself on the land."[38]

Locusts can thus be seen in part as a social phenomenon within the colonial process. Locusts traveled across a landscape transformed by the economics of colonialism and its system of cleared agricultural fields and roads, devouring export, provisionary, and tributary agricultural goods. As locusts swarmed and ate their way across colonial Guatemala during an infestation, they became historical agents in their own right, in much the way that the smallpox virus also became an important historical agent in the colonial Americas.[39] While I argue throughout this chapter that locusts acted as historical agents in colonial Guatemala, this does not mean that I attribute agency or sentience to their behavior. It does mean that agents of the colonial state and inhabitants of Guatemala had to react to the periodic locust swarms in identifiable and measurable ways. By destroying a wide range of crops, locust activity had the potential to cause poor harvests and played a significant role in shaping the development of future agricultural production. Locusts attacked and consumed crops grown on both small farms and large plantations, as Gage describes during a seventeenth-century infestation: "The farmers towards the south sea coast cried out for their indigo which was then in grass was like to be eaten up; from the Ingenios of sugar [sugar plantations], the like moan was made, that the young and tender sugar canes would be destroyed, but above all grievous was the cry of the husbandmen of the valley where I lived, who feared that their corn would be swallowed up by that devouring legion."[40] As Gage notes, the locusts repeatedly attacked some of the mainstays of the colonial economy: sugar and indigo, both key export goods.[41] Locusts also attacked corn, a Mesoamerican dietary staple, and wheat, an Iberian dietary staple. The historian William Cronon has linked increased population numbers of some insect species native to the New World with the arrival of European colonialism as they fed on colonial gardens and orchards.[42] While more research needs to be undertaken on this issue regarding locusts in the New World, it may also be the case that the changes in the cultivation of crops under Spanish colonialism, both food crops used to supply growing urban populations and export crops destined for European markets, encouraged the frequency of locust plagues and the size of the swarms in Guatemala.

Locust actions not only had direct consequences for the agricultural economy, but also affected the health of colonial populations. Locust plagues could reoccur over consecutive years, as did the widespread infestations of 1660–63 that hit the valley towns around the capital city of Santiago de Guatemala.[43] Such lengthy infestations across multiple growing seasons drove food prices higher, and precipitated shortages and even famine. Eyewitnesses to the 1660s locust infestations described how the insects had destroyed indigenous agricultural production there, especially the cornfields. This caused corn prices to double. *Corregidores* and *alcaldes mayores*, local and regional political officials, distributed corn and vegetables to Indian communities in stricken areas to prevent famine.[44]

Colonial authorities reveal a sense of desperation in the reports they filed concerning the widespread locust infestations, linking them not only to food shortages and the threat of famine, but also to epidemic disease outbreaks. Writing in July 1660, Martín Carlos de Mencos, president of the Audiencia of Guatemala, described the chaotic atmosphere of the start of his administration, caused in part by locusts.

> A year and a half since I have arrived [from Spain] and taken possession of this office [of president of the Audiencia of Guatemala], these provinces have suffered from three contagious illnesses, called here *pestes* (plagues; epidemics). And because of the devastation and mortality they have caused, it seems that one was smallpox and the other measles. These resulted in high mortality among the Indians, the majority of whom were very young. And more deaths occurred when some volcanoes exploded, hurling much fire and ash [into the air], causing significant damage to the towns closest to them, to a distance of twelve to fourteen *leguas* on all sides. Now these provinces are flooded with locusts, a difficult plague to treat.[45]

Antonio Molina, a Dominican friar and historical chronicler who also witnessed the locust plagues of the early 1660s, noted the simultaneous arrival of the locust swarm with a smallpox epidemic as well.[46] Others saw direct causal connections between locusts, famine, and epidemic disease in humans, as in this report from 1706: "The locust causes woeful widespread damages, such as famine. [The locusts] take over the land, destroying the fields and preventing sowing. Famine is usually followed by an epidemic [outbreak]."[47] One colonial official warned that even in death, locusts caused illness in humans as their rotting exoskeletons released disease-causing fumes: "I have discovered that [the locusts] have been carelessly left to rot. Their fetidness can cause putrid illnesses [*enfermedades pútridas*] in nearby towns, and can promote a general epidemic."[48] To avoid illness from the rotting insect bodies, he advised, community members were to bury the "dead *chapulín*."[49] An early nineteenth-century handbook on how to exterminate locusts warned its readers to beware of the "fetid exhalations of locust cadavers."[50]

This is a common observation after locust plagues—the enormous piles of dead locusts. Saint Jerome, in a commentary on the locust plague described in Amos, reported that in his time, the shores of the Mediterranean and Dead Sea "were filled with heaps of locusts that the waters had cast up, their stench and putrefaction was so noxious as to corrupt the air, so that a pestilence was produced among both beasts and men."[51] This associative link to sacred and medieval commentaries is important in the Spanish imperial context. A major goal that worked to justify colonialism and colonial expansion was the conversion of native peoples to Christianity, an ongoing project over the course of Spanish rule in Guatemala. There is also a long-established association between bad air and disease going back to Galen, and this association operated in colonial Guatemala as well. Galenic medical culture saw the body as controlled by humors, such that decay and bad air could penetrate the human body. Commonly known in Guatemala as *mal aire*, the connection between humoral conceptions of "bad airs" and their links to human illness interacted with Mesoamerican medical cultures that also considered certain kinds of noxious smells as disease-causing.[52]

Most likely the associations that colonial peoples made between locusts and disease in humans contributed to the extensive use of medical metaphors in the sources.[53] Reports from the field described locust infestations as epidemics, plagues, and contagion (e.g., *epidemia de langosta, plaga de langosta,* and *el contagio de dicho langosta*), and called for *remedios* (cures) to fight against their spread.[54] Juan López de Azpestia, an eyewitness to a locust plague in 1707, wrote that "the locusts have consumed so much, and little remains in the affected area, and because it is a mountainous [region], they could not apply the *remedio* (cure)."[55]

These remarks and others that characterized locusts as unsettling, devouring, alien, and disease-spreading provide a context for explaining and historicizing why humans put so much effort into killing locusts in colonial Central America. Locusts posed periodic threats to colonial Guatemala's economy, both to export products and to products for the internal market, such as domestic foodstuffs. These infestations also disturbed the equilibrium between colonial administrators and the governed, and formed one in a spectrum of threats to tributary Indian populations and the poor who had been rocked by the great epidemics that followed the first European–Indian encounters and that had continued through the colonial period.

Locust-induced famine and the perceived links of locust infestations to epidemics created anxiety at all social levels. Then there was the collective psychological repulsion at the sheer, unstoppable voraciousness of the seemingly endless number of insects, which may have found an outlet in a desire to wipe them out as enemies of both the sacred and social order. In 1797 officials of the city of Granada, Nicaragua, reported that a locust swarm there covered an area of twenty leguas (some sixty square miles) in and around the city, and that the insect's destruction of indigo, cacao, and food crops "threatened a terrible famine within a few months, which has unnerved the spirit of the residents."[56]

Early Colonial Locust-Killing Campaigns

Who killed locusts, and how did they kill them? Before the beginning of the eighteenth century, locust-killing campaigns tended to be local, community-organized events coordinated at the town level and made up of two intersecting strategies, the religious and the secular. Male religious leaders, through their parish churches, directed religious-based forms of locust extermination through prayers, processions, and fasts, and called for the assistance of specific saints thought to have particular locust-killing and -repelling abilities. Political leaders in individual communities organized their own citizenry—men, women, and children from afflicted towns, along with the labor of their domestic animals—to repel and kill locusts. Colonial officials' and residents' lived experiences with locusts had shown that both types of strategies needed to be used together to confound the swarms and kill the insects.

Specific saints could be called on to repel and kill locusts in colonial Central America. Known as "abogados contra la langosta" (advocates against the locust), these saints and their power were invoked by religious authorities and the Catholic faithful through prayer, processions, and other rituals.[57] Colonial officials, farmers, and their families turned to the supernatural because they recognized locusts as powerful adversaries that threatened more than the secular order. Locusts could not be defeated by earthly human power alone, because of their sheer numbers and the famine and epidemics that often accompanied their arrival. The justification for these strategies could be found in the Bible and the traditions of the Church as well.[58]

In 1685 Tomás Delgado de Najera, leading a locust-killing campaign in the town of Sacatepéquez, reported to Santiago de Guatemala's *cabildo* (town government) that he had used every strategy possible to kill the locust hoards over the first fifteen days of the infestation, but he lamented: "I have recognized that human forces are not enough, and because [the locusts] are so abundant, [and the people] are starving."[59] He then advocated calling on *los divinos auxilios* (divine aid) to help defeat the locusts.[60] In response, the cabildo consulted with the Bishop of Guatemala to organize a procession to pray for the destruction of the locusts by carrying the images of Nuestra Señora de la Merced (Our Lady of Mercy), San Agustín (Saint Augustine), and San Nicolás de Tolentino (Saint Nicolas of Tolentino) to the cathedral, then leaving the statues there for eight days.[61]

San Nicolás de Tolentino in particular has historically been regarded as the most important locust-killing saint in Guatemala.[62] In the 1640s residents of the primarily indigenous Maya towns of Mixco and Pinula organized a series of locust-repelling rituals that included special masses and processions of people carrying the images of San Nicolás de Tolentino and the Virgin Mary. Farmers and their families brought holy wafers stamped with an image of San Nicolás de Tolentino to the religious services; priests blessed the wafers, and the farmers took the wafers back to their *milpas* (cornfields), "some [farmers] casting them into their corn, some burying them in their hedges and fences, strongly trusting in Saint Nicholas [i.e., San Nicolás de Tolentino] that this bread would have the power to keep the locust out of their fields."[63] Historically, some non-Christian cultures have also used prayer and called on supernatural beings to halt or deflect locust swarms. Starting in the sixteenth century, the Chinese made offerings to locusts in response to infestations, and between 1500 and 1900 built some 870 locust temples to worship the insects and to carry out rituals to repel or lessen the plagues.[64]

Religious officials and the faithful in Guatemala continued to venerate San Nicolás de Tolentino as an important intercessor during locust plagues into the eighteenth century. San Agustín and the Virgin Mary also played key roles as antilocust patrons. In 1706 the president of the Audiencia reported that church officials organized a series of religious processions "to liberate this city from the locust and the plagues that they cause, praying for intercession from San Agustín and San Nicolás de Tolentino, advocates against the locust. And we will hold processions for three days, leaving from the San Agustín convent."[65] The costs for the antilocust religious rituals were borne first by local residents via alms donated by the faithful, supplemented when necessary by town government funds.[66]

Once the locusts had arrived, some priests and friars resorted to exorcism to expel them, as they would expel the demons that possessed a human subject.[67] The Dominican friar Antonio Molina, writing about a mid-seventeenth-century locust plague, described the goal of the rituals in this way: "They conducted [religious] processions and *rogativas* (ritual prayers) to exorcise (*conjurar*) the locusts. Bishop Payo de Rivera led the religious procession into the countryside to exorcize the locusts [from the fields]."[68] The verb *conjurar* in the colonial period was used to describe the formal Catholic ritual conducted by a priest to drive out demons using specific prayers.[69] Recourse to the language of exorcism, with its association with repelling or driving out demons, provides more evidence for just how locusts figured in the colonial semiotic space. Locusts possessed a double aspect as both natural entities and supernatural signs. In the latter aspect, they could not be defeated by mortal means alone.[70] Based on both tradition and an ordinary experience of the extraordinary power of the locust swarm, Guatemalans made alliances with powerful saints that lasted for decades.

Locust-killing campaigns in Guatemala utilized secular strategies in conjunction with religious strategies. Local colonial officials drew on the tributary labor system to organize Indian workers at the town level to kill locusts. This colonial labor force applied the available technologies to kill the locusts: fire to burn the insects and agricultural tools to crush and bury them. A focus on secular campaigns to kill locusts highlights the contradictions in the colonial dynamic, as Spanish administrators, who represented colonial authority and power, uprooted villagers and farmers to fight and kill locusts. Villagers became caught between locusts and the colonial state, represented by local authorities, who pooled community resources through cooperation and through coercive colonial labor institutions to kill the insects.[71]

Spaniards, in general, did not directly kill the locusts. Instead, Spanish colonial officials organized and managed subordinates—mostly Indians, both adults and children—to kill locusts. Colonial officials put Spaniards seen as experienced hands in charge of organizing and directing systematic locust killing and other antilocust measures during infestations. In 1685 Tomás Delgado reported to Santiago de Guatemala's town government that he had organized "all of the people from the surrounding towns, who have descended on the valley of Sacatepéquez" to kill locusts there.[72] And, once the locust plague arrived, residents used a variety of methods, including noise, to influence the swarms. Gage picturesquely describes local officials ordering Indians to the cornfields with their musical instruments to repel the invading insects: "The care of the magistrate was that the towns of

Indians should all go out into the fields with trumpets, and what other instruments they had to make a noise, and so affright them from those places which were most considerable and profitable to the commonwealth. And strange it was to see how the loud noise of the Indians and the foundling of the trumpets defended the fields from the fear and danger of them."[73]

The majority of firsthand accounts of antilocust measures focus on the arsenal of weapons—especially fire, agricultural implements, and suffocation—that were applied to the task of destroying the locusts quickly and efficiently. Eyewitnesses used the language of annihilation, measures that aimed to "kill and destroy the langosta," and described the killing event as a "massacre" (*matanza*).[74] Locust-killers targeted the young locusts with undeveloped wings as the cornerstone of the killing.[75] Residents drove nonflying locusts into pits, where they were crushed, burned, or buried alive. The historian and naturalist Francisco Ximénez explained that locusts had to be killed before they developed wings: "First [the locusts] eat up everything, later they begin to procreate their *hijuelos* (little children) and send them forth, and they cannot fly because they have not yet grown wings. And the [people] drive them so that they fall together into pits, and then burn them."[76] Gage, too, described the use of collective community labor to drive the young locusts into ditches, crush them with agricultural tools, and bury them alive in insect mass graves: "Where they lighted in the Mountains and High wayes, there they left behind them their young ones, which were found creeping upon the ground ready to threaten with a second years plagues if not prevented wherefore all the Towns were called with Spades, Mattocks and Shovels to dig long Trenches and therein bury all the young ones."[77] Guatemalans saw the young locusts as particularly threatening because they represented the possibility that they would swarm again in the near future.[78] One official warned, "This is an opportune moment to stop them by killing *la pequeña semilla* [lit., 'the little seed,' i.e., the young locusts] before they grow the wings that they need to fly and destroy the fields."[79]

By the late seventeenth century, the practice of using collective community labor, primarily that of Indian residents organized by local Spanish political and religious officials through elite Indian intermediaries, took hold as the operating model for locust-killing campaigns and continued throughout the colonial period. One morning at 6 AM during the 1707 locust plague, one friar García led his scribe, two ministers of the Indian governor, the mayors and other justices of the town of San Cristóbal Amatitán, and the town's resident Indians to a nearby hamlet called Pazón. Officials reported, "On a plot of corn there we discovered another large cluster of small chapulín, and the Indians startled them and drove them toward a trench and killed them."[80] Residents with sticks (*ramas*) crushed to death those young chapulines that fell into the ditches, and started fires to burn the adults who attempted to fly out of the field.[81]

This drafting of the Indian population is reminiscent of other labor mobilizations, for instance tributary Indian labor used in mines and on cacao, indigo, and sugar plantations. There are also resonances with earlier Spanish mobilizations of tributary Indians to participate in wars of conquest against other indigenous ethnic groups, and with colonial mobilizations of tributary Indians along with free and enslaved blacks to put down rebellions in other parts of the Audiencia, as in the wars of conquest against the Itzá Maya in the Petén at the beginning of the eighteenth century, and in the 1712 Maya rebellion in Chiapas.[82] Colonial political officials' enactment of a labor draft to combat the locusts rested firmly within established patterns of colonial social control developed by the Spanish to govern and exploit Indian populations in the New World.

Organized groups of residents who killed locusts not only protected their own fields, but also traveled to other nearby towns and hamlets to help as well. San Cristobal Amatitán had a particularly well-organized and efficient group of locust workers that included not only local Spanish and Indian colonial officials, but also adults and children over ten years old. They traveled one day to Pazón and another day to the hamlet of Pacún, both times leaving San Cristobal Amatitán at six in the morning with "the adults carr[ying] large hoes (*azadones*) and dry straw" used to crush and burn the chapulines.[83] Locust workers left early in the morning because experience had demonstrated that overnight the locusts "soaked up the dew of the atmosphere" and were largely at rest. The best times to kill locusts were at dawn, on rainy days, and on "humid nights of the full moon," as the moisture impeded their flying ability.[84]

Killing locusts could take up the entire day, with the implication that Indian laborers had to remain until those who officials who managed them allowed them to stop for the day. Juan López de Azpestia, working in and around San Cristobal Amatitán during the 1707 infestation, commented, "[We] worked hard [killing locusts] for more than twelve hours until we quit work for the day."[85] Friar García organized another group of Indian residents and led them to the hamlet Rincón Grande del Mattadero, where they were joined by *maestro del campo* Juan Lucas de Hurtarte and his group of residents. "We worked at killing [chapulines] all afternoon," López de Azpestia reported. After more than six hours, García, the priest, had everyone stop work for the day, and to rest to continue killing chapulines the next day."[86] Locust-killing labor thus took away from the other farming and family responsibilities of the Indians.

Eyewitnesses often interspersed descriptions of long days spent killing locusts with imagery of locust-filled graves. In the aftermath of locust-killing campaigns, masses of dead locust bodies carpeted the ground and floated on the surfaces of streams, lakes, and the ocean. The decay of these locust bodies filled the countryside with a pervasive and, it was thought, pestilential smell. Fray Antonio Molina wrote in the mid-seventeenth century about the attempt to forestall the noxious effect of the odor: "In the countryside they made great pits and graves (*sepulturas*) where they buried [the langosta] after they killed them."[87] Thomas Gage also described vast locust graves, but raised the depressing possibility of the resurrection of the locust foe, which could never be truly defeated and could return from the grave: "Thus with much trouble to the poor *Indians*, and their great pains (yet after much hurt and losse in many places) was that flying Pestilence chased away out of the Countrey to the South Sea, where it was thought to be consumed by the Ocean, and to have found a grave in the waters, whilst the young ones found it in the Land. Yet they were not all so buried, but that shortly some appeared, which not being so many in number as before were with the former diligence soon overcome."[88]

The consideration of both religious and secular campaigns to kill locusts suggests that Guatemalans saw locusts as an ambiguous species within the natural world. The metaphoric here locates the locusts in two spaces: on the one hand as a plague, and thus as nonhuman; and on the other hand as an enemy, and thus human. And if locusts, as well as people, could be exorcised, then in the Christian semiotic, the locust was not the opposite of human but instead had the humanlike quality of susceptibility to demonic possession. Locusts became enemies not just of man but also of trees, of corn, of other organisms that formed the colonial dominion. And locusts became enemies of Christianity as their repeated swarms questioned the Christianizing mission of Spanish colonial rule.

Colonialism and Locust Extermination in the Eighteenth Century

Entomologists and historians have tended to analyze insect-extermination campaigns as modern phenomena, framed by technological advances and the professionalization of entomology. As a result, historians date the beginning of sustained insect-extermination campaigns to the late nineteenth century and early twentieth. This is also the case for much of the literature on locust extermination.[89] But technological development is not the only way to periodize human relationships with insects. In this final section I examine the rise of increasingly centralized, uniform, state-directed extermination campaigns over the long eighteenth century in Guatemala. This shift did not rest on new technological innovation or scientific advancement. Instead, it corresponded to the changing needs of Spanish colonialism and colonial rule in the eighteenth century. For colonies like Guatemala to remain so in the Age of Revolution, political elites, both in Spain and Guatemala, turned to a reinvigoration of colonial export economies and the restructuring of colonial bureaucracies to rebind Spain's colonies to the imperium.[90] Changes in locust-killing campaigns can be viewed in this context. Community-level religious and secular strategies were absorbed into a new kind of locust-killing campaign that was increasingly regionally organized and directed at the Audiencia and archbishopric level, streamlined to promote a more uniform response to the threat of locust plagues across Guatemala. These regionally coordinated campaigns also began to include preventive elements, rather than simply reactive ones. This shift seems to stem from Guatemalan Audiencia officials and other colonial elites, however, rather than from any specific antilocust directives from Spain itself.

The beginning of the eighteenth century also signals the beginning of a new kind of locust expert—Spanish men who tended to be political officeholders in agriculturally productive areas and who reported to the president of the Audiencia.[91] These men had practical experience in local locust-killing campaigns in rural, primarily Maya towns. Colonial officials and local communities saw these men as having a particular skill for killing locusts, helping to lessen their effects, to deflect the insects from their area, and to prevent their reoccurrence. They also had personal and professional relationships with bilingual Indian elites in the towns of their jurisdiction, or they were bilingual themselves, and these connections led to their viability as locust experts. As officials slowly developed consistent, regionally applicable antilocust campaigns, they argued for the use of locust "experts." Evidence for the development of locust expertise can be seen in reports written from the field to political officials in the capital city. While officials felt that these experts and their antilocust strategies did not fully halt locust invasions, it seemed to them that they did help lessen the locusts' destruction during infestations.

The development of locust expertise did not suddenly emerge in the eighteenth century, but as this evidence shows, instead had roots in the earlier campaigns of the seventeenth century. In 1665 officials in Guatemala reported to royal authorities in Spain, "We have felt [the langosta plague] less in this capital [of Santiago de Guatemala] and towns of this valley and district [because] of the great care that the government took to prevent the langosta plague by organizing residents to repel them and kill them."[92] In 1707 the president of the Audiencia, Toribio de Cosio, warned that locust swarms had reached the nearby towns of San Juan Amatitán, San Cristobal Amatitán, and Petapa and threatened the valley towns in and around the capital Santiago de Guatemala.[93] Audiencia and town officials called on Tomás de Arrevillaga to go to Petapa to organize the town's residents to kill the locusts there. Arrevillaga held a number of political positions over the course of his career, including *alferez mayor* (militia commandant), *alcalde ordinario* (mayor) of Santiago de Guatemala, and *corregidor* (magistrate) of its valley, and had developed a reputation as a particularly seasoned veteran of antilocust campaigns, who used creative strategies to kill the insects. This was just the kind of person officials wanted to face the locust swarm in Petapa: "[Arrevillaga] acquired experience in the past through careful and persistent efforts and because of the zeal that he has shown for royal service and in [promoting] the public good."[94] Corregidores like Arrevillaga knew how to use their local knowledge as well as their connections with the Indian elites of the towns in their jurisdiction to organize locust-killing efforts. Officials told Arrevillaga to gather the Indian *justicias* of the towns threatened by locusts and inform them that they were to be "at his orders."[95] As in the Conquest period, the Spanish used Indian auxiliaries, coordinated via elite Indian intermediaries, to create an army, although this particular army was "battling" swarms of invading locusts.

The practice of using collective community labor at the local level that began in the seventeenth century continued to be the model for organizing locust-killing campaigns throughout the colonial period. What changed, however, is that this model became more formalized and began to be organized as an Audiencia-wide response, rather than as a series of responses of individual towns according to need. Workers continued to be Indians but they were now called by the formal title of *indios operarios* (Indian workers) and could receive a small salary for their labor. During this same 1706 locust plague, *oidor* (judge) Pedro de Ozaeta y Oro raised 1,175 *pesos* by personally going to the homes of prominent residents in the capital and asking them to contribute money to "set in motion the destruction of the langosta."[96] Some of these funds went to pay a small salary to the Indian workers. Officials were also encouraged to offer "prizes" or money to the locust workers who gathered the most locust egg sacs (*cañutillos*) once they were located in the surrounding countryside.[97] Presumably, however, locust killing remained undesirable labor despite the inclusion in some cases of small salaries.

Eighteenth-century campaigns also focused on identifying and forming ties to acculturated Indian elites in key agricultural towns to coordinate locust-killing campaigns. During a 1707 infestation, Audiencia officials named Juan López de Azpestia, also an alcalde ordinario and corregidor, to head to the primarily Indian town of San Cristobal Amatitán to organize residents to kill locusts there. In a letter dated 7 July 1707, he reported to the president of the Audiencia that he had arrived at the town of San Cristobal Amatitán at eight in the morning, and gathered together all of the Indian justices in the town, including the governor Don Cristobal Saguah, two mayors, five aldermen, and the town scribe. He assured the president that "all are *ladinos* in the Spanish language, and they are ready to obey your orders."[98] Colonial officials described Indians as *ladinos* or *indios ladinos* to indicate that they were acculturated Indians who spoke and understood Spanish, as well as one of the Maya languages and perhaps Nahuatl. Colonial officials also used ladino Indians to kill locusts in the town of San Juan Amatitán during the same locust infestation. Azpestia organized a troop consisting of the *primer alcalde* (first mayor) Don Francisco Chrisostomo, Domingo Alonzo, Chrisostomo's *compañero* (signifies a close ritual relationship), the chief constable Nicolás García, the town scribe Miguel de Ozma, seven of the town's Indian aldermen, and other Indian *principales* (elites). Colonial officials described all these Indian men as ladinos as well. These officials then organized other nonelite Indians in their communities, men, women, and children.[99]

How residents and operarios repelled or destroyed locusts also became more systematized and elaborate. No longer was using musical instruments or simply hacking at locusts with hoes enough. One strategy antilocust campaigns used was encircling and beating locusts to death. In this method, Spanish officials organized "as many indios operarios as necessary [to] encircle the infested terrain, barricading the insect[s] and driving them to the center. They beat them with poles, and then they burned the heaps of dead insects or buried them in trenches."[100] Coordinated groups of locust exterminators also lit fires in an attempt to catch the locusts in the flames. Officials cautioned that this method worked best when the wind blew in the right direction and only as long as the locusts swarmed far enough away from towns and homes to make the use of fire safe. The Indian recruits would first gather dry shrubs and underbrush, piling them into huge pyres of combustible debris that surrounded the locusts in the infested area. Then they lit the pyres so that the locusts in the middle would be incinerated, and those trying to fly away would be scorched.[101]

In the eighteenth century colonial officials began to recruit other human and nonhuman animals, in addition to Indian workers, as locust killers. Now "vagabonds," presumably coerced by local officials, were employed as workers to destroy locusts during infestations.[102] Colonial officials ordered residents in areas adjacent to infested areas to let loose their domesticated animals—pigs, flocks of sheep, or other animals—to eat the insects and trample and crush their egg sacs. To make sure that all the egg sacs had been destroyed, residents were then to plow the infested ground.[103]

Evidence for the centralization of locust extermination in the eighteenth century can also be seen in the systematic use by Spanish officials of local residents for regular reconnaissance in the countryside and agricultural fields, followed by inspections by locust experts. Colonial officials and residents under their control made a habit of conducting postswarm inspections to check for any signs of survival, as when, after the 1706 swarm, colonial officials, Indian elites and "all the Indians [from San Cristobal Amatitlán], adults and children more than ten years old" left town and walked along the Camino Real (Royal Road) toward the capital, inspecting fields along the way.[104] During the same infestation, colonial officials ordered the [Indian] governor and aldermen from San Cristobal Amatitán to gather the town's residents to "search for [the locusts] in the area." Chief Constable Azpestia notified authorities that he, along with his scribe José de Meza and other officials, planned to leave that day for the town of Bachia, located near San Cristobal Amatitán, to inspect a small cluster of chapulines that residents there had found and killed.[105]

In addition to programmatic and organizational changes, colonial officials and local scientists published handbooks to spread information on how to kill locusts in general, and in relation to the cultivation of specific crops. In 1799 José Mariano Moziño, a Mexican botanist then living in Guatemala, published a handbook for indigo growers designed to maximize production of the crop.[106] Over the course of the eighteenth century indigo became one of Guatemala's most dynamic export crops. Yet insects, caterpillars, and especially locusts repeatedly threatened its cultivation. Locusts attacked indigo farms in 1702, 1723, 1732, 1769, 1771, 1772–75, 1798, and 1799–1805.[107] As a warning to indigo growers, Moziño compared the relative destructive capabilities of locusts to termites based on the speed at which each insect could consume a tree: "We are accustomed to [thinking of] the termite (*el comején*) as one of the most destructive insects, who attack the roots and travel through the heart of the [tree] trunk and dry it out. The locust, commonly know as *chapulín*, can devour [the tree] in [just] a few hours."[108]

Starting in the eighteenth century, handbooks that focused exclusively on locust extermination were published in Guatemala and circulated there, among them José de Valle's "Instructions about the Locust Plague, and Methods to Exterminate It" (1804).[109] The target readership was "los magistrados y jueces subalternos," Spanish colonial officials and Indian town officials. These handbooks described the insect's reproduction and physical development, the most current killing methods, and how to organize Indian *operario* and domestic animal labor, and provided tips for the conservation of grains to prevent famine during infestations. The handbook also included new scientific information from Guatemalan scientists and growers in the eighteenth century, who successfully experimented with rice and *yuca dulce* (sweet cassava) as locust-resistant alternative food crops.[110]

Finally, in the eighteenth century, locust-killing handbooks and eyewitness accounts introduced the systematic use of the term *extermination* in reference to locusts, a word that we associate with mass insect killing today. The antilocust handbook that circulated in Guatemala contained the word (*exterminio*) in its title.[111] The same language can be found in field reports as well. In 1706 the president of the Audiencia called on alcalde ordinario Arrevillaga "to exterminate by the most effective method the locusts who have infested all kinds of *milpas* (agricultural fields) in the valley of this city."[112]

Even with more systematized Audiencia-directed campaigns, locusts indeed proved hard to kill off completely with the technology and manpower available to colonial Guatemalan society, which did not have recourse to a chemical arsenal of toxins or other modern-era methods. Throughout the colonial period, eyewitnesses noted the ever-present danger that the antilocust measures were unable to entirely eliminate, and they feared that the insects would emerge again later that year, or the following year. This experience with the impressive regenerative power of locusts underscored colonial policies that mandated that locusts be exterminated, which mandate, in the eighteenth century, was framed by Audiencia officials as a "public good" that the colonial government provided to the Guatemalan people. As one official reported, "I have ordered the [Indian] governor, alcaldes, and other justices in this pueblo of San Juan Amatitán to be sure to inspect the boundaries of this town to see if there are any langosta anywhere, so that then they can be killed for the public good."[113]

Despite all the locust killing and postkilling inspections, however, and even with the more organized and regionally implemented locust-extermination campaigns, Guatemalans could not completely eradicate the insects. Their efforts only temporarily diverted the swarms until the next infestation arrived. Thus, even by the end of the eighteenth century, written reports for areas after extermination campaigns include resurrection imagery: "The langosta de chapulín that have been born all over these lands have also [emerged] in Amatitán. They thought [that] they had killed [all of] them last year, but this is not the case."[114]

Conclusion: Locusts and Colonial Latin American History

Locusts proved ambiguous beings. Residents of colonial Guatemala saw the locust as a kind of supernatural messenger, evidence of divine disfavor and an apocalyptic symbol. Locusts also seemed to have special powers of survival that human intervention alone could not control. Moreover, locusts were not the right kind of animals, in that they could not be domesticated. Guatemalans could not contain them, prevent their movement, or harness their production as they could with the more orderly domestic animals: sheep, cattle, pigs, honeybees, and cochineal bugs. Nor, at least from the Spanish point of view, did locusts produce anything useful, though they did remain a food source for Amerindian peoples in the form of toasted chapulines.

In other ways locusts acted like humans—swarming and flying together across the land, acting as a kind of community, and eating human, not insect, food.[115] Mary Fissell has argued that early modern European conceptions of vermin as competitors for human food sources made it socially acceptable to kill them.[116] Locusts, however, did not confine themselves to human foodstuffs. They also ate lucrative colonial export products. And they depleted other colonial sources of wealth, in particular Indian labor. As locust swarms threatened colonial farmlands and plantations, tributary Indian labor activities had to be rechanneled to combat the insects. Hunger and famine frequently followed in their wake, weakening Indian residents and making them susceptible to epidemic illness. Thus in some ways it can be said that while swarming, locusts acted as a kind of conquering "army," though the goal was simply to eat and reproduce, not to establish a colony.

Colonial locust-extermination programs can thus be seen as rooted in imperial attempts to impose control and order over the environment and its animal populations, and as akin to efforts of imposing control over subject human populations in Guatemala. To keep Guatemala safe for Spanish colonialism, the Crown had to protect subject populations and maintain a healthy agricultural sector. This was carried out by local and regional elites who were experienced in locust-killing campaigns and who had a political or economic stake in the system. While more research needs to be conducted, Guatemala's locust-extermination campaigns suggest that colonialism is not just about controlling a geographic place and its subject peoples and extracting resources. It is also about controlling the environment and the species that populate that environment.[117] This includes harnessing productive species like domestic animals and productive insects, not only for labor and food, but also to kill or eat unproductive or threatening species like locusts. And colonialism also included killing locusts and other animals whose actions threatened agricultural production and public health in the colony.

Notes

I thank Roger Gathman, Zeb Tortorici, and Adam Warren for their insightful comments on earlier drafts of this chapter. I also received helpful feedback from the First Fridays Seminar in the History Department at the University of Arizona, and from the participants in a panel on animal histories held at the American Historical Association meeting in January 2011, especially from the commentator Harriet Ritvo. I have updated spelling and grammar in directly quoted Spanish-language materials and their translations.

1. For gypsy moths, see Robert J. Spear, *The Great Gypsy Moth War* (Amherst: University of Massachusetts Press, 2005). For the Mexican campaign against mosquitoes, see Marcos Cueto, "Appropriation and Resistance: Local Responses to Malaria Eradication in Mexico, 1955–1970," *Journal of Latin American Studies* 37 (2005): 533–59. For the medfly, see "Malathion and the Medfly," *Science News* 120.3 (18 July 1981): 35.

2. For an interesting study of the intertwined processes of scientific professionalization and insecticide use, see Charlotte Sleigh, "Inside Out: The Unsettling Nature of Insects," in *Insect Poetics*, ed. Eric C. Brown (Minneapolis: University of Minnesota Press, 2006), 281–97. Sleigh argues that in the early twentieth century, professional entomologists reworked insects as "modernist bogeys" whose presence required "professional intervention" ("Inside Out," 281–82).

3. Under Spanish colonial rule, this region fell under the jurisdiction of the Audiencia of Guatemala, with its capital first in Santiago de Guatemala, then, from 1773 on, in what is today Guatemala City. The Audiencia of Guatemala formed roughly the southern third of the viceroyalty of New Spain, with its capital in Mexico City. In this chapter, I use the terms "Guatemala" and "colonial Guatemala" to refer to geographic area encompassed by the Audiencia, which is much larger than the modern nation-state of Guatemala.

4. For locusts as part of agricultural history, see Edward Hyams, *Soil and Civilization* (New York: Harper Row, 1976). For locusts as part of environmental history, see William Cronon, ed., *Uncommon Ground: Toward Reinventing Nature* (New York: W. W. Norton, 1995). For an exception to this trend, see Jeffrey A. Lockwood's pathbreaking work *Locust: The Devastating Rise and Mysterious Disappearance of the Insect That Shaped the American Frontier* (New York: Basic Books, 2004).

5. Eric C. Brown, "Introduction: Reading the Insect," in *Insect Poetics*, ed. Eric C. Brown (Minneapolis: University of Minnesota Press, 2006), ix–xxiii. See also the online journal *Cultural Entomology Digest*, http://www.insects.org/ced, which published four issues devoted to insects between June 1993 and June 1997. Brown notes that works important to establishing the field of animal studies left insects out.

6. *Antennae*, http://www.antennae.org.uk, currently has three thematic issues that focus on insects.

7. May R. Berenbaum, *Bugs in the System: Insects and Their Impact on Human Affairs* (New York: Basic Books, 1995), 114.

8. Berenbaum, *Bugs in the System*, 114.

9. Olfat S. El-Mallakh and Rif S. El-Mallakh, "Insects of the Qur'an," *American Entomologist* 40.2 (1994): 82–84.

10. "And the locusts came up over all the land of Egypt, such a dense swarm of locusts as had never been before, nor ever shall be again. For they covered the face of the whole land, so that the land was darkened, and they ate all the plants in the land and all the fruit of the trees which the hail had left; not a green thing remained, neither tree nor plant of the field, through all the land of Egypt." Quoted in Lockwood, *Locust*, 20.

11. Quoted in Gene Kritsky and Ron Cherry, *Insect Mythology* (New York: Writer's Club Press, 2000), 76.

12. See E. P. Evans, *The Criminal Prosecution and Capital Punishment of Animals* (New York: E. P. Dutton, 1906). Evans refers to a fourteenth-century excommunication of locusts during an infestation in Tyrol, and another locust excommunication in the fifteenth century in northern Italy. See fn70 for works examining locust exorcisms in sixteenth-century Spain.

13. For historical changes in the use of the term *predator* and U.S. government policies regarding coyotes in the American West, see Thomas R. Dunlap, "American Wildlife Policy and Environmental Ideology: Poisoning Coyotes, 1939–1972," *Pacific Historical Review* 55.3 (August 1986): 345–69.

14. For example, Erica Fudge in her influential essay "A Left-Handed Blow" does not include insects as a distinct part of her call for activist animal-rights perspectives within the emerging field of human-animal studies. See Fudge, "A Left-Handed Blow: Writing the History of Animals," in *Representing Animals*, ed. Nigel Rothfels (Bloomington: University of Indiana Press, 2002), 3–18. Nor does the Animal Studies Group address insects in its consideration of animal slaughter through hunting, human food preparation, and in animal shelters in their influential *Killing Animals* (Urbana: University of Illinois Press, 2006).

15. An exception is the professional entomologist Jeffrey A. Lockwood, whose career has focused on grasshopper extermination in the American West. He considers these issues in his memoir, written as a series of linked essays. Lockwood likens his career as an entomologist developing and experimenting with pesticides used to kill grasshoppers in the West's rangelands to his father's career as a nuclear physicist, helping to develop and test atomic bombs in the Nevada desert. He argues that both pesticide and atomic-bomb development are "weapons of mass destruction in desolate places" (*Grasshopper Dreaming: Reflections on Killing and Loving* [Boston: Skinner House Books, 2002], 75). See also Jeffrey A. Lockwood, "Not to Harm a Fly: Our Ethical Obligations to Insects," *Between the Species: An Online Journal for the Study of Philosophy and Animals* 4.3 (1998): 204–11, http://digital commons.calpoly.edu/bts/vol4/iss3/12.

16. For more on methodologies for using travel writing as historical sources, see Mary Louise Pratt, *Imperial Eyes: Travel Writing and Transculturation* (New York: Routledge, 1992). See also Alfred Crosby's *Ecological Imperialism: The Biological Expansion of Europe, 900–1900* (New York: Cambridge University Press, 1986), about the European tendency to overemphasize New World species as monstrous or spectacular in comparison to the same species in Europe.

17. *Langosta* can also refer to the marine lobster. It is unclear if the term *cigarra* came to Spain with tobacco, a New World product, and so was a new term coming into currency in Spain. For example, the entry "cigarra" in *Diccionario de las Autoridades* notes that the name comes from the sound that grasshoppers made, "cig, cig" (*Diccionario de las Autoridades: Edición Facsímil* [Madrid: Editorial Gredos, 1964], 1:348). Locusts and cicadas may have been conflated in Spain in this period. Gene Kritsky notes that what settlers in eastern colonial America in the seventeenth century called locusts were really periodical cicadas ("Periodic Revolutions and the Early History of the 'Locust,'" *American Entomologist* 47.3 [2001]: 186–88).

18. *Diccionario histórico biográfico de Guatemala* (Guatemala City: Associación de Amigos del País, 2004), 258. There is an interesting transcultural resonance of locusts and devouring in Mesoamerican cultures, with the locust as a devouring apocalyptic symbol in Christianity, featured prominently in Revelation: "When the fifth seal is broken, a swarm of monstrous locusts arises from the bottomless pit, led by a king named Abaddon, or Devourer" (see Revelation 9:1–11).

19. "Nombre de pájaros en lengua quic[he]," ca. 1750, Newberry Library (NL), Ayer Manuscript 1587, f. 8.

20. Ross Hassig, *Time, History, and Belief in Aztec and Colonial Mexico* (Austin: University of Texas Press, 2001), 60. I thank Roger Gathman for pointing me to this source.

21. Raúl MacGregor-Loaeza, "Los insectos y las antiguas culturas mexicanas: Un ensayo etnoentomológico," *Revista de la Universidad de México* 29.6–7 (1975): 8–13.

22. Julieta Ramos-Elorduy, "Insectos comestibles," *Arqueología Mexicana* 6.35 (1999): 68–73.

23. Quoted in Gene Kritsky and Ron Cherry, *Insect Mythology* (New York: Writer's Club Press, 2000), 75.

24. Quoted in Kritsky and Cherry, *Insect Mythology*, 75.

25. See, for example, this government publication: Guatemala, Secretaría de Agricultura, *Destrucción de la langosta o chapulín* (Guatemala: Tipografía Nacional, 1935).

26. Francisco Ximénez, *Historia natural del Reino de Guatemala* (Guatemala City: Sociedad de Geografía e Historia de Guatemala, Editorial "José de Pineda Ibarra," [1722] 1967), 227.

27. Ximénez, *Historia natural del Reino de Guatemala*, 227.

28. Thomas Gage, *A New Survey of the West-India's or the English American His Travail* (London, [1648] 1655), 163. The Dominican priest Thomas Gage spent twelve years in Chiapas and Guatemala, from 1625 to 1637.

29. This stands in contrast to Brown's argument that humans see bugs as parts and segments. See his essay "Introduction: Reading the Insect," in *Insect Poetics*, ed. Eric C. Brown (Minneapolis: University of Minnesota Press, 2006), ix–xxiii.

30. See the *Real Academia Española* website, http://buscon.rae.es, entry "langosta."

31. Gage, *A New Survey of the West-India's*, 163.

32. Berenbaum, *Bugs in the System*, 111.

33. Archivo General de Indias (AGI), Guatemala 219, marzo 1706, "El oidor D. Pedro de Ozaeta y Oro acompaña certificación por donde parece que habiendose experimentado en diferentes territorios de aquellas provincias por marzo del año de 1706 epidemia de langosta," f. 1v.

34. Archivo General de Centro América (AGCA), A1.2.5-3099-29833, 1706, "Exposición del ayuntamiento de Guatemala elevada al superior gobierno sobre ayuda para el extermino de la langosta," f. 1.

35. NL, *Gazeta de Guatemala*, 1 October 1804, 455.

36. Gage, *A New Survey of the West-India's*, 163.

37. Gage, *A New Survey of the West-India's*, 163.

38. Antonio de Molina, *Antigua Guatemala: Memorias del M. R. P. maestro Fray Antonio de Molina continuadas y marginadas por Fray Agustín Cano y Fray Francisco Ximénez, de la Orden de Santo Domingo* (Guatemala City: Unión Tipográfica, 1943), 106.

39. See the introduction to this volume for more on our position on the issues of animal agency and posthumanism. For more on smallpox as a historical agent during the continent-wide epidemic during the 1770s to early 1780s, see Elizabeth Fenn, *Pox Americana: The Great Smallpox Epidemic of 1775–1782* (New York: Hill and Wang, 2001).

40. Gage, *A New Survey of the West-India's*, 163.

41. For corn and "planted fields of all kinds," see also AGI, Guatemala, carta 3 July 1665, oficiales reales a su Magestad, n.p. For more on the locust threat to indigo production, see José Mariano Mozino, *Tratado del xiquilite y añil de Guatemala* (Guatemala City: Ministerio de Educación, Dirección de Publicaciones, 1976), 66.

42. William Cronon, *Changes in the Land: Indians, Colonists, and the Ecology of New England* (New York: Hill and Wang, 1983), 154. Cited in Eric C. Brown, "Insects, Colonies, and Idealization in the Early Americas," *Utopian Studies* 13.2 (2002): 20–37, 34fn10.

43. AGI, Guatemala 47, carta 3 July 1665, oficiales reales, Guatemala, a su Magestad, n.p.

44. AGI, Guatemala 21, carta 25 February 1662, D. Pedro Frasson, Guatemala, a su Magestad, f. 1.

45. AGI, Guatemala 20, 25 July 1660, carta de D. Martín Carlos de Mencos a su Magestad, f. 1. Mencos was president of the Audiencia of Guatemala from 1659–88. A *legua* is a colonial unit of measurement of about three miles.

46. Molina, *Antigua Guatemala*, 106. Antonio Molina (1628–83) was a creole (born in the Americas) Dominican friar and chronicler.

47. AGCA, A1.2.5-3099-29833, 1706, "Exposición del ayuntamiento de Guatemala elevada al Superior Gobierno sobre ayuda para el extermino de la langosta," ff. 1–1v.

48. AGCA, A1.24-6091-55306 [n.d.], f. 118.

49. AGCA, A1.24-6091-55306 [n.d.], f. 118v.

50. José de Valle, *Instrucción sobre la plaga de langosta, medios de exterminarla, o de disminuir sus efectos, y de precaber la escasez de comestibles* (Nueva Guatemala: Ignacio Beteta, 1804), 4.

51. Cited by Thomas Findley in *Joel, Amos, Obadiah: An Exegetical Commentary* (Chicago: Moody Press, 1990), 611. I thank Roger Gathman for this reference.

52. For medieval theories, see J. N. Hays, *The Burdens of Disease: Epidemics and Human Response in Western History*, rev. edn. (New Brunswick: Rutgers University Press, 2010).

53. At this point I have not been able to locate sources that allow me to make a distinction, say, between Indian and Spanish metaphoric uses, or cultural distinctions among different social, racial, and ethnic groups.

54. For references to *epidemia de langosta*, *plaga de langosta*, and *el contagio de dicho langosta*, see, respectively, AGI, Guatemala 219, March 1706, "El oidor D. Pedro de Ozaeta y Oro acompaña certificación por donde parece que habiendose experimentado en diferentes territorios de aquellas provincias por marzo del año de 1706 epidemia de langosta," f. 1; AGCA, A1-1783-11777, Libro de Cabildo, 6 October 1665, ff. 113–113v; and AGI, Guatemala 219, March 1706 "El oidor D. Pedro de Ozaeta," f. 1v, and Molina, *Antigua Guatemala*, 106.

55. AGCA, A1.22.8-151-2955, "Providencias tomadas por el corregimiento del valle, capitán don Juan López de Azpestia para destruir la plaga de chapulín que ha invadido varios sectores del valle," 8 July 1707, n.p.

56. NL, *Gazeta de Guatemala*, 4 September 1791, 247.

57. AGCA, A1.2.5-3099-29833, "Exposición del ayuntamiento de Guatemala elevada al Superior Gobierno sobre ayuda para el extermino de la langosta," 1706, f. 1v.

58. Specific information regarding precolonial Mesoamerican responses to locust

infestations, aside from state-funded food distribution during fifteenth-century infestations in the Mexica empire, for now remains elusive in the sources.

59. AGCA, A1-1783-11777, Libro de Cabildos, 5 October 1685, ff. 111–112v.

60. AGCA, A1-1783-11777, Libro de Cabildos, 5 October 1685, ff. 111–112v.

61. AGCA, A1-1783-11777, Libro de Cabildos, 5 October 1685, ff. 111–112v. The Italian Nicholas of Tolentino, known as San Nicolás de Tolentino in Spanish, was born circa 1246 and died in 1305. He was a member of the Augustinian order and known as the "Patron of Holy Souls."

62. It is unclear whether San Nicolás de Tolentino's locust-killing and -repelling specialty extends past the Audiencia of Guatemala in the colonial period.

63. Gage, *A New Survey of the West-India's*, 164.

64. Shin-Yi Hsu, "The Cultural Ecology of the Locust in Traditional China," *Annals of the Association of American Geographers* 59.4 (December 1969): 731–52, 738.

65. AGCA, A1.2.5-3099-29833, ff. 1–1v.

66. AGCA, A1-1781-11781, Libro de cabildos, 2 March 1706, ff. 9v–10.

67. Alexandra Parma Cook and Noble David Cook describe locust exorcisms in the towns around Seville, Spain, in the late sixteenth century as part of the religious response to locust swarms there (*The Plague Files: Crisis Management in Sixteenth-Century Seville* [Baton Rouge: Louisiana State University Press, 2009], 21). For other accounts of locust exorcisms in sixteenth-century Spain, see also William A. Christian Jr., *Local Religion in Sixteenth-Century Spain* (Princeton: Princeton University Press, 1989); and Jody Bilinkoff, *The Avila of Santa Teresa: Religious Reform in a Sixteenth-Century City* (Ithaca: Cornell University Press, 1989). Thanks to David Cook and Marcy Norton for referring me to these works. For a fascinating and detailed analysis of a locust trial in an ecclesiastical court in seventeenth-century Spain, see Georgina Dopico Black, "The Ban and the Bull: Cultural Studies, Animal Studies, and Spain," *Journal of Spanish Cultural Studies* 11.3–4 (2010): 235–49. Thanks to Zeb Tortorici for bringing this article to my attention.

68. Molina, *Antigua Guatemala*, 106.

69. *Diccionario de las autoridades*, entry "conjurar," 516.

70. Church tradition preserves strong imagery connecting locusts to the diabolical, as in Revelations 9:1–12, where locusts emerged out of the bottomless pit, portending the end of the world.

71. Cook and Cook describe a locust-killing campaign in late sixteenth-century Seville, Spain, where city authorities recruited community members to collect tubes of locust eggs (*cañutillos*) by offering to pay one *real* for each *almud* collected (an almud equaled roughly 27.75 liters). Officials would then burn the eggs. If an inadequate number of paid volunteers offered to perform this labor, city officials would compel each of Seville's parishes to contribute a certain number of locust-egg collectors. See Cook and Cook, *The Plague Files*, 19–21.

72. AGCA, A1-1783-11777, Libro de cabildo, 5 October 1685, f. 112.

73. Gage, *A New Survey of the West-India's*, 163.

74. Respectively, AGI, Guatemala 219, March 1706, "El oidor D. Pedro de Ozaeta," ff. 1–1v; AGCA, A1-1781-11781, Libro de cabildos, 2 March 1706, ff. 9v–10.

75. Gage, *A New Survey of the West-India's*, 163.

76. Ximénez, *Historia natural del Reino de Guatemala*, 227.

77. Gage, *A New Survey of the West-India's*, 163.

78. The longest locust plague I have found documented lasted three years (1660–62), but there may be other, longer-lasting infestations. AGI, Guatemala, carta 3 July 1665, oficiales reales a su Magestad.

79. AGCA, A1.2.5-1772-25239, 1772, f. 3

80. AGCA, A1.22.8-151-2955, 1707, n.p., letter dated 8 July 1707.

81. AGCA, A1.22.8-151-2955, 1707, n.p., letter dated 8 July 1707.

82. See Kevin Gosner, *Soldiers of the Virgin: The Moral Economy of a Colonial Maya Rebellion* (Tucson: University of Arizona Press, 2002); and Grant D. Jones, *The Conquest of the Last Maya Kingdom* (Stanford: Stanford University Press, 1998).

83. AGCA, A1.22.8-151-2955, 1707, n.p., letter dated 9 July 1707; Valle, *Instrucción sobre la plaga de langosta*, 4.

84. Valle, *Instrucción sobre la plaga de langosta*, 5.

85. AGCA, A1.22.8-151-2955, 1707, n.p., "Providencias tomadas por el corregimiento del valle," n.p., letter dated 9 July 1707.

86. AGCA, A1.22.8-151-2955, 1707, n.p., "Providencias tomadas por el corregimiento del valle," n.p., letter dated 9 July 1707.

87. Molina, *Antigua Guatemala*.

88. Gage, *A New Survey of the West-Indians*, 163, emphasis in original.

89. See Lockwood, *Locust*. Lockwood, an entomologist working in the American West, dates the first historical era of locust extermination to the Industrial Revolution and the introduction of mechanization to farming. The Riley Locust Catcher, developed in 1878, is an example of technological innovation applied to kill locusts. See David Serlin, "Days of the Locust: An Interview with Jeffrey Lockwood," *Cabinet* 25 (spring 2007): 57–61. Another aspect of the technological transformations in insect extermination came out of the application of chemistry to human warfare during the First and Second World Wars. Military historians have noted the related developments of the science and technology of chemical warfare and nerve gases with synthetic insecticides, such as DDT (dichlorodiphenyltrichloroethane). For more on this, see Edmund P. Russell, "'Speaking of Annihilation': Mobilizing for War against Human and Insect Enemies, 1914–1945," *Journal of American History* 82.4 (March 1996): 1505–29. Starting in the late twentieth century, scientists began using satellite and remote-sensing technologies to identify the emergence of locust swarms. This technology could then be used in conjunction with "natural" pathogens known as bioinsecticides such as *Nosema locustae*, placed in bait, for locust extermination. For more on this, see Jeffrey A. Lockwood, Charles R. Bomar, and Al B. Ewen, "The History of Biological Control with *Nosema Locustae*: Lessons for Locust Management," *Insect Science and Its Application* 19 (1999): 333–50. See also Berenbaum, *Bugs in the System*,

204; D. A. Streett, "Grasshoppers: Their Biology, Identification, and Management," United States Department of Agriculture Agricultural Research Service, n.d., http://www.sidney.ars.usda.gov/grasshopper/Handbook/I/i_2.htm; and Environmental Protection Agency, *Nosema locustae* fact sheet, http://www.epa.gov/pesticides/chem.../red_PC-117001_1-Sep-92.pdf.

90. These eighteenth-century political and economic reforms were collectively known as the Bourbon Reforms. Changes in locust-killing campaigns that concern this chapter did not come directly out of the Bourbon Reforms, but certainly fit within its rubric and goals. For more on the Age of Revolution (ca. 1760–1830) and threats to absolutist European powers and their colonies, see, for example, Eric Hobsbawm, *The Age of Revolution: 1789–1848* (New York: Vintage, 1999); Catherine Macaulay and Mercy Otis Warren, *The Revolutionary Atlantic and the Politics of Gender* (Oxford: Oxford University Press, 2005); and Cathy Matson, "The Atlantic Economy in an Era of Revolutions: An Introduction," *William and Mary Quarterly* (July 2005): 357–64.

91. The viceroyalty of New Spain was subdivided into smaller administrative units that included the Audiencia of Guatemala. A Crown-appointed Audiencia judge, who also held the title of president and captain general, administered it.

92. AGI, Guatemala, carta 3 July 1665, oficiales reales a su Magestad.

93. AGCA, A1.22.8-151-2955, 1707, n.p.

94. AGCA, A1.22.8-151-2955, 1707, n.p.

95. AGCA, A1.22.8-151-2955, 1707, n.p.

96. AGI, Guatemala 219, March 1706, "El oidor D. Pedro de Ozaeta y Oro acompaña certificación por donde parece que habiendose experimentado en diferentes territorios de aquellas provincias por marzo del año de 1706 epidemia de langosta," ff. 1–1v.

97. Valle, *Instrucción sobre la plaga de langosta*, 3, 4. In *The Plague Files* Cook and Cook note that during a late sixteenth-century locust plague in Seville, Spain, officials there paid residents to collect locust egg sacs, which would then be burned (19–20).

98. AGCA, A1.22.8-151-2955, 1707, n.p.

99. AGCA, A1.22.8-151-2955, 1707, n.p., letter dated 15 July 1707.

100. Valle, *Instrucción sobre la plaga de langosta*, 4–5.

101. Valle, *Instrucción sobre la plaga de langosta*, 4.

102. Valle, *Instrucción sobre la plaga de langosta*, 7.

103. Valle, *Instrucción sobre la plaga de langosta*, 3; 4.

104. AGCA, A1.22.8-151-2955, 1707, n.p., letter dated 15 July 1707.

105. All information in this paragraph on the locust plague of 1707 comes from AGCA, A1.22.8-151-2955, 1707, n.p.

106. John Carter Brown Library, Brown University (JCBL), José Mariano Moziño, *Tratado del xiquilite y añil de Guatemala, con notas puestas por el socio mencionado Dr. Fr. José Antonio Goycoechea* (Guatemala: Ignacio Betata, 1799). Moziño was a naturalist in the Real Expedición de Nueva España.

107. Horacio Cabezas Carcache, "Producción Agropecuaria," in *Historia general de Guatemala*, ed. Jorge Luján Muñoz (Guatemala City: Asociación de Amigos del País, Fundación para la Cultura y el Desarrollo, 1995), 3:294–303, 3:295.

108. Moziño, *Tratado del xiquilite y añil de Guatemala*, 66.

109. Valle, *Instrucción sobre la plaga de langosta*, title page.

110. Valle, *Instrucción sobre la plaga de langosta*, 3.

111. AGCA, A1.2.5-3099-29833, 1706, n.p.; and AGCA, A1.24-6091-55306, n.d., f. 118.

112. AGI, Guatemala 219, March 1706, "El oidor D. Pedro de Ozaeta y Oro acompaña certificación por donde parece que habiendose experimentado en diferentes territorios de aquellas provincias por marzo del año de 1706 epidemia de langosta," ff. 1–1v.

113. AGCA, A1.22.8-151-2955, 1707, n.p., letter dated 15 July 1707.

114. AGCA, A1.2.5-1772-25239, 1772, "Cartas dando informes sobre que en jurisdicción de Amatitán ha aprecido langosta," f. 1.

115. Mary Fissell, "Imagining Vermin in Early Modern England," *History Workshop Journal* 47 (1999): 1–29.

116. Though Fissell considered animal vermin and not insects in her essay, locusts fit the conception of vermin as well.

117. This can be seen in other colonial and postcolonial settings as well, a point made especially in environmental histories. See, for example, Alfred W. Crosby, *The Columbian Exchange: Biological and Cultural Consequences of 1492* (Westport: Greenwood Press, 1972), especially chap. 3, "Old World Plants and Animals in the New World"; Alfred W. Crosby, *Ecological Imperialism: The Biological Expansion of Europe, 900–1900* (New York: Cambridge University Press, 1986); Elinor G. K. Melville, *A Plague of Sheep: Environmental Consequences of the Conquest of Mexico* (Cambridge: Cambridge University Press, 1994); Andrew Sluyter, "The Ecological Origins and Consequences of Cattle Ranching in Sixteenth-Century New Spain," *Geographical Review* 86.2 (1996): 161–77; Jon T. Coleman, *Vicious: Wolves and Men in America* (New Haven: Yale University Press, 2004); and Lockwood, *Locust*. Historical connections between specific animals and state formation can benefit from a longue durée approach, as in J. R. McNeill's fascinating work *Mosquito Empires: Ecology and War in the Greater Caribbean, 1620–1914* (Cambridge: Cambridge University Press, 2010).

3

"In the Name of the Father and the Mother of All Dogs"

Canine Baptisms, Weddings, and Funerals in Bourbon Mexico

ZEB TORTORICI

In 1770 an odd festive occurrence involving two dogs caught the attention of the Mexican Inquisition due to its heretical nature and accompanying sacramental desecration.[1] According to the voluntary denunciation of twenty-eight-year-old don Juan Antonio López de la Paliza, and those of other witnesses gathered by the Inquisition, in a party given in a house on San Ramón Street in Mexico City owned by a woman known as La Panchita and her husband, don Francisco, the following prank occurred: "Many people of both sexes and all classes [frequented] this profane, sacrilegious, scandalous, and superstitious function of dance and diversion in which a marriage between [two] dogs was performed with the utmost formality. . . . Disorder was precipitated through the abominable excess of solemnizing the said matrimony with such seriousness and formality *as if it were contracted between Christians and able persons.*"[2] Oddly enough, the priest responsible for this profane marriage was genuine. His name was Thoribio Basterrechea, a twenty-six-year-old priest who had studied theology in Mexico City at the colleges of San Pablo and San Pedro, and he functioned as a regular priest and confessor in a church in the town of Guachinango. According to witnesses, in the midst of this drunken party, the priest was presented with a little dog (*una perrita*) dressed as the bride and another dog (*un perrito*) dressed as the bridegroom. They were then "mutually asked about their consent to wed [by Basterrechea], and consent was given by those present through the responses of those that held the dogs in their hands [a man and a woman who were named the dogs' godparents]. This defendant [Basterrechea] then solemnized and formalized the said wedding between the two little animals [*animalejos*]."[3]

According to everyone questioned by inquisitors, this canine wedding took place in the midst of jokes, drinking, and music. Numerous witnesses made reference to the lively dances and fandangos in which partygoers participated both before and after the dogs were wed.[4] One witness, Mariano Correa, related that he saw the dogs dressed up and placed in a miniature matrimonial bed before the fandango began.[5] However, inquisitors were suspicious that the jocose nature of the marriage disguised more profound levels of heresy and sacrilege, and were in particular interested in uncovering whether or not the sacrament of marriage had been formally profaned. To this end, the inquisitors interrogated numerous witnesses about the priest and the motives of those who had attended the mock wedding, yet they always received similar responses. According to one Spaniard, don Joseph González, the wedding was performed as a "joke and as entertainment" (*bulla y diversión*) because they had done nothing fun on his son's birthday a few days earlier. However, González unequivocally asserted that nothing was done to desecrate the religious sacraments and, in a statement corroborated by the other testimonies, that the wedding was seen by all "merely as a joke and pastime" (*sólo por chiste y pasatiempo*).[6]

Despite assertions that the canine marriage was altogether a joke, the Inquisition proceeded with the case as though it were a grave matter and transferred Father Basterrechea to the secret prisons of the Inquisition for further interrogation.[7] There he remained for the duration of his trial, during which time he was threatened with instruments of torture—though never actually tortured—in order to make him confess his crimes. In the initial denunciation that brought this potentially heretical case to the attention of the Mexican Inquisition, López de la Paliza specifically noted that in the dog marriage ceremony the priest observed many of the formalities that marked a marriage between two humans. Two canine *padrinos*, or godparents, were named—a woman named doña Juana and Tranquilo, the eighteen-year-old son of the household servant—and they held the dogs and, speaking for them, gave the dogs' consent to be wed. What most perturbed ecclesiastical authorities about Basterrechea, however, was the denouncer's claim that the priest had married the dogs *in the name of the Father, the Mother, and the Son*, a phrase disconcertingly similar to the phrase "in the name of the Father, the Son, and the Holy Spirit" regularly professed by priests at Mass and during the consecration of marriage.[8] This did smack of heresy, or at least of blasphemy.

As was the procedure with all individuals charged with heresy and brought before the Inquisition for questioning, the inquisitors informed Basterrechea neither of the charges against him nor of the person who had accused him. When asked during the interrogation if he knew why he had been imprisoned in the secret jails of the Inquisition, he replied that he did not know. Only later, when questioned again, Basterrechea asserted that it might have been because of his role and participation in the marriage of two dogs (*un casamiento de dos perritos*); however, according to him, this had been done without any heretical intention, but had merely been a form of diversion (*diversión*); if any disrespect had thereby been shown to the holy sacrament of marriage, it had been done inadvertently. He defended his actions by pointing out that during the mock wedding he did not don clerical attire, and he had avoided uttering the phrase "in the name of the Father, the Son, and the Holy Spirit." Instead, Basterrechea had jokingly married the dogs "in the name of the Father and the Mother of all dogs" (*en el nombre del Padre, y de la Madre de los perros os caso*), which, it was true, may have been misheard by those present at the drunken festivities.[9]

Inquisitors scrutinized the syntax and word choice of Basterrechea at the exact moment he solemnized the marriage between two dogs in order to see if he had breached the line that held between abjuration *de vehementi* (of strong suspicion of heresy) and abjuration *de levi* (of slight suspicion). Had he consecrated the marriage of the two dogs "in the name of the Father, the Son, and the Holy Spirit" as opposed to "in the name of the Father and the Mother of all dogs," the punishment he merited would have been more severe and likely would have included fines, spiritual penance, and stern admonishments. In his defense, four witnesses, when called again for questioning, affirmed that he had indeed said the latter. Of special interest here is that inquisitors appear to have been so concerned with Basterrechea's statements, confirming Stuart Schwartz's conclusions that inquisitors tended to focus much more on statements than on acts or internal belief structures, because "an element of heretical content could be found in any statements or thoughts that contested dogma, even when they were uttered in anger, ignorance, or jest."[10] Basterrechea's statements having been clearly uttered in jest, what made all the difference for Basterrechea was the semantic and linguistic line between the species—a line between the Father and Mother of all humans and the Father and Mother of all canines. In the end, after weeks in prison, Basterrechea was simply admonished for his actions and warned that he would be severely punished in the event of future sacramental misbehavior. A contrite Basterrechea acknowledged his errors, and inquisitors found him not guilty of heresy, which comes as no surprise given that "heresy lay not necessarily in the doubting of dogma or in statements at variance with the Church's position, but rather in the refusal to accept correction and in the stubborn persistence of error."[11]

Animals, Sacramental Desecration, and Changing Sensibilities

Throughout early modern Europe, inquisitorial and ecclesiastical author-ities were adamant about keeping religious sacraments undefiled by ani-mals. That dogs were especially deprecatory to the sacraments in the early Church can be gleaned from a number of historical contexts. The early Irish penitentials, for example, stipulated that inadvertently vomiting the Eucha-rist could require up to forty days of penance and that penance for this act would be increased to one hundred days if a dog were to lick up the vomit containing the Holy Eucharist.[12] E. P. Evans in his phenomenal book on medieval ecclesiastical and criminal trials against insects and animals, *The Criminal Prosecution and Capital Punishment of Animals*, mentions one case from 1394 in which authorities hanged a pig at Montaigne, France, for having sacrilegiously eaten consecrated Eucharist.[13] The Inquisition histo-rian Henry Charles Lea cites a 1460 French Inquisition case in which a man supposedly baptized a toad, fed it with the consecrated Host, and then killed the toad, using it to poison an entire family.[14] Adding to this bizarre list of examples, David Cressy writes about a number of instances in which dogs, cats, horses, pigs and calves were baptized and christened, and also touches on a "profane marriage of a goose and a gander" in Tudor and Stuart En-gland.[15] Though he asserts that "baptisms of beasts are puzzling phenom-ena, and it is by no means clear how they fit into our larger understanding of history," he advances our comprehension of such mock ceremonies by look-ing closely at historical context, religious sectarianism, spontaneous the-atricality, the production of rituals and counterrituals, and the role that alcohol may have played in irreligious acts.[16] These diverse chronological and geographical examples from the early modern European context help to frame the theological context in which the Church viewed animals mixing with holy sacraments and sacrosanct rituals as defiling and contaminating, and therefore worthy of punishment.

The Church's interest in stamping out heresy and superstitious beliefs, as evidenced by the founding of medieval inquisitorial tribunals between 1184 and the 1230s, partially signals the penetration of the Church into popular religious culture, practices, and beliefs. In late medieval Europe, the Church grew more aggressive in asserting its monopoly over the sacred, persecuting the "superstition" of the populace—the use of holy water on crops, for example. In the context of Spain, Ferdinand II of Aragon and Isabella I of Castile set up the Spanish Inquisition in 1478 with the approval of Pope Sixtus IV. The Inquisition operated in Spain and in its colonies and territorial possessions, at least until independent tribunals were established in Lima (1569), Mexico City (1570), and Cartagena de Indias (1610). Creating a backdrop for the transatlantic context of the suppression of heresy and superstition was the Protestant Reformation and the changes instated by the Council of Trent (1545–63). In the context of a radically changed religious climate and challenges to Catholic orthodoxy, the tribunals of the Inquisition in the Americas sought to regulate popular and folk religious beliefs and practices throughout Spain's colonies. Even in the New World, reports of animal baptisms and weddings, though rare, can be found. One newly converted New Christian named Fernão Pires, for example, was denounced in late sixteenth-century Bahia during the Portuguese Inquisition's first visitation to Brazil for having baptized dogs and given them Christian names.[17]

The scattered reports of animal baptisms and weddings in parts of Catholic and Protestant Europe and in colonial Latin America demonstrate that the colonial Mexican cases analyzed here are not completely isolated cases, and our scholarly interest in them is not a matter of digging up curiosities, but of bringing into focus a very thin but persistent parameter from which we can get a better sense of human-animal interactions in the transatlantic world of the time. These exemplary cases mark broader, less visible trends on the part of Mexican religious authorities in determining and justifying the subordinate place of nonhuman animals within a sacred geography that was determined by the Church's rules about orthodoxy, heresy, and the uniqueness of the human soul. Even as late as the end of the eighteenth century in Mexico, the mixing of animals and sacraments still elicited the charge of heresy among Church officials and inquisitors.

The Catholic Church's official abhorrence of actions and beliefs that hinted at animals having souls and that, as a consequence, recognized their right to partake of sacraments tells only part of the long and contested history of Christian religious beliefs regarding animals and their proper place in relation to humans.[18] From around 900 AD running through the thirteenth century, there were persistent reports of a Christian sect with strong Gnostic elements, called the Bogomils (in Bosnia) and the Cathars (in southern France and northern Spain), who defined their relationships with animals in terms that were different than what eventually became hegemonic Christian theology. Members of these groups espoused vegetarianism on the ground that killing an animal was the same as killing a man, as animal souls were likewise immortal.[19] Saint Augustine (354–430) and, centuries later, Thomas Aquinas (1225–74) among others disdainfully rejected the Manichaean belief that man had no right to kill other animals, echoing the Old Testament suspicion that the refusal to kill animals could be a sign of animal worship or idolatry. Thus, the moral relationship between animals and humans was read under the sign of the relationship between Christianity and other religions—a template that was already in place when the Spanish arrived in the Americas and that determined the reactions of many Spanish priests and missionaries on encountering Mesoamerican religions in which "idolatry" was seen everywhere, alongside the supposed deification of certain Mesoamerican animals.

Many central early Christian figures, however, including Saint Jerome, Saint Aventine, Saint Anselm, and Saint Isidore, emphasized compassion toward animals.[20] Running counter to the tendency of the Church fathers to conflate the kindly treatment of animals with worship of them was the valorization of compassion itself, which for some medieval Christians, most notably Saint Francis, was extended to beasts. This theme emphasizes another Old Testament narrative, that of the peaceful companionship of all living creatures in the Garden of Eden. It was largely due to Saint Francis—who was proclaimed by Pope John Paul II as the patron saint of ecology on Easter Sunday, 1980—and his positive reconceptualization of nature and animals in the thirteenth century that some of the clergy came to perceive animals as worthy of sacramental recognition and blessings.[21] That animals, largely household pets, today are regularly brought into public spaces and sometimes churches to be blessed with holy water by priests around Easter and on 4 October, the feast day of Saint Francis, in cities around the world potentially dates back to the sixteenth century, when farmers brought their animals to churches to be blessed for fertility and health.[22]

Ultimately, the ambivalent traditions governing interactions between animals and Christian sacraments in the early modern European context provide the framework within which ecclesiastical authority in late colonial Mexico perceived and reacted to canine weddings. This discussion of human-animal interactions in the context of sacraments and sacrilege, however, only shows part of the cultural and religious backdrop against which, as outlined by Keith Thomas in *Man and the Natural World*, a significant change in sensibilities toward animals occurred in the early modern European context and spread across the Atlantic as colonists arrived in the

Americas. Thomas notes that for the most part in early modern England, the "working dogs" commonly used in war, transport, farming, meat production, and in the protection of sheep and households "seem to have been regarded unsentimentally."[23] It was, however, in the sixteenth and seventeenth centuries that pets—the category comprising certain "privileged species," including dogs, cats, and birds—"really established themselves as a normal feature of the middle-class household, especially in the towns, where animals were less likely to be functional necessities and where an increasing number of people could afford to support creatures lacking any productive value."[24] While the historical sources on which this paper is based come from a very different cultural and historical context, Spain's colonies in the Americas were also affected by the sensibilities of the bourgeoisie in Europe and by the humanism that expressed itself in civil usage of and bonds of affection with "pets."

Baptisms of Dogs (and Dolls) in Mexico

As unusual as the 1770 canine marriage case might seem to us, don Joseph González, a thirty-one-year-old Spaniard residing in Mexico City, disconcertingly assured inquisitors that weddings between dogs "were a very common thing in Mexico" and then related how a woman named doña María Agustín had spent a considerable amount of money buying a bellyband (*un hombliguero*) and other ornaments for a dog that was to be wed with another.[25] This aspect of the canine weddings in Mexico shares something in common with later French fin-de-siècle canine weddings in which, according to Alfred Bonnardot's *Des petits chiens de dames* (1856), one man in attendance reported that it was customary for the "husband to present a marvelous set of wedding presents, for everyone to lunch, then for guests to leave the [canine] newlyweds 'finally alone.'"[26] Though it is impossible to determine the frequency of dog weddings in colonial Mexico, we do find other cases in the historical archives in which animals were sacrilegiously and mockingly granted the sacraments.

In 1770, in an attempt to cheer up the young daughter of don Balthasar Mendieta, who had fallen ill, don Antonio Balbuena and others held a large party where they performed the baptism of two newborn puppies that were the daughter's pets.[27] Nearly a year later, this "scandalously jocose" event was brought to the attention of the Mexican Inquisition in a denunciation that named Balbuena (who had baptized the dogs), doña María Agustina Parteiro (who had played godmother of the dogs), and Padre Anjoso (a Franciscan friar who had lent Balbuena his priestly garb). Inquisitors, who interrogated some twenty witnesses, quickly seized on the comparison between the baptism of an "irrational animal such as a little dog" (*animal yrrasional como perrito*) to those of dolls, inanimate objects, and images in order to serve blasphemous and heretical ends.[28] One Spaniard present at the festivities noted that Antonio Balbuena had dressed up like a cleric, had tossed water from a baptismal font upon the small dogs during the fandango, and, in a phrase reminiscent of Basterrechea's lexical play in the 1770 canine wedding, had announced that they had been baptized "in the name of the cock and of the hen" (*en el nombre del gallo i de la gallina*).[29] In the words of that same witness, "The little dogs were dressed like infants, and the puppies' mother rested in the [miniature] bed, with its canopy [suspended over the bed], as if pretending to be human" (*como fingiendo ser gente*).[30] Despite numerous eyewitness testimonies against Balbuena, he and his wife, along with their daughter, denied the charges when interrogated by inquisitors. The incomplete Inquisition case unfortunately tells us nothing about the fates of those charged.

In a later Inquisition case, more thoroughly investigated than Balbuena's case, the Spanish tailor Joseph Armas jokingly administered the sacrament of baptism to two dogs, in 1780. This case caught the attention of the Mexican Inquisition through the denunciation of María Manuela Vázquez, a thirty-three-year-old *doncella* who came forward with her accusation on 7 August 1780 on the advice of her confessor.[31] According to her story, a mock ceremony of baptism had been elaborately prepared, with lit candles placed on the dogs' paws, salt put in their mouths, and water in a large bowl, as though it were holy water in a baptismal font, in which to immerse the animals while an invited crowd of bystanders looked on.[32] During the event, the owner of the dogs, María Dolores, became concerned that they would be harmed if they were doused in the makeshift baptismal font, so the "baptism" of the dogs was never actually carried out. Interestingly, Vázquez in her initial denunciation also made reference to another bystander, a certain María Rodríguez, who claimed having once seen a horse being baptized with *pulque* (a fermented alcoholic beverage made from the maguey plant) by a secular priest, and who also recounted a story about sacraments being administered to a doll.[33]

Though Vázquez's assertion that the candles were put on top of the dogs' paws and the salt placed in their mouths was denied by other witnesses, other facts in her story were corroborated on many counts. After investigating some fourteen participants and witnesses, inquisitors concluded that in fact two separate baptisms of individual dogs had taken place on two consecutive days. They concluded that on the first day the following took place:

> In the presence of many neighbors who attended, the baptisms of two little dogs were performed by a secular tailor in the following manner: dressed as a priest, as if his clothing were that of an ecclesiastical minister, he [Armas] adorned himself with a cassock, surplice, cincture, and a cloak.... Nearby he had a bowl that served as the baptismal font, salt on a small plate, and a lit candle; the "godmother" [María Guadalupe de Cuebas] neared the dog, and the minister pretended to read from a large book in front of him. He then took the salt as if to throw it at the dog, and made a similar motion to anoint the dog with water, dipping a conch into the large bowl of water three times, without ever touching the dog so that, according to the dog's owner, the dog would not die. Armas also took a feather from a flask and anointed the dog, being held by its "godfather," three times [with oil]. A boy holding the candle then cleaned the dog with cotton cloth.... The godfather gave one *real* to the minister and half a *real* to the boy holding the candle, they returned the dog back to its owner, and with this the ceremony was concluded.[34]

The second baptism was performed similarly, though the inquisitors gathered fewer details about it. To make matters worse in the eyes of the Inquisition, one of the dogs mockingly baptized had died only a few days after the ceremony and was given a funeral. That dog was shrouded, as if it were human, with a remnant piece of cloth from the Carmelite habit, and those present at the burial, including the dog's owner, María Dolores de Cuebas, among others, placed palms, garlands, and flowers on the canine corpse. They then placed the dog in a small box, made "in the shape of a coffin," with its paws crossed (*con las manos cruzadas*), and buried the dog under the patio of her home.

Some witnesses spoke of irreverence not simply for the sacraments, but also for the clergy. The "idiot brother" of Joseph Armas had, for example, been arrayed like a bishop for the occasion. No doubt, the tailoring supplies of Armas had helped, here, to produce clothing to that effect. Miguel Cervantes, a witness, said that though he could not recall specific occurrences, he believed that fake baptisms, especially of dogs, were very common in Mexico. While the weddings and baptisms were clearly performed for novelty, curiosity, and fun (though the grief that might drive a dog's human companion to bury her dog with ceremony might be genuine enough), the inquisitors characterized these events as "such perverse and diabolical diversion" (*tan perversa y diabólica diversión*).[35] After gathering information on the case of Joseph Armas, inquisitors ultimately charged him with the "extremely irreligious and suspicious act of having baptized two dogs, with grave scandal to the participants and in notorious deprecation of the ecclesiastical rites and ceremonies of baptism, the most necessary of all sacraments."[36]

Despite the inquisitors' recognition that this was merely a "burlesque mass celebrated for diversion," as in the earlier case of the canine wedding, the case was treated seriously and as one with heretical potential. The prosecution requested that Joseph Armas be temporarily placed in the secret prisons of the Inquisition and that his goods be confiscated for the duration of the trial. Ultimately, however, as with Father Basterrechea in 1770, Joseph Armas was not found guilty of heresy. Both Basterrechea and Armas were merely given serious warnings by ecclesiastical authorities that they would be severely punished in the event of future sacramental misconduct.[37] Given these cases, I can agree with David Cressy, though he is writing about the early modern English historical context, that "the baptisms [and marriages] of beasts were minor scandals, soon suppressed though often remembered."[38] Still however, the Church treated such cases seriously because of the "insulting and incongruous manner" in which common domesticated animals were brought together with the holiest of rituals.[39] In canine weddings and baptisms, carnivalesque religious rituals veered dangerously close to heresy, but the larger issue at stake was that, at least in the eyes of ecclesiastical authorities, these acts challenged the divinely ordained natural and social orders.

The Death of Pamela, the Noblest of Dogs

Though meant to be jocular spectacles for the most part, events such as canine weddings, baptisms, and funerals might also be seen as elaborate displays of affection and attention bestowed on middle- and upper-class pets. This was not lost on one Mexican intellectual who anonymously published a brilliant late eighteenth-century satirical piece dedicated to a fictitious dog named Pamela and directed against antiquated, baroque funerary pomp.[40] The so-called "Honras fúnebres a la perra Pamela" (Funerary honors for a dog named Pamela) tells of the extraordinary life, tragic death, and funerary honors accorded to a little dog named Pamela. Is it possible that stories of canine weddings, baptisms, and funerals among Mexico's elite might have reached the ears of this author? Can his text be seen not only as a diatribe against funeral splendor and luxury, but also partially as commentary on the putatively profligate nature of bourgeois pet-keeping culture?[41] The author begins his text by questioning the scholarly category under which such writings about Pamela would fall. He offers the following choices: dogology (*perrología*), dogosophy (*perrosofía*), dogomancy (*perromancia*), and dogometry (*perrometría*). Yet, finding them all inadequate to properly extol the virtues of Pamela, the author settles simply on "Honors" (*Honras*).

Honras begins with an epitaph, written in both Latin and Spanish, to be inscribed on Pamela's tombstone:

A PAMELA

Perrita. Finísima

Descendiente. De. Abuelos. De. La. Mejor. Raza.

Nacida. En. Puebla.

Criada. En. Acaxete.

Admirada. En. México. Por. Sus. Esclarecidos. Hechos.

Y. Allimismo. Con. Universal. Sentimiento.

Arrebatada. Por. Una. Temprana. Muerte.

Acabando el Siglo VXIII.

Su. Casa.

Ocupada. De. La. Mayor. Tristeza.

Para. Prueba. Perpetua. De. Su. Amor.

La. Erigió. Este. Magnífico. Mausoleo.[42]

TO PAMELA

Noblest. Dog.

Descendent. Of. Grandparents. Of. The. Best. Breed.

Born. In. Puebla.

Raised. In. Acaxete.

Admired. Throughout Mexico. For. Her. Enlightened. Deeds.

And. There. With. Universal. Sentiment.

Seized. By. An. Early. Death.

The Eighteenth Century Drawing to a Close.

Her. Home.

Occupied. By. The. Greatest. Sadness.

As. Perpetual. Proof. Of. Their. Love.

This. Magnificent. Mausoleum. Was. Erected.

This parody of an epitaph is followed by a number of sonnets and octets dedicated to the dog's life, tragically cut short by sickness, and "the bitter death of the very noble, exquisite, and very fine little dog, *doña* Pamela."[43] The author further laments, "We had all promised ourselves, and not without reason, that one day, arriving at an adult age, Pamela would know how to sit, stand on two legs, join her paws when begging, jump for a piece of bread and open her mouth to catch it, and play dead, among a number of other tricks that are recommended of her species."[44] The references to dog tricks can be interpreted at multiple levels. On the one hand, they act as a humorous jab at the frivolous amusements (and pet-keeping practices) of Mexico's upper classes. On the other, if we take Pamela as a metaphor for Mexico City elites themselves, then Pamela's skillful yet ultimately meaningless tricks evoke the perhaps equally meaningless accomplishments and feats—so devotedly recounted at their funerals—of deceased elites. Pamela's "tricks" *were* the tricks of the elites, often performed at the expense of the disenfranchised masses that made up most of colonial society.

The sardonic poetry highlights such values as Pamela's loyalty, chastity, humility, obedience, and docility, as well as her only defect—the ominous and impending marriage with a male dog "of inferior birth," not suited for her dignified class, social status, and lineage (*las sospechas de que pretendía casarse con un perrillo de inferior nacimiento*).[45] The author concludes his panegyric by hypothesizing that, after death, Pamela will rise to the heavens and become one with the celestial constellations. In his penultimate paragraph, he broaches the possibility of Pamela's corporeal relics being divided among a group of elite women: "How sweet it would have been, had you left your neck to doña Fepita, your teeth to doña Joseta, your little festive tail to doña Guadalupe, and your eyes with your entire stomach lining stuffed [*menudo entero relleno*] to doña Teresa!"[46]

In this short, sarcastic, and impressive tract, the author literally and sarcastically reads Pamela into nearly every aspect of human life. Pamela is used by the author to criticize elite extravagance (in funerary processions and in pet-keeping practices) and to exemplify and poke fun at human divisions among social strata regarding marriage, social constructions of honor and nobility, emotions about and reactions to death, religious beliefs regarding relics and the afterlife, and even culinary practices (by mention of Pamela's menudo entero relleno). Through the device of this panegyric to a fictitious little dog, the author sees into the inauthenticity of the professed values and moral standards of late eighteenth-century colonial Mexican society, which was so entrenched in codes of honor and so divided along the lines of gender, social class, ethnicity, *and* species. For the purpose of our investigation, the importance of "Honras fúnebres a la perra Pamela" is not so much in the author's intent to express a satirical and hyperbolic critique of human society, but rather in how it indicates the multiple ways that humans imbued nonhuman animals (especially dogs) with all sorts of socially and historically specific meanings. This text is thus like a gloss on the Inquisition cases dealing with canine weddings and dog baptisms, giving us a sense of the broader meanings at work in late colonial society.

The Social Meanings of Pets and Animal Spectacles

How do we begin to synthesize these unique historical evidences and literary narratives of canine spectacles in colonial Mexico? As Donna Haraway notes, dogs are and always have been bound with humans as "companion species" in what she terms "significant otherness."[47] In the historical and literary texts analyzed here, such "otherness" becomes evident as the human historical actors and writers repeatedly use canines to poke fun at human institutions, such as sacraments and the Church, or at the pomp of funerals for the rich. Haraway writes, however, that dogs are not merely "a projection, nor the realization of an intention, nor the telos of anything. They are dogs; i.e., a species in obligatory, constitutive, historical, protean relationship with human beings."[48] Part of the project of "centering animals" in Latin American history is to shift the focus from the discourses on and about animals to the actual histories of those animals in order to better understand their mutable relationships with the humans around them. This is methodologically more difficult than it might at first sound. What does this exercise in colonial Mexican perrología (dogology) actually tell us, in general and in particular, about dogs and their changing relationships with humans in colonial Mexico? From them, what might we learn from human cultural perceptions and theological conceptions of nonhuman animals in New Spain? Can an analysis of the weddings and baptisms of dogs—ultimately involving the use of dogs as a form of entertainment—allow us to lay bare some aspects of the anthropocentrism stitched deep within the fabric of colonial Mexican society?

At first glance, these cases provide us with a window into pet-keeping as a cultural practice among different classes in late colonial Mexico, through which we can glean the colonizing logic and the cultural binaries that governed the ideas surrounding pets and affections bestowed on them: domestic/savage, pet/stray, purebred/mixed, leisurely/utilitarian, and private/public. Kathleen Kete's assertion that in nineteenth-century Paris "pet-keeping culture existed quietly within the history of private life on an everyday level" may be applied to colonial Mexico as well.[49] As in other historical contexts, there was a definite class aspect to pet ownership in Mexico.[50] That the majority of human participants in the 1770 wedding and the 1770 baptism were referred to with the honorific titles *doña* or *don* before their name (including doña María Agustín, don Antonio Balbuena, don Joseph González, etc.), which conveyed honor, social status, and a sense of privilege, suggests that the pet owners in these cases were of a certain class.[51] The author of "Honras fúnebres" even sarcastically refers to the exquisite *perrita* as *doña* Pamela. The 1780 baptism involving Joseph Armas, a tailor, and a number of people lacking honorific titles indicates that companion dogs, less likely to be purebred, were also common among Mexico City's artisans, craftsmen, and day laborers. Clear indications of ownership appear in some of the documents; for instance, María Dolores de Cuebas is repeatedly referred to as the "owner" (*dueña*) of the two dogs in the 1780 baptism.

The ubiquity of dogs not only as pets among all social classes, but also as strays should come as no surprise, especially given a 1728 Mexico City ordinance that "ordered that no person shall throw into the streets, plazas, or canals, dogs nor horses, nor other dead animals."[52] Among the bourgeois and upper classes, however, many pets were ultimately seen as sources of companionship, affection, and entertainment for their "owners," and, especially in the case of purebred dogs and exotic pets, they were seen as luxuries and important markers of social status.[53] Importantly, in these historical documents, no reference is made to the utilitarian nature of *these* particular dogs. The dogs that were mockingly baptized and wed or grievously mourned were bestowed with a certain level of affection by their human companions. They were clearly *not* the working dogs that were used by hunters, shepherds, butchers, bullfighters, criminal authorities, and soldiers. Rather, they were leisurely dogs, individually named and cared for, and indicative of the social world that was built up around the domestic dog in late colonial society.

In order to further access the social meanings behind narratives of canine marriages, baptisms, and funerals in eighteenth-century Mexico, we must first theoretically locate these events within the carnivalesque tradition, through which all sorts of social hierarchies were inverted, in a social order that was replete with mundane religious rituals. At the same time, however, we must recognize that the tradition of carnivalesque animal mockery is juxtaposed with the notion that leisurely, domestic dogs as pets were entitled to human affections. Here we might take hints from Robert Darnton's essay "Workers Revolt: The Great Cat Massacre of the Rue Saint-

Séverin," which traces the meanings behind a group of apprentices in a Paris printing shop who in the 1730s staged mock trials for a number of neighborhood cats and subsequently hanged them. They evidently found the whole process hilarious. He asks, "But why cats? And why was the killing so funny?"[54] Working in a similar vein, we might ask here, why were dogs chosen in these particular cases? And what was so funny about married and baptized dogs? Indeed, it was often the hilarity of the events that, through hearsay and rumor, initially drew onlookers and participants. Others were attracted to the festive events simply in passing by or by overhearing riotous laughter on the streets. Attesting to the spontaneity of such events, María Guadalupe de Cuebas, who acted as the dog's godmother in the 1780 baptism, was drawn to the dog baptism on hearing *carcajadas de risa*, roaring laughter, as she passed the building where the dog baptism was about to take place. The dogs—dressed as human beings, offered a matrimonial bed, assigned godparents, and conferred the holiest of sacraments—essentially served as the entertainment centerpiece of such parties. Clearly, partygoers and participants found the idea of parodying central religious rituals with dogs to be not only funny, but also a distinct reason to give a party in the first place. The meanings behind this deserve scrutiny.

While onlookers saw the (unnatural?) mimicry of human religious rituals using dogs as actors as humorously absurd, part of the humor, at least in the wedding, was expressed in the ways that those present at the event could project human gender roles and social relations onto their pets (much in the same way that the author of "Honras fúnebres" projected *limpieza de sangre*, social status, honor, and chastity onto the beloved fictitious Pamela). The mock imposition of human institutions like matrimony on nonhuman animals that could not and would not remain "faithful" as husband and wife added to that humor. The idea of making dogs the object of a ceremony that implied heterosexual romance, courting, and consummation of marriage (through the small matrimonial bed provided for the newlyweds) playfully mocked human institutions by temporarily casting animals in gendered human roles. That female dogs typically went into heat a couple of times per year and that otherwise "promiscuous," uncastrated male dogs had seemingly indiscriminate sexual relations with female (and male) dogs, as well as with a number of other animate and inanimate objects, would certainly have been known to any urban inhabitant of Mexico City who ventured out into the streets and encountered stray dogs—dogs whose sexuality had *not* been contained by humans through spaying, neutering, or through selective breeding practices.[55] Given the festive context of drinking and dancing, the spectators must have found the spectacle of a marriage contract which bound the dogs to sexual fidelity and monogamy hilarious.

In all these cases, the dogs were enlisted as participants in jocular spectacles and mimicry of human institutions and religious rituals that they would otherwise have no reason to be a part of. Norine Dresser, in an essay on the horse and cat bar mitzvah, a dog wedding, and a lobster wedding in the late twentieth-century United States, poses interesting questions about the deeper affective meanings behind such celebratory events.[56] The affective meanings behind dog weddings and canine baptisms are also evident in the colonial Mexican cases. Dresser, however, also raises the notion of "human imperialism" in tandem with the question of why some people see the need to impose human celebrations on other species. Whereas Cressy concludes that in the early modern period, "by sprinkling the horses and cats with water and giving them names in the parody of the ritual by which babies become Christians, they profaned the sacred ceremony and blurred the boundary between humans and beasts," I would argue that, on one level, such parodic religious rites actually reified the human-canine boundary in the minds of onlookers.[57] Human spectators present at the parties and fandangos likely perceived the holy sacraments bestowed on dogs as humorous precisely because they saw a clear boundary between human and nonhuman animals. To administer religious sacraments to canines was considered ludicrous, farcical, and therefore comic and entertaining. Yet, on the other hand, the dogs were involved in such mock rituals and celebrations precisely because of the strong affective ties between humans and their canine companions—affective ties that may be said to blur species boundaries in a particular way. The complexity of human-animal interactions is such that the funeral of María Dolores de Cuebas's dog seems to have expressed authentic grief. Obviously, the ideological and theological boundaries between humans and animals were in flux at the time. That the Mexican Inquisition dealt with these doctrinal infractions so seriously signals its deeply rooted historical anxieties about the borders between humans and animals as well as anxieties about the very state of the "human."[58]

No matter how we analyze the playful yet mischievous administration of sacraments to dogs, these carnivalesque weddings and baptisms challenged church dogma in a very different way than did the seventeenth- and eighteenth-century opponents of Cartesianism, who by conceding to animals the "powers of perception, memory and reflection . . . were implicitly attributing to animals all the ingredients of an immortal soul."[59] Despite the fact that the ecclesiastical authorities recognized and confirmed that everyone present at these events knew that the baptism or wedding was merely a joke, a form of entertainment, the worrisome thing for the Church was the doctrinal challenge to the concept of the immortal souls of humans. After all, according to inquisitors, the sacraments were granted as if to "Christians and able persons."[60] Given that the Protestant Reformation and Martin Luther's challenge to Catholic rituals and sacraments remained pervasive threats to the Church throughout the sixteenth, seventeenth, and eighteenth centuries, the tremendous lengths to which the Inquisition went to uncover and extirpate the potential heresies involved in administering sacraments to dogs make sense. Furthermore, the ways in which humans used animals to mock church dogma and the sanctity of religious rituals greatly concerned the Holy Office. One gets the sense, however, that, at the dawn of the nineteenth century, the rigid and increasingly obsolete institution of the Mexican Inquisition was applying certain measures that lagged behind the emotional mood of the time. At the same time, it is significant that the Inquisition in all of the cases cited above failed to mete out significant penalties to the humans responsible for the mock administration of the sacraments. All of these signs point to the new marks of sensibility that extended to some privileged (non-utilitarian) animals in Bourbon Mexico. As we can glean from late-colonial cases of dog baptisms, weddings, and funerals, popular and elite segments of society experienced a rise in consciousness about the sentiments that could be bestowed on domestic creatures and household pets.[61]

Anthropomorphism, Anthropocentrism, and Animal Anxieties

In the archival cases cited above, anthropocentrism led directly to anthropomorphic gestures and to a sort of transspecies sartorial drag through which dogs were made to wear human clothing—dressed up like brides, grooms, and soon-to-be-baptized infants—and placed in miniature matrimonial beds or alongside baptismal fonts. I acknowledge that the invocation of a term as contentious as *anthropomorphism* might be problematic, but it is not without merits in understanding human treatment of other animals. For, as Tom Tyler critically comments, "Anthropomorphism, both as term and concept, imprudently starts with the human, even though the whole question of the nature of the human has yet to be determined."[62] Rather than rely here on that term in an ahistorical manner to think about humanity first, I want to highlight specific anthropomorphic moments and gestures as a way to look at animal anxieties and to acknowledge certain historical shifts in anthropomorphism and in human-animal relationships in Bourbon Mexico.[63] Dogs, it is clear, did not mock the sacraments of baptism and marriage of their own accord, but rather were manipulated by humans to dress like humans and to receive the simulated waters of baptism and marital vows. It was the human-centered desire to use animals as entertainment that led to the anthropomorphic and apparently humorous spectacles of dogs being married and baptized. Most remarkable in these cases was the historically determined tenderness and sensibilities shown toward these particular dogs. We might even read María Dolores's apprehension that her dogs might be harmed, were they actually to be submerged in the makeshift baptismal font, as indicative of shifting apprehensions and anxieties about cruelty to animals, which led to the passage of a number of laws in Europe and the Americas in the eighteenth and nineteenth centuries aimed at protecting domestic and domesticated animals.[64] Humans and dogs have been shaping their relationships with one another since the domestication of the canine species; yet here we find humans gently coercing other creatures in ways radically different than in dogfights, bullfights, cockfights, or hunts. The changing relationships between humans and dogs indicate historical shifts in human sensibilities about domestic dogs as well as changes in anthropomorphic practices and anthropocentric attitudes. Late eighteenth-century humans coerced their canine pets in different ways than they did the working dogs found throughout the early modern world.

Significantly, nothing in the records of these cases intimates how the dogs themselves experienced these events. While those present at the events would most likely have been able to read the dogs' bodily expressions, vocalizations, and other communicative acts, we learn nothing about them from the reports. Witnesses and participants did not comment on these aspects, and inquisitors were not interested. And while we have no indication of the animals' feelings or experiences, we can postulate that the dogs, secured by humans for the duration of the mock ceremonies and encircled by loud music and even louder laughter, might have been terrified. On the other hand, they might have relished the attention and affection bestowed on them by the human bystanders. Or they might simply have been accustomed to such surroundings and therefore might have had little reaction. In any case, no testimony was elicited about the behavior of the individual dogs, and thus the dogs, though vital to the narratives, essentially "disappear" from the transcripts. While the dogs were placed in central positions, humans ultimately thrust the dogs into human roles as a way of providing themselves with the laughter necessary to sustain a festive and jocular atmosphere. It has been suggested that "animals somehow transgressing the conceived natural order between man and beast are regarded as impure," yet here the anthropomorphized animals evoke (at least among human participants) notions of the carnivalesque, rather than ideas about contamination and pollution.[65] This amounts to a rescripting of traditional notions of the carnivalesque, wherein inherently unclean animals were temporarily elevated in status. The dogs here were pampered household pets, already regarded by the human bystanders as clean creatures in comparison to the untamed stray dogs that did not, even in a carnivalesque climate, enter the (elite) human domestic sphere from the street.

While the nonhuman animals do drop out of the transcript in these Inquisition cases, they do not always drop out of the judicial system. In colonial Latin America and early modern Europe, for example, animals were regularly killed in bestiality cases (which did tend to raise the specter of corporeal pollution).[66] Those medieval and early modern animals suspected of being witches' familiars, those pests exorcised by ecclesiastical authorities, or those roving creatures tried by secular authorities for a variety of crimes (like eating infants or degrading the sacraments) could also be publicly put to death. In these eighteenth-century cases, however, neither the onus nor the punishment was placed on the dogs themselves. Animals were central to these cases and to the arguments put forth by ecclesiastical authorities, yet at the same time the dogs themselves, as well as notions of animal agency, disappear from inquisitorial debates on the root causes of heresy. In these cases, humans, and never the dogs themselves, are in danger of committing acts of heresy and being punished for them. The actions of Father Basterrechea, for example, "put his soul at grave risk," according to the inquisitors presiding over his case. In contrast to this explicit reference to the souls of humans, we only have a few scattered references to the dog as an *animal yrrasional*, or an "irrational animal." In what might be described as a type of theological colonization, animals are consistently denied spiritual existence, to the extent that dog burials also largely disappear from the transcript. Finally, we notice that in Bourbon inquisitorial accounts of dog baptisms and marriages, there is absolutely no suggestion of animals being suspected of demonic possession, which indicates that at the official level, too, attitudes toward nonhuman animals implicated in heresy had largely shifted by the late eighteenth century. Changing official and popular sensibilities about dogs as symbols of domesticity were not untroubling for the Inquisition, which, as the colonial period drew to a close, uncovered ample evidence of disrespect being shown to the sacraments though new types of human-animal interaction.

Linda A. Curcio-Nagy notes that the large-scale spectacles characteristic of late medieval and early modern Europe served a number of functions: "They were hegemonic tools of the State, they articulated or sustained community or group identity, and they defined human-divine relations."[67] Might it therefore be said that the small-scale spectacles like canine weddings and baptisms, with animals as their focal points, both defined human-animal relations and sustained group identity among humans? The weddings and baptisms of dogs took on a ludic and carnivalesque nature, but what, then, do we make of the burial of María Dolores de Cuebas's dog? Clearly different from the fictitious extravagance of Pamela's funeral, the quiet nature of this burial also differentiated itself from the ludic spectacles of the weddings and the baptisms. It hints at some affective elements of colonial Mexican society that, given the nature of historical documentation, are difficult to access. The burial of Dolores de Cuebas's dog, with its paws crossed in a miniature coffin, offers a unique glimpse at human emotional attachment to a canine companion. The funeral afforded to Pamela also reflects, albeit sardonically, the affective dimension of pet-keeping among the eighteenth-century Mexican bourgeoisie. Jon T. Coleman has observed that dogs "mongrelize species categories, and their impurity threatens to collapse other boundaries."[68] While on the one hand the dogs in these cases may be seen as living, breathing props in staged spectacles of entertainment for humans, on the other hand the affective ties between humans and their lapdogs did indeed confound—and simultaneously reify—species categories. While dogfights, for instance, may be said to increase the ontological distance between humans and nonhuman animals, in the cases discussed here the clear concern for the well-being of the dogs expressed by some participants mitigates the human-canine ontological distance. Though who is to say how the dogs themselves may have viewed the events in which they participated?

These are not only cases of, as Kathleen Kete puts it, "bourgeois history as it unfolded in the history of the dog," but also cases of inquisitorial, theological, and literary history as manifested through the histories of a few very specific dogs.[69] While acknowledging the importance of historical works that have offered histories of dogs, I have steered away from the aggregate, abstract category of "dogs" by focusing on the experiences of seven late eighteenth-century colonial Mexican dogs (and one literary dog, Pamela). Paradoxically, while I have attempted a history of animals, in the

Inquisition cases, as opposed to in "Honras fúnebres," surprisingly little detail is provided about the dogs themselves, aside from what they were wearing. What we can understand about these dogs and the uses to which they were put by humans clearly broadens our understanding of the polysemous nature of nonhuman animals in colonial Mexico. If the Inquisition cases dealing with animals are never really about the animals themselves, they are, rather, about the transgressions of humans in relation to nonhuman animals. Herein lays the fundamental methodological problem of writing histories of animals: How do we come to terms with the anthropocentric nature of historical sources?

While it is undeniably difficult to "center animals" within our historical analyses of the past, it is neither a fruitless nor an impossible task. Ultimately, the project of centering animals stems from the recognition that humans and *other* animals exist on the same ontological continuum. The cases I have presented came into existence at the intersection of an Inquisition that was intent on protecting the dignity and sanctity of the sacraments, and a lay population that was in some degree touched by a new sensibility about domestic dogs even while enacting carnivalesque spectacles of animal mockery that had deep historical roots. While mock baptisms and weddings involving dogs reified the division between the species, the softening of attitudes toward privileged species and the increase in affective ties with pets together broke down species barriers. In this foray into colonial Mexican perrología, I have set out neither to reify the human-animal divide nor to celebrate a fluid and porous human-animal boundary. Rather, I have sought to show (through a limited set of human-animal interactions) that colonialism was relational and highly dependent on alterities not only of language, race, class, gender, and sex, but of species as well.

Notes

1. Bancroft Library (BANC) MSS 96/95m, 10 (1771).
2. BANC MSS 96/95m, 10: "Muchas personas de ambos sexos, y de todas calidades [asistieron] a una profana sacrilega, escandalosa, supersticiosa función de vaile y diversión que con la maior formalidad se havia dispuesto para celebrar un Matrim° entre perros. Y deviendo por su estado reprender, y abominar una profanación tan sacrilega, y heretical, bien lejos de haverlo asi executado; se havia precipitado al desorden, y exceso abominable de solemnizar dho Matrimonio, con tanta seriedad, y formalidad, como se fuese contrahido entre Christianos, y personas aviles." Emphasis added.
3. BANC MSS 96/95m, 10: "Se le havia presentado un perrito vestido de hombre por uno de este sexo, y una perrita vestida de muger por una de esto otro, y pregun-

tados mutuam^te sobre sus consentim^tos entendido de ellos por las respuestas de los que los tenian en las manos, havia solemnizado, y formalizado este reo el referido matrimonio entre los dos animalejos."

4. The term *fandango* popularly referred to a lively dance or a form of festive entertainment. The Italian friar Ilarione da Bergamo, who traveled through Mexico between 1761 and 1768, noted that in colonial Mexico, the fandango was "the most universal one [dance] among the common people and where, for the most part, they perform the dances they call the *chuchumbe, bamba,* and *guesito,* which are all quite indecent." See Ilarione da Bergamo, *Daily Life in Colonial Mexico: The Journey of Friar Ilarione da Bergamo, 1761–1768,* ed. Robert Ryan Miller (Norman: University of Oklahoma Press: 2000), 116.

5. BANC MSS 96/95m, 11: "Antes de empezarse el fandango vio unos perritos compuestos y una camita que le tenían puesto a los novios."

6. BANC MSS 96/95m, 15.

7. This in itself marked the gravity of the case, for as Charles Nunn has noted, in the eighteenth century the cost of supporting prisoners was high enough to prevent the Inquisition from using incarceration in frivolous cases. See Charles Nunn, *Foreign Immigrants in Early Bourbon Mexico, 1700–1760* (Cambridge: Cambridge University Press, 2003), 59.

8. BANC MSS 96/95m, 2.

9. BANC MSS 96/95m, 57: "Nunca llegó a proferir las palabras; Yo os caso en el nombre del Padre, y del hijo, y del Espiritu Santo, y solam[en]te dixo, en el nombre del Padre, y de la Madre de [todos] los perros os caso, lo que pudieron entender mal los que se hallaron presentes."

10. Stuart Schwartz, *All Can Be Saved: Religious Tolerance and Salvation in the Iberian Atlantic World* (New Haven: Yale University Press, 2008), 19.

11. Schwartz, *All Can Be Saved,* 18.

12. Rob Meens, "Eating Animals in the Early Middle Ages: Classifying the Animal World and Building Group Identities," in *The Animal/Human Boundary: Historical Perspectives,* ed. Angela N. H. Creager and William Chester Jordan (Rochester: University of Rochester Press, 2003), 13. This of course raises a number of other questions. Joyce E. Salisbury writes: "One cannot help but wonder whether this last instance was a regularly occurring problem. It seems that the combination of circumstances that would place a vomiting individual in the same location as a dog after mass would not occur often enough to call forth legislation against it. It does, however, express the very real desire to preserve the one purely human food [the Eucharist], spiritual food, away from [nonhuman] animals" (*The Beast Within: Animals in the Middle Ages* [New York: Routledge, 1994], 65).

13. E. P. Evans, *The Criminal Prosecution and Capital Punishment of Animals* (London: Faber and Faber, 1987), 156.

14. James A. Serpell, "Guardian Spirits or Demonic Pets: The Concept of the Witch's Familiar in Early Modern England, 1530–1712," in *The Animal/Human Boundary:*

Historical Perspectives, ed. Angela N. H. Creager and William Chester Jordan (Rochester: University of Rochester Press, 2003), 171.

15. David Cressy, *Agnes Bowler's Cat: Travesties and Transgressions in Tudor and Stuart England* (New York: Oxford University Press, 2000), especially chap. 11, "Baptized Beasts and Other Travesties: Affronts to Rites of Passage."

16. Cressy, *Agnes Bowler's Cat*, 174.

17. Laura de Mello e Souza, *The Devil and the Land of the Holy Cross: Witchcraft, Slavery, and Popular Religion in Colonial Brazil* (Austin: University of Texas Press, 2004), 73.

18. The strong condemnation of animal worship and idolatry is a recurring theme in the Old Testament, the prime example being the story in Exodus 32 in which the Israelites built and worshiped a golden calf while Moses was on Mount Sinai communing with God.

19. Colin Spencer, *The Heretic's Feast: A History of Vegetarianism* (Lebanon, N.H.: University Press of New England, 1995).

20. Rod Preece, *Brute Souls, Happy Beasts, and Evolution: The Historical Status of Animals* (Vancouver: University of British Colombia Press, 2005), 131.

21. Keith Thomas, *Man and the Natural World: Changing Attitudes in England, 1500–1800* (New York: Penguin, 1983), 23.

22. Norine Dresser, "The Horse *Bar Mitzvah*: A Celebratory Exploration of the Human-Animal Bond," in *Companion Animals and Us: Exploring the Relationships between People and Pets*, ed. Anthony L. Podberscek, Elizabeth S. Paul, and James A. Serpell (Cambridge: Cambridge University Press, 2005), 98.

23. Thomas, *Man and the Natural World*, 102.

24. Thomas, *Man and the Natural World*, 110.

25. BANC MSS 96/95m, 18. For her part, María Agustín denied having attended any canine wedding, but admitted to having bought the ornaments for a dog. Because of conflicting testimonies, it is uncertain whether or not this canine wedding took place, but given that Agustín knew about the case against Father Basterrechea, it is possible that she denied that this other canine wedding took place in order to avoid being implicated.

26. Cited in Kathleen Kete, *The Beast in the Boudoir: Petkeeping in Nineteenth-Century Paris* (Berkeley: University of California Press, 1994), 94.

27. Archivo General de la Nación, Mexico (AGN), Inquisición 1241, exp. 4, ff. 64–102.

28. AGN, Inquisición 1241, exp. 4, f. 66v. In colonial Mexico, there are a number of cases in which individuals were punished by the Inquisition for having baptized, married, or held funerals for dolls. See AGN, Inquisición 594, exp. 5, ff. 491–517, for the case of Fray Miguel, in New Mexico, who found himself in trouble with the Inquisition for heretical acts including having held a funeral for a doll, acted irreverently toward the Eucharist, and solicited women in the confessional. Fray Miguel hanged himself in 1663. See Zeb Tortorici, "Reading the (Dead) Body: Histories of Suicide in New Spain," in *Death and Dying in Colonial Spanish*

America, ed. Martina Will de Chaparro and Miruna Achim (Tucson: University of Arizona Press, 2011), 53–77. See also AGN, Inquisición 731, exp. 30, ffs. 391–401, "Autos sobre un bautismo de muñecos que se celebró en el pueblo de San Juan del Rio, 1707"; AGN, Inquisición 1051, exp. 14, "Denuncia por haberse celebrado un bautizo de muñecos y una misa nueva y sermon en la casa de Juan de Figuredo, 1717"; AGN, Inquisición 777, exp. 63, ff. 472–86, "Autos sobre unos bautismos y casamientos de muñecas efectuados en la ciudad de Zacatecas, 1719"; AGN, Inquisición 872, exp. 27, ff. 395–404, "El Señor Inquisidor contra Manuel de Cordova, official de carpintero y demás complíces en el bautismo de ciertos muñecos, Guadalajara, 1735"; and AGN, Inquisición 753, exp. 2, "Certificación de los autos contra Manuel de Cordova, official de carpintero, y demás complices, por un bautismo de muñecos, 1736."

29. AGN, Inquisición 1241, exp. 4, f. 72.
30. AGN, Inquisición 1241, exp. 4, f. 73: "Y tambien vio que los perritos estavan vestidos como criaturas, y la perrita madre acostada en su cama con su pabellon como fingiendo ser gente."
31. AGN, Inquisición 1535, exp. 5, 175.
32. According to the Council of Trent, the placement of salt into the mouth of the person to be baptized signifies a deliverance from the corruption of sin, the desire to do good works, and the gift of divine wisdom.
33. Pulque is a thick, fermented, alcoholic beverage made in Mexico from various species of the maguey or agave plant. It is defined in the 1737 *Diccionario de la lengua Castellana* by the Real Academia as "the juice or liquor of the maguey made by cutting its trunk when it is ready to be opened and then leaving a large cavity where it is then distilled. This drink is highly esteemed in New Spain where they are used to adding certain ingredients in order to give it a greater punch" (*Diccionario de la lengua Castellana* [Madrid: Real Academia Española, 1737], 430).
34. AGN, Inquisición 1535, exp. 5: "A presencia de muchos de vecinos que concurrieron, se procedió al bautismo de los perrillos que practicó un secular de oficio sastre en el modo y forma siguiente. Vestido al modo de sacerdote, haciendo de sus ropas las que correspondian a un Miñtro Ecco como fueron sotana, sobrepelliz, estola, cingulo, y capa; la sotana de la capa que usaba; la sobrepelliz de un desaville; la estola de un zeñidor, y la capa plubial de um pano cuapascle de rebozo; y com bonete q formó de um sombrero; y en esta disposicion teniendo a la mano agua en un vaso q servía de pila, sal, y vela encendida, acercandose con uno de los perros lo que hacía de madrina, el Ministro fingia que leia en un libro que tambien tenian dispuesto; y a breve rato hizo la demostracion de tomar la sal, para echarla al perrito: que asi mismo hizo la demostracion de echar la agua al perrito, sacandola con una concha del vaso grande por tres veces, y dejandola caer otras tantas ocasiones en el mismo vaso sinque tocase al perro, porque no se muriese, segun se produjo la dueña de ellos. Que tambien el que hizo de Ministro, saco una pluma de una redomita, y untó por tres veces al perrito, que tenia en sus manos el padrino, y otras tantas hizo que le limpiaba con los algodones un mucha-

cho que tenia en las manos una vela encendida. . . . Dando un real el padrino al Miñtro, y médio a que tenia la vela, y entregando el perro su dueña: con lo que se concluio el acto."

35. AGN, Inquisición 1535, exp. 5, f. 207.

36. AGN, Inquisición 1535, exp. 5, f. 173: "Por el irreligiosísimo y sospechoso hecho de haber bautizado dos perrillos con grave escándalo de los concurrentes, y notorio desprecio virtual de los ritos y ceremonias eclesiásticos del más necesário de los sacramentos qual es el bautismo."

37. Given that legal precedence was important in colonial Spanish America, it is also very likely that don Antonio Balbuena would not have been found guilty of heresy by inquisitors for his dog baptism in 1770.

38. Cressy, *Agnes Bowler's Cat*, 184.

39. Cressy, *Agnes Bowler's Cat*, 184.

40. See "Honras fúnebres a la perra Pamela," transcribed by Edmundo O'Gorman, *Boletín del Archivo General de la Nación* 15.3 (1944): 525–44. In 1818, this text was also reproduced in its entirety (with some variation in the Latin segments) by José Joaquín Fernández de Lizardi, in *La Quijotita y su prima* (Mexico City: Editorial Porrúa, 1979), 193–97. It is uncertain whether Lizardi was the original author of "Honras fúnebres," although some have concluded that this was the case. For commentaries on "Honras fúnebres," see Juan Pedro Viqueira, "El sentimiento de la muerte en el México ilustrado del siglo XVIII a través de dos textos de la época," *Relaciones* 2.5 (1981): 27–62; María Isabel Terán E., "Dos sátiras del siglo XVIII contra las actitudes funerarias barrocas," in *Las dimensiones del arte emblemático*, ed. Bárbara Skinfill Nogal and Eloy Gómez Bravo (Zamora: El Colegio de Michoacán, 2002), 247–62; and Claudio Lomnitz, *Death and the Idea of Mexico* (New York: Zone Books, 2008), 346.

41. For a wonderful analysis of the movement against late colonial funerary extravagance, see Pamela Voekel, *Alone before God: The Religious Origins of Modernity in Mexico* (Durham: Duke University Press, 2002). To get an idea of the type of funeral splendor that the author of "Honras fúnebres" was critical of, Voekel offers this description of elite funerals and public funerary processions: "Funerals also provided the Church with an opportunity to sanctify elites' lofty social position. Elaborate funeral corteges were a common sight on Veracruz and Mexico City streets until the waning years of the colonial period. Although twelve was the usual number, hundreds and even thousands of clergy regularly escorted the dead [elite] to the church and prayed at burial ceremonies. With clergy clogging the streets, hundreds of flickering candles, bright cloths, ornate carriages, musical accompaniment, and paid mourners, these funerals stunned the senses of urban crowds" (*Alone before God*, 6).

42. O'Gorman, "Honras fúnebres a la perra Pamela," 528.

43. O'Gorman, "Honras fúnebres a la perra Pamela," 534: "¿No deberíamos usar de otras mayores para llorar la muerte que aún no podemos olvidar, la amarga muerte de la muy noble, muy exquisita, y muy fina perrita doña Pamela?"

44. O'Gorman, "Honras fúnebres a la perra Pamela," 537: "Todos nos prometíamos, y no sin fundamento, que llegando a una edad adulta, debería sentarse, pararse en dos pies, juntar las manos como quien pide, brincar para alcanzar un pedacillo de pan, abrir la boca para acertar el que le tirasen, hacer el muerto y otras gracias que recomiendan a los de su especie."

45. O'Gorman, "Honras fúnebres a la perra Pamela," 539.

46. O'Gorman, "Honras fúnebres a la perra Pamela," 543: "¡Qué dulce os hubiera sido, que hubiese dejado su pescuezo a doña Fepita, sus dientes a doña Joseta, su colita fiestera a doña Guadalupe, y sus ojos con su menudo entero relleno a doña Teresa!"

47. Donna Haraway, *The Companion Species Manifesto: Dogs, People, and Significant Otherness* (Chicago: Prickly Paradigm Press, 2003).

48. Haraway, *The Companion Species Manifesto*, 11.

49. Kete, *The Beast in the Boudoir*, 49.

50. For a variety of creative essays on the culturally contingent nature of canine companionship and pet ownership, see Lynn Festa, "Person, Animal, Thing: The 1796 Dog Tax and the Right to Superfluous Things," *Eighteenth-Century Life* 33.2 (2009): 1–44; Alma Gottlieb, "Dog: Ally or Traitor? Mythology, Cosmology, and Society among the Beng of Ivory Coast," *American Ethnologist* 13.3 (1986): 477–88; Eduardo Kohn, "How Dogs Dream: Amazonian Natures and the Politics of Transspecies Engagement," *American Ethnologist* 34.1 (2007): 3–24; and Aaron Skabelund, "Can the Subaltern Bark? Imperialism, Civilization, and Canine Cultures in Nineteenth-Century Japan," in *JAPANimals: History and Culture in Japan's Animal Life*, ed. Gregory M. Pflugfelder and Brett L. Walker (Ann Arbor: Center for Japanese Studies, University of Michigan, 2005), 195–243.

51. Kenneth J. Andrien, *The Human Tradition in Latin America* (Wilmington, Del.: SR Books, 2002), vix.

52. Cited in Jay Kinsbruner, *The Colonial Spanish-American City: Urban Life in the Age of Atlantic Capitalism* (Austin: University of Texas Press, 2005), 60.

53. Dogs in pre-Conquest Mesoamerica and in colonial Mexico were, of course, imbued with a wide variety of historically and economically contingent meanings. See Marion Schwartz, *A History of Dogs in the Early Americas* (New Haven: Yale University Press, 1997), for a discussion of the ways that some representations of dogs attained cosmological significance among indigenous inhabitants of the Americas, while other breeds became culinary delicacies, herding dogs, and beasts of burden. Dogs were also used as vicious animal conquistadors by the Spanish. See John Grier and Jeannette J. Varner, *Dogs of Conquest* (Norman: University of Oklahoma Press, 1983).

54. Robert Darnton, *The Great Cat Massacre and Other Episodes in French Cultural History* (New York: Basic Books, 1984), 82.

55. See Kete, *The Beast in the Boudoir*, 92–114, for an erudite discussion of canine sexuality in nineteenth-century French pet-keeping culture.

56. Dresser, "The Horse *Bar Mitzvah*."

57. Cressy, *Agnes Bowler's Cat*, 182.

58. Kete, *The Beast in the Boudoir*, 57. See also Erica Fudge, *Brutal Reasoning: Animals, Rationality, and Humanity in Early Modern England* (Ithaca: Cornell University Press, 2006).

59. Keith Thomas, *Man and the Natural World: Changing Attitudes in England, 1500–1800* (New York: Penguin, 1983), 34.

60. BANC MSS 96/95m, 10.

61. This is of course not to say that prior to the eighteenth century no human shared an affective bond with a canine companion. In 1626, for example, sailors on a frigate navigating between the Philippines, California, and Acapulco denounced Pedro de Valle for witchcraft, because he had exchanged amorous words with Francisco de Barrios and regularly kissed and shared a bed with his pet dog (*besaba un perro y dormía con él*). See AGN, Inquisición 356, exp. 117, ff. 224–30. There were no insinuations of sexual activity between Pedro de Valle and his dog, and it thus appears that the explicit bonds of affection between the man and his dog were reason enough to cause a scandal in the early seventeenth-century maritime world.

62. Tom Tyler, "If Horses Had Hands . . . ," in *Animal Encounters*, ed. Tom Tyler and Manuela S. Rossini (Leiden: Brill Academic Publishers, 2009), 23.

63. We need not look far for evidence that the "humanity" of humans is historically, culturally, and theologically determined. Indeed, it was only in 1537—almost five decades after Europe's initial contact with the indigenous inhabitants of the Caribbean and some fifteen years after Cortés and his allies invaded the Mesoamerican center of Tenochititlán—that Pope Paul III's papal bull, *Sublimis Deus*, declared the "Indians" to be "truly men" and thus capable of Christianization.

64. See Reinaldo Funes Monzote's essay in this volume.

65. Meens, "Eating Animals in the Early Middle Ages," 15.

66. On bestiality in colonial Mexico, see Lee Penyak, "Criminal Sexuality in Central Mexico, 1750–1850" (PhD diss., University of Connecticut, 1993); Mílada Bazant, "Bestialismo, el delito nefando, 1800–1856," *Documentos de Investigación 66* (2002): 1–22; and Zeb Tortorici, "Contra Natura: Sin, Crime, and 'Unnatural' Sexuality in Colonial Mexico, 1530–1821" (PhD diss., University of California, Los Angeles, 2010).

67. Linda A. Curcio-Nagy, "Introduction: Spectacle in Colonial México," *The Americas* 52.3 (1996): 275.

68. Jon T. Coleman, "Two by Two: Bringing Animals into American History," *Reviews in American History* 33 (2005): 491.

69. Kete, *The Beast in the Boudoir*, 53.

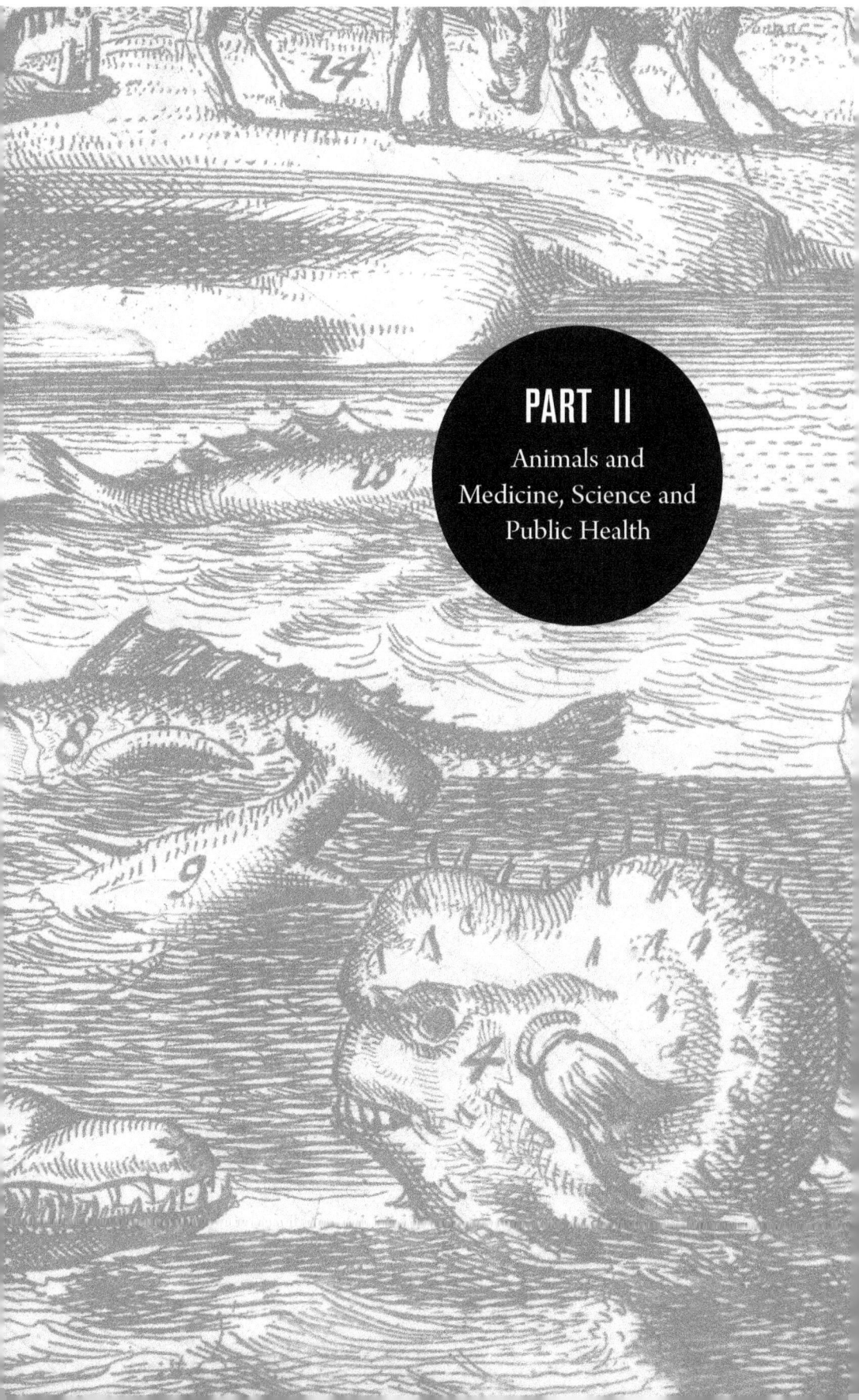

PART II

Animals and
Medicine, Science and
Public Health

4

From Natural History to Popular Remedy

Animals and Their Medicinal Applications among
the Kallawaya in Colonial Peru

ADAM WARREN

> "Chinchillas": This is the name the Naturals [Amerindians] use for a
> little animal like the rabbits of Castille. The wool of this animal, when
> cut and mixed with "restrictive powders" and egg whites, is good for
> stopping the flow of blood from wounds. The meat of this animal is
> indigestible and heavy, which is why I have seen those who eat it suffer
> great afflictions at night, and retching as if they had eaten a poisonous
> delicacy. Its temperament is cold.
> —MARTÍN DELGAR, *RECETARIO EFICAZ PARA LAS FAMILIAS:*
> *MEDICAMENTOS CASEROS*

In recent years, scholars of Latin America have produced a large number of
historical works that examine medicinal uses of plants in the context of a
colonial science of botany, which linked New World intellectuals to their
counterparts in Spain.[1] Few studies, however, have paid much attention to
zoology or the exchange of animals between colony and metropole. Even
fewer still have problematized medicinal uses of animals among different
sectors of colonial society itself.[2] This is in many ways surprising. Animals
often appeared as part of natural histories of the New World, they were included in shipments to Spain's Royal Botanical Gardens, and studies of animals informed research on both botany and medicine.[3] Dissection of animals,
especially dogs, became a popular way to teach anatomy in eighteenth-
century Lima, and animal vivisection and dissection had been practiced for

much longer in Spain.[4] Perhaps more important, however, animals formed a key part of indigenous ethnopharmaceutical practices both in Mesoamerica and the Andes. By the eighteenth century, in fact, such local practices involving animal-based treatments had spread beyond indigenous communities and had come to constitute a key feature of popular home medical guides known as *recetarios* in Peru.[5] Recetarios provided lists and instructions for the preparation and application of treatments, which were often made using common household ingredients or items acquired from nature. The description of the chinchilla in the epigraph, taken from the late colonial work *Recetario eficáz para las familias: Medicamentos caseros*, is just one such example of this process of knowledge transference.[6] In this way, indigenous Andean medical practices dependent on local fauna sometimes became part of a parallel Spanish tradition of cataloging and utilizing animals for healing.[7]

In this chapter I focus on how one eighteenth-century surgeon and author of recetarios, Martín Delgar, translated and integrated indigenous Andean medical knowledge about the uses of animals to create and catalog new folk medical treatments. In doing so, I explore how indigenous populations and other groups converted animals and animal parts into medicinal substances and explained their healing properties, situating Delgar's work within a longer history of information gathering on indigenous medical practices in the Andes. I examine links between four copies of Delgar's recetario and the Jesuit Bernabé Cobo's natural history, *Historia del Nuevo Mundo*, which was completed sometime in the early 1650s.[8] I then compare Cobo's natural history and Delgar's catalogs of treatments to ethnographic descriptions of the medical practices and knowledge of the modern Kallawaya, a population from the southern Andes who served as official healers and diviners to the Incas, and who for centuries have been famous for their cures. Historically the Kallawaya have traveled extensively beyond their communities. Kallawaya practices incorporate many widely shared indigenous Andean beliefs about health and the body, and they were almost certainly the basis for Delgar's and Cobo's works. They thus provide a window into broader Andean uses of nonhuman animals in treating the human body and its ailments.

The comparison of Delgar's work with texts completed in different time periods is both important and useful because Delgar embraced indigenous Andean medicine at a time when most prominent, formally trained doctors in Peru did not. Delgar's proposals and approaches to healing were thus in many respects exceptional and unlike other published works in the second half of the eighteenth century. Instead of calling for the triumph and widespread dissemination of formal Spanish medical knowledge, Delgar argued the sick would do best to employ more mundane indigenous and popular folk remedies from the Andean highlands, many of which were based on the perceived medicinal properties of animals and plants. Following the lead of writers in New Spain, who since the sixteenth century had published recetarios based on New World flora and fauna, Delgar believed knowledge of Andean treatments should be spread widely and integrated with Spanish practices for the benefit of the colony's poor. Producing recetario manuals based in part on Cobo's earlier work and the observation of indigenous Andean healers, Delgar wrote at a time in the eighteenth century when self-consciously professionalizing doctors sought to persecute folk practitioners and displace popular medical knowledge.[9] Dismissing such efforts, he compiled elaborate recipes and instructions for converting Andean animals, animal products, plants, rocks, soils, and waters into substances with known palliative or curative properties. His work, perhaps more than any other produced in the Bourbon period, therefore sheds light on common medical practices and uses of animals and plants in the Andes.

In order to gain access to indigenous beliefs in the past, I focus on how descriptions of remedies in Delgar's texts reflect the influence of indigenous Andean and Spanish humoral systems, in which health depended in distinct ways on the management of fluids or humors thought to be in flux within the body. By juxtaposing the work of anthropologists, in particular, with Delgar's earlier texts and those of Cobo, I problematize the complex Spanish and indigenous Andean belief systems about the body and disease that underlay Delgar's explanations of treatments, their applications, and their healing powers. I use ethnographic scholarship on the Kallawaya, arguably the most important practitioners of indigenous Andean healing traditions, to question whether humoral analyses and understandings of flora and fauna in fact differed from one another at distinct historical moments among Andean peoples, and why. Despite differences in descriptions of plants and animals evident in Delgar's work—differences that corresponded to Spanish stylistic and organizational traditions—indigenous populations such as the Kallawaya understood the medicinal applications of animals in much the same way that they thought about plants. Both functioned according to broader indigenous Andean notions of the circulation and expulsion of humors from the body to preserve or restore health. Given that many of these beliefs persist in Kallawaya medicine today, the practice of tracing back or "upstreaming" from the ethnographic present is both appropriate and useful.[10]

In addressing these questions, I ultimately ask how animals may have come to be used as part of popular, everyday colonial medical practices in the late colonial period. Indigenous populations obviously made widespread use of animals for healing, transforming their respective parts into medicinal items that could be exchanged, manipulated, and consumed. As Delgar makes clear, however, animals at times also constituted "crossover" commodities of sorts that bridged the division between predominantly indigenous societies and those regions of Peru with more diverse colonial populations. In this way, I look beyond mere questions of the "history of human attitudes toward animals," which have tended to dominate the historiography of animals in the past.[11] I also move past the notions that animals were primarily important in indigenous Andean belief systems for their symbolic and cosmological values, and that they were not particularly important at all in other belief systems. I aim instead to build on Erica Fudge's call that we not just examine representations of animals in human thought, but also problematize human uses of animals in daily life.[12] I thus shed light on the complex ways colonial subjects of various kinds may have exchanged and put such animals to use to care for themselves and cure their ills.

Observers, Indigenous Peoples, and Animals

How did Martín Delgar come to know about the animal-based treatments indigenous people employed in the Andean highlands? Traveling, reading, and writing in the eighteenth century, the author in fact drew on and appropriated many of the animal and plant descriptions Father Bernabé Cobo had included in his *Historia del Nuevo Mundo*. In doing so, Delgar transformed a work from the genre of natural history into a piece belonging to the tradition of popular medical recetarios.[13] Using Cobo's work as the basis for his own, Delgar expanded the curative applications of animals and plants the priest had described, and he added knowledge he had learned from indigenous healers, including the Kallawaya, from various parts of the southern Peruvian highlands. By building on Cobo's history and advocating remedies made by merging Spanish and indigenous Andean ingredients and treatments, he compiled a medical guide that effectively negotiated, integrated, and reconciled divergent medical systems. As a work influenced by Spanish medicine and concepts of nature, it reflects at one level an observation noted in Philippe Descola and Gísli Pálsson's edited volume, *Nature and Society*: that the idea of nature "as an abstract inventory of things, distinguished by a small number of features" is more apparent in natural history museums and other kinds of Western cataloging practices "than in the lived culture of indigenous peoples."[14] At the same time, Delgar's work clearly reflects the complex concepts and theories of animal- and plant-based healing central to southern Andean indigenous medical thought, furthering their transfer into popular colonial medical practice.

Despite the richness of his work, we know less about the life of Martín Delgar than we do about that of the Jesuit natural historian Bernabé Cobo. Born in Lopera, Spain, in about 1580, Cobo received very little education as a young man. In 1596, however, he left the peninsula in search of adventure, setting sail for the Caribbean. Eventually his travels took him to Peru, in 1599, where he gradually became an expert in science and trained as a Jesuit priest. He first enrolled in Lima's Jesuit school, the Colegio Real de San Martín, and spent approximately two years studying humanities there.[15] He continued to study in other schools in Lima until 1609, at which point he traveled to Cuzco and spent about four years in the Andes. In his biography of Cobo, Father Francisco Mateos notes that the young man traveled to Cuzco to study theology at a Jesuit institution there, but he clearly traveled to La Paz as well, spending considerable time in the village of Tihuanaco on the southern shores of Lake Titicaca.[16] His early journeys to these regions likely placed him in contact with the Kallawaya, a population who traveled long distances as popular healers in the seventeenth century. The Kallawaya resided in villages to the east of the lake, where they had developed a complex medical repertoire of rituals, diagnoses, treatments, and practices. They were already notable healers at this time, having previously served the Inca royal family as curers and religious specialists. Guaman Poma even mentions their favored position in the Inca royal court and depicted them carrying the Inca on his litter, a clear sign of prestige.[17] Given the Kallawaya's expertise, status, and popularity as healers, the Spanish Jesuit Cobo would certainly have learned of their medical knowledge about animals and plants in his various journeys.[18]

Cobo's contact with indigenous herbalists and healers resumed after he studied theology in Lima, from 1613 to 1615. In late 1615 he traveled to the highland town and Jesuit mission of Juli, located near the shores of Lake Titicaca. There, he appears to have entered another Jesuit school, studied Quechua and Aymara (indigenous languages spoken widely across the Andean highlands and in the southern Andean highlands, respectively), and carried out missionary work in several villages in a region traversed by the Kallawaya.[19] He also traveled and worked extensively in the province of

Chucuito, which pertained to Juli, and spent an additional two years, in 1617 and 1618, traveling through the region known as the Collao, which now forms part of the Bolivian highlands.

It appears that during this time as a missionary Cobo gathered extensive information about societies and regions where the Kallawaya practiced medicine. This contact enabled him to compile elaborate descriptions of animals and plants, explaining how indigenous people used such items. Mateos notes that this work is made evident through "the interactions with Indians of which he gives testimony in his writings, often taking advantage for scientific goals of learning about their antiquities."[20] Cobo spent considerable time in Oruro but also visited other mining areas in Pacajes, Cochabamba, and possibly Charcas and Potosí. These regions were home to populations that sought the expertise of the Kallawaya. According to Mateos, in these journeys "the scientific ends were by no means secondary" for Cobo.[21] This is evident in the content of Cobo's *Historia del Nuevo Mundo*, much of which was written after a third journey to the highlands in 1626 and a trip to New Spain in 1643. The text drew on names of animals and plants in Quechua and Aymara (as well as in the Mesoamerican language Nahuatl), and it provided elaborate descriptions of their properties and curative powers.[22]

Although by comparison little is known about Martín Delgar's life, I have been able to piece together some of the eighteenth-century author's travels and activities through the anecdotes of doctors and students who knew him or knew of him.[23] Trained as a doctor and surgeon, Delgar practiced in both France and Spain before traveling to Peru in 1744 to work in mining communities. It is unclear how his travel to Peru came about, but we do know that he spent much of his time in Upper Peru (now Bolivia), working for many years at a hospital in the large silver mining city of Potosí. Delgar traveled to Potosí to study mine-worker diseases and to improve their health and life expectancy. However, existing sources do not make clear whether he was sent there on royal orders or traveled on his own volition. In any case, once he was there his contact with indigenous mine workers exposed him to native Andean medical concepts, both local ones and those brought to Potosí by migrants from other parts of the highlands. Although lifted in part from Cobo's *Historia del Nuevo Mundo*, lists of plant and animal remedies in Delgar's recetarios clearly reflect this contact. Among other things, Delgar expanded on the curative properties of various highland items based on his own communication with residents. For example, he added to Cobo's discussion of *añas* or *sorrino*, which was most likely a kind of Andean skunk. Cobo described it as a small creature related to foxes, and he called it "the animal with the most profound and pestiferous stench in the world."[24] Suggesting that it was indeed "a foul-smelling animal," Delgar added that in his experience a man by the surname of Mexía used the animal medicinally and "had been healed of severe mercury poisoning, which he had acquired in benefit of the metals in the Villa of Potosí."[25]

Doctors at the end of the eighteenth century referred to Delgar as having traveled widely in Peru, especially in the highlands, and he appears to have been well known in Lima as well. In his inauguration of the colony's first dissection amphitheater in 1792, Lima's most prominent physician, Hipólito Unanue, credited Delgar with bringing new surgical knowledge and techniques to Peru. Unanue claimed Delgar gained fame in the colony's interior provinces, "where people traveled great distances to consult him about their suffering."[26] In early nineteenth-century medical theses from Lima, students cited Delgar's popularity as a healer in the Andes, and they acknowledged that his treatments proved useful and successful.[27] Delgar's lists of cures, moreover, indicate that he knew the region well, whether through travel, reading the works of others such as Cobo, or hearsay. They include references to conversations he had with "Indians" from different regions, who shared with him their cures.

Delgar's reliance on Cobo's work was nevertheless substantial. His catalog of animals and plants represents an eighteenth-century appropriation and refashioning of earlier Spanish medical and scientific genres. It results from the transformation of a natural history written for a peninsular Spanish audience into a popular medical guide, one aimed at an audience of colonial subjects. In directly crediting indigenous Andean people for providing him with treatment knowledge and instructions, however, Delgar largely neglected to acknowledge Cobo's role in compiling such information in its basic form a century earlier. Claiming to have worked with indigenous experts primarily in Charcas, the surgeon described the remedies he listed as "acquired with the reason of the most expert and practiced in their knowledge of plants, trees, roots, stones, flowers, birds, animals, lakes, springs, fish, and other things, which the infinite piety of the Almighty created in this New World of Peru."[28] Delgar thus advocated a local science of sorts grounded in indigenous understandings of the surrounding animal and plant environment in the Potosí and Charcas regions.

At the same time, by engaging contemporary indigenous Andean beliefs about bodies, diseases, and cures, the surgeon's work suggests that boundaries between culturally distinct medical practices and medical ideologies may have blurred substantially in the cities, towns, and villages of the southern Andes a century after Cobo finished his opus. Drawing heavily on indigenous medical practices, Delgar wrote at a time when many doctors and scientists still regarded New World animals and plants as scientific curiosities to be studied for their unusual features, for the knowledge their bodies could convey about faraway lands, and for what they could tell us as a reflection of God's plan.[29] Delgar, however, articulated a more regionally oriented, applied vision in which such organisms possessed curative properties that should be understood and utilized locally. His work thus differs from many texts produced in Lima during the Bourbon period, which sought to create more unified practices by dismissing most forms of indigenous medical knowledge. Delgar's work instead reflects a long process of day-to-day indigenous contact with missionaries such as Cobo, as well as with doctors. Through these individuals, indigenous Andean medical knowledge had come to inform Spanish medical thought among a culturally diverse population in the colonial Peruvian highlands.

By the start of the nineteenth century various pieces of Delgar's writings (and those of Cobo) on animals and plants had become part of home medical manuals found in Lima, Arequipa, and Upper Peru. Two of these manuals are directly attributed to Delgar and are arbitrarily dated 1800 and 1836, which probably correspond to moments when additional catalogs and extracts of texts were bound to them; one of them has an inscription of ownership dating from 1797, suggesting it must have been completed earlier. A third manual is archived in Bolivia, while a fourth one entitled *Recetario eficáz para las familias*, mentioned at the start of this essay, was found in Arequipa and appears to be a near verbatim match to the largest section of the other two. Although it bears a different title from Delgar's other works and nowhere is attributed directly to Delgar himself, the Arequipa volume appears to have been his work and likely lost its title page sometime in the nineteenth century. Composed in a late eighteenth-century script, it joins Delgar's writings with a set of other catalogs bound together. Taken as a whole, however, Delgar's recetarios all reflect the influence of Cobo's work, leading one to question why little attention has been paid to the relatively continuous use in Peru of the medical knowledge about animals and plants that Cobo documented.[30]

Beyond Delgar's own claims and the links to Cobo's writings, I have found some overlap between the eighteenth-century author's lists of herbal remedies and the published lists of nineteenth- and twentieth-century medicinal substances employed by the Kallawaya, who are still much celebrated for their knowledge of plant-based cures. Thirty-one out of ninety-seven plants from Delgar's "Libro de medicina y cirugía para el uso de los pobres" correspond by name and by descriptions of curative properties to those listed as samples in inventories of Kallawaya medicinal plants, which the Bolivian government displayed at the Paris Exposition in 1889 with the hope of attracting European chemists and pharmacists to the country.[31] Numerous other herbs match lists of plants that Joseph Bastien compiled during fieldwork among the Kallawaya in the 1970s and 1980s.[32] While one should be careful in using materials from the ethnographic present and the late nineteenth century to make claims about the colonial period, this overlap certainly suggests that both Cobo and Delgar had access to knowledge and treatment practices common among Kallawaya herbalists. This claim becomes even more plausible when one considers the extent to which Kallawaya healing practices had spread throughout the Andes during the precolonial and colonial periods, transforming local indigenous beliefs. As a favored population under the Inkas, some of the Kallawaya were relocated to Cuzco, while others migrated as healers and practiced their animal- and plant-based cures on various populations. Such practices continued under Spanish rule. Many of the treatments Delgar and Cobo describe must thus pertain either to a Kallawaya medical system or to a system influenced by Kallawaya medical thought.

Analyzing Animals and Humors in Delgar's Work

How did Delgar organize the inclusion of animals in his medical guide, and how did his understandings correspond to those of Cobo and those of the modern Kallawaya? To answer these questions, we must first examine the history of Spanish cataloging practices in Mesoamerica and the Andes, and the degree to which they can convey non-Spanish medical beliefs. Traditionally, authors of New World recetarios cataloged plants and, to a lesser extent, animals by describing them within Spanish humoral frameworks that emphasized and measured qualities of heat, cold, moisture, and dryness. This is certainly true of recetario manuals produced in New Spain. Although recetarios were created and distributed throughout much of Latin America, the most famous home medical guides were those published in Mexico City, which cataloged the healing properties of New Spain's flora and fauna. Perhaps the most famous of these was the Moravian Jesuit missionary Juan de Esteyneffer's *Florilegio medicinal*, first published in Mexico City in 1712, then republished in Amsterdam in 1729, in Madrid in 1729 and 1755, and again in Mexico City in 1887. According to the late George Foster, Esteyneffer's lengthy guide was the most influential of its kind in the New

World and was clearly intended for home medical use, providing "a wide variety of herbal, animal, and mineral remedies, including many drawn from indigenous pharmacopoeias."[33] Perhaps overestimating the popular circulation of Esteyneffer's text and its importance in transforming popular beliefs throughout the colony, Foster claimed that ever since he first examined it he felt "it *must* have played a major role in making humoral medicine common knowledge in Latin America."[34]

In Peru little evidence of medical recetarios exists prior to the seventeenth century, but some works were completed within decades of Delgar's recetarios, drawing on comparable cataloging frameworks. In particular, a 1777 manuscript titled *El médico verdadero* offered various animal- and plant-based cures. According to Foster, a large section had been lifted from classical works, and this likely explains the text's emphasis on nonindigenous animals. Other recetarios were imported from Buenos Aires at the start of the nineteenth century, and travelers in late colonial and early republican Peru also frequently wrote about local diseases and remedies they encountered on their journeys.[35] In general, these texts, much like the manuals from New Spain, reflected a shared belief that the natural features of the New World's environments possessed treasures useful for treating disease and ill health among the poor. In this way, they fomented notions of locally specific medicine, science, and illness while leaving questions about the value of indigenous medicine itself unresolved. They therefore provide few clues as to the origins of indigenous humoral medical knowledge.

Unlike these contemporary recetarios and much like Cobo's natural history completed a century earlier, Delgar's writings attempted to explain indigenous Andean knowledge of plant and animal remedies within a humoral medical framework. The analysis of these kinds of frameworks, however, has proven complicated elsewhere in the Americas, as has been made clear by Foster's debates with various scholars.[36] It is certainly no less complicated here, especially with respect to understanding the medicinal uses of animals. For one thing, although humors can tell us much about indigenous explanations of remedies, we do not actually know when humoral concepts entered indigenous Andean pharmacopoeia and medical thought. Moreover, there is substantial disagreement as to whether such concepts originated in the New World independently of Old World humoral concepts, or whether the Spanish brought such views with them. Nevertheless, scholarship on this matter does prove helpful for understanding the intersection of indigenous and Spanish medical views, and should therefore be examined here at length. For example, Foster argued humoral theories had entered indigenous medical practices and pharmacopoeia after the Spanish Conquest.[37] Challenging Andeanists such as Joseph Bastien and Meso-americanist scholars such as Bernard Ortíz de Montellano and Alfredo López Austin, who argued for the New World origins of humoral models among indigenous people, Foster proposed a "filtering down" model instead.[38] He suggested colonial institutions and practices facilitated the spread of humoral concepts between Spanish conquerors and indigenous people. Educated Spanish residents of the colonies, physicians, and members of religious orders such as the Jesuits figured prominently in disseminating these views due to their contact and evangelizing work among indigenous communities. Often charged with attending to the bodies of indigenous peoples along with their souls, the colony's religious, in particular, rendered such views pervasive, according to Foster.[39]

Cobo's missionary work fits well within Foster's "filtering down" model, perhaps explaining the predominance of humoral thought among the Kallawaya by the eighteenth century, when Delgar reached the Andes. Foster identified four institutions and practices that served as vehicles for the transfer of humoral medical concepts. He considered hospitals and pharmacies to be key sites where indigenous and Spanish beliefs and remedies competed with one another and at times merged. Beyond these urban institutions, his list emphasized the importance of missions and recetarios. He suggested the work of missionary orders extended humoral medical concepts into more remote regions because missionaries often engaged in medical care with indigenous people, who drew on local plants and animals. This was certainly true of Cobo, and it is likely true of other Jesuit missionaries who worked in the Lake Titicaca region, where the Kallawaya resided and practiced medicine among neighboring groups. Foster also identified recetarios such as Delgar's as a parallel written channel for the dissemination of humoral models.[40] Assuming that the distribution of home medical guides was widespread and perhaps overestimating literacy rates and the frequency of access to such knowledge, Foster claimed recetarios "constituted one important avenue of transmission of humoral theory and therapy to New World populations."[41] Although not technically considered a recetario, Cobo's *Historia del Nuevo Mundo* may also have fulfilled this purpose if portions were circulated among evangelizing priests in the highlands.

Foster only knew vaguely about one copy of Delgar's work, and he knew nothing about its author. If he had had a greater sense of the person behind the work and the larger historical context, he would have likely seen Delgar's recetarios not just as tools for the dissemination of a Spanish humoral model, but also as texts that reflected hybrid medical perceptions of treatments, healing, the body, animals, and plants in the eighteenth-century southern Andes. These perceptions were formed as ideas spread between indigenous Andean groups and other sectors of colonial society. The descriptions of medical treatments and beliefs about cures implicit in Delgar's and Cobo's studies, moreover, resemble concepts the Kallawaya articulated to anthropologists in the twentieth century. Joseph Bastien, an outspoken opponent of Foster on these issues, has argued convincingly that many of these concepts had longer, precolonial roots.[42] The modern work of medical anthropologists can thus assist us in gaining a greater historical sense of these indigenous treatment processes and beliefs about humors, provided that we are mindful of the possibility that medical beliefs changed substantially over time. In upstreaming we must also consider the methodological risks of using contemporary studies to gain insight into the past.

Although he looked exclusively at plants and largely ignored indigenous medicinal uses of animals, Joseph Bastien's work on Spanish, Kallawaya, and broader indigenous Andean humoral systems sheds light on how Delgar and others in the Charcas region likely understood the medicinal effects of both plants and animals on the human body. Trained as a medical anthropologist, Joseph Bastien rightly characterized humoral medicine in Latin America as less homogeneous than Foster had described.[43] Having conducted fieldwork among the Kallawaya in the 1970s and 1980s, Bastien argued that Kallawaya healers saw health as dependent on the proper cycle or flow of fluids and semifluids in the body, which included water, air, blood, and food. The body distilled these materials into secondary fluids and semifluids, among them mucus, bile, sweat, urine, gas, milk, semen, feces, and fat. While the Kallawaya saw fat as a marker of good health and a source of energy in the second half of the twentieth century, they believed all other secondary fluids and semifluids needed to be eliminated regularly from the body in order to preserve health. If not removed, they could become toxic. Disease therefore resulted from stopping the cycle of circulating blood, distilling fluids, and eliminating waste products.[44]

This modern Kallawaya understanding of the human body in many ways bears relation to the traditional Greek humoral model, which Foster believed the Spanish brought with them to the Andes, and which doubtlessly formed part of Delgar's medical training. However, Kallawaya notions of humors also differ considerably in ways that reflect indigenous Andean influences. These influences are evident in Delgar's descriptions of animals.

In the Greek or Spanish model, good health was achieved by maintaining the proper balance of four humors, disease was caused when this balance was disturbed, and medicine cured disease by restoring proper balance. The Kallawaya, on the other hand, emphasized the circulation and transformation of fluids, not their equilibrium. They understood the body as shaped by a set of hydraulic dynamics, in which centripetal forces within the body distilled liquids and centrifugal forces dispersed them to the periphery. Bastien claimed that for the Kallawaya specifically, and for indigenous Andean herbalists more generally, the idea of balance so fundamental to Greek humoral theory was irrelevant. This was the case because the basic humors in the Andean humoral system—air, blood, and fat—were in a constant process of movement, concentration, and dispersal that generated fluctuations in quantity. Health was created through the maintenance of this circulation and the process of converting fluids, while disease was caused by the disruption of this cycle through the loss of fluids or the inability to expel fluids. Indigenous Andean herbalists in Bastien's analysis were thus concerned with cleansing, repairing, and maintaining the conduits through which bile, feces, gas, phlegm, semen, sweat, and urine flowed.[45] These practices are repeatedly emphasized in the work of Delgar and, to a much lesser degree, of Cobo.

Despite these fundamental differences, indigenous Andean humoral systems resemble their Spanish and Greek counterparts in that both systems focus on qualities of heat, cold, moisture, and dryness for understanding and treating health problems. Although Bastien disagreed with Foster about the origins of indigenous humoral systems, he joined him in attributing this particular feature to the presence of missionaries in Upper Peru such as Cobo, who traveled with herbal manuals during the early colonial period and taught indigenous populations to classify plants in this manner. Bastien also claimed, however, that the uses of these categories differed from other Greek-based systems. The Greeks saw applying remedies according to their properties of heat, cold, moisture, and dryness as a necessary procedure for restoring the balance of humors. The Kallawaya and other indigenous populations in the Andes, on the other hand, saw these qualities as crucial for restoring the fluidity of what Bastien described as the body's "cyclical hydraulic system." According to Bastien, indigenous Andean peoples understood that the cyclical processes within the body depended on an asymmetry of heat, cold, moisture, and dryness that varied in a pendulum-like movement over time, facilitating the movement of liquids. Excessive levels of these properties could cause disease by disrupting the circulation or exacerbating the elimination of fluids. These problems could be resolved, however, through the application of medicinal remedies known to possess the opposite properties. Kallawaya medicine thus drew on multiple ideas of humors from both Spanish and indigenous Andean cultures.[46]

It is important to consider the subtleties of these humoral systems because both Cobo's descriptions of indigenous practices involving animals and Delgar's descriptions of animal-based cures reflected the influence of Kallawaya (and broader indigenous Andean) assumptions about the workings of the human body. However, the authors also reconfigured such understandings within traditional Spanish concepts of humoral systems, and they employed Spanish stylistic formats used for cataloging and describing remedies in recetario manuals. At one level, the merging of these concepts and practices may be the result of considerations about the desired audience on the part of both authors. Cobo intended through his work to convey knowledge about New World flora, fauna, and minerals to an audience in Spain curious about how to make sense of this newly discovered environment. He thus sought to "translate" indigenous medical knowledge in a way intelligible to an elite Spanish reader. On the other hand, Delgar's recetario reflected an effort among Spanish settlers in the Americas to make practical sense of the environment around them. Delgar employed multiple medical frameworks to create what one might describe as a locally specific science of sorts, which could be employed by everyday colonial subjects of diverse backgrounds. In this way, the work of Delgar drew on knowledge of natural history but had a different set of goals. He sought to foment local knowledge of the uses of animals and plants in order to improve health conditions and provide everyday subjects with the means to overcome illness. By transforming Cobo's natural history for this purpose, Delgar intended to popularize the understanding and use of Andean animals, animal parts, and plants in healing.

The Workings of Animal-Based Treatments

In order to understand how Delgar and indigenous populations like the colonial Kallawaya explained the healing properties of animals, we must now compare Delgar's work both to Cobo's *Historia del Nuevo Mundo* and to Bastien's ethnographic findings. In doing so, we must focus especially on how Delgar cataloged animals in relation to plants. Delgar categorized and described plants according to whether they had intrinsically warm or cold properties, and whether they were inherently moist or dry. He described these main humoral properties as "the qualities philosophers describe as primary."[47] His work mimicked Cobo's writings by employing the same categories and the language of degrees, and it generally reproduced the strategies found within broader Spanish genres of medical and botanical writing. However, Delgar's narrative descriptions of remedies also reflected the understanding of humors Bastien described for the Kallawaya. Furthermore, his approach to discussing the humors and properties of remedies appears to have served as a conscious teaching device for explaining both of these systems to everyday people.

Outlining his method of describing medicinal items found in nature, Delgar wrote that "in order to know them, [each one] will be categorized within four degrees, which rational philosophy explains, leaving aside what is considered [neutral] temperament, which is that which is neither cold, nor hot, nor dry, nor wet."[48] Drawing on Cobo's catalog of the natural world and selectively incorporating different sections, Delgar then listed animals and plants one by one. He described where they were found, what they looked like, and how indigenous Andean peoples used their specific parts as curative medicines. Basing these descriptions on Spanish and local humoral models, Delgar claimed indigenous plant remedies worked because they counteracted extremes of heat, cold, moisture, and dryness, which impeded or otherwise affected the circulation of fluids in the human body. He thus argued that plants defined as intrinsically warm could be used to cure ill health caused by the presence of cold. Conversely, plants defined as intrinsically cold could be used in the treatment of fevers. Both would adjust the expulsion or retention of humors. In this way, his descriptions merged the indigenous models Bastien analyzed for the Kallawaya with Spanish humoral systems, emphasizing both the balance and circulation of humors.

Animals, animal parts, and items produced by animals often served as mediums or vehicles that facilitated the application or consumption of herbs and other plants with curative powers. These vehicles most commonly were milk, honey, and animal fat, but they could also include items animals made, such as nests. In many cases, however, animal parts and products also served as substances with palliative or curative powers in their own right. Much like plants, they could be made into compound treatments through burning, drying, mashing, grinding, or mixing with other ingredients.[49] Nevertheless, in his catalogs of remedies Delgar described animal- and plant-based treatments in different stylistic fashions. On the one hand, it is clear from both Cobo's and Delgar's works that plant remedies in the seventeenth and eighteenth centuries were classified in varying degrees of heat, cold, moisture, and dryness, much as they are today among the Kallawaya.

Their palliative and curative powers, moreover, were understood as counteracting temperature and moisture levels in the body, which obstructed the circulation of elements. On the other hand, Delgar discussed animals in most cases without complete or systematic reference to their intrinsic heat, cold, moisture, or dryness, and he focused much more heavily on methods for converting animals and animal parts into medicinal treatments. Such cataloging practices thus require explication, especially with regard to indigenous beliefs.

Delgar's failure to catalog animals in a manner similar to plants may be the result of a number of factors. For one thing, Spanish herbalists and botanists appear to have traditionally thought of animal substances as neutral vehicles for applying cures, rather than as substances with their own curative and humoral properties. As a result, while they included with classifications of plants a formulaic description of their properties of heat, cold, moisture, or dryness, they did not employ a strong parallel practice for listing animals. In addition, unlike plants, animals were thought of as more complex sets of parts with qualities that varied and fluctuated. The various pieces of each animal's body—whether mammal, insect, bird, or fish—likely differed in terms of their humoral properties, and they certainly differed in terms of their applications. Each part thus served to counteract a variety of disruptions to the circulation of humors caused by irregularities of heat, cold, moisture, and dryness in the body and on its surface. Finally, many of the animals were native to the Andes and specific to particular regions, so Delgar may not have observed them firsthand or known how to classify them. However, descriptions of how animal remedies worked in Delgar's manual certainly did correspond to Kallawaya and broader indigenous Andean beliefs about the role of temperature and moisture as properties that restored and mediated the cycle of fluid circulation. Plants and animals thus worked in similar ways in terms of functions and cures.

The only animals that Delgar and Cobo categorized systematically according to degrees of heat, cold, moisture, and dryness were two kinds of worm, both of which they considered highly caustic. Descriptions of their medicinal uses, moreover, supported an indigenous Kallawaya understanding of health problems caused by the loss of fluids and the blockage of fluids. The worm *mupullo* or *musullo*, which was found in trees and which Delgar and Cobo classified as hot in the fourth degree and dry in the third, could heal and close infected cuts if doctors toasted its flesh and mashed it into a dry powder, which they would then apply topically. Providing the initial description of this item, Cobo offered less elaborate instructions than did

Delgar, but described indigenous Andean people provocatively as using the worm "for their sensualities."[50] More important, however, Delgar claimed it also cured difficulties with urination when applied directly to the patients' genitalia, provided one had previously coated that region with oil. The resulting irritation of the genitals, Delgar suggested, led the patient to "urinate with great force," expelling fluids trapped in the body and restoring the proper internal cycle of fluid movement. In order to prevent excessive irritation and loss of fluids, which would ultimately worsen health, healers could apply fat from cows or pigs to the irritated regions, neutralizing the caustic properties of the worm.[51]

Delgar also classified the small worm *chuqui chuqui* as hot in the fourth degree and dry in the third degree, and he claimed it was useful for several kinds of treatments. Its body could be used to remove superfluous, infected flesh from cuts if first toasted and ground into a powder, then applied directly to the patient. This enabled the cut to heal, protecting the patient from further loss of blood, a substance the modern Kallawaya have long seen as precious and finite in quantity within the human body, according to Bastien.[52] The worms' inner fluids, moreover, could cause warts to dry up and disappear if applied properly.[53]

In other cases Delgar described animals by using the language of humors, but without using consistently the categories and degrees of heat, cold, moisture, and dryness. For example, Delgar described the paste from a catfish—known here as *suches*—as particularly hot, but he gave no indication of a precise degree, nor did he speculate as to its qualities of moisture or dryness. Drawing on Cobo's description in which the freshwater fish was listed by its other name, *bagre*, he noted that it could be used to treat "hard and rebellious abscesses," and that if mixed with another kind of fish, it could resolve hard tumors, sebaceous cysts, and hardenings of the kidney and bladder. He also suggested that consumption of the fish could inflame the throat and cause gumboils and cuts.[54] In another case, he described the chinchilla rabbit as being of a cold temperament, but he did not describe it in terms of degrees and again neglected categories of moisture and dryness.[55] The meat of the chinchilla was absolutely useless and consuming it provoked responses similar to ingesting poison, yet according to Delgar one could make a compress to stop bleeding in wounds by mixing the animal's abundant fur with egg whites and other powders. This was something Cobo also acknowledged.[56] Likewise, Delgar described the vicuña, a wild camelid prized for its wool in the Andes, as "an animal of cold temperament, both in its meat and its wool." While Cobo believed the vicuña's meat was useful for reducing swelling around the eyes, Delgar argued it was particularly effective at reducing swelling around the testicles caused by hernias. It could be applied warm or cold "according to the state of the patient."[57] Delgar, however, did not propose any uses for the animal's wool. While most animals were considered to have different curative properties through their multiple body parts, the vicuña was not alone in being identified as having just one body part with medicinal value.

On other occasions, Delgar employed the language of degrees while drawing on new categories absent from traditional humoral classifications. For example, Delgar described an animal known as *vinco vinco* as being of "a poisonous temperament that reaches the fourth degree."[58] Providing a description of his own that bears no relation to those found in Cobo's *Historia del Nuevo Mundo*, Delgar noted that when the substance of the vinco vinco is spread on hemorrhoids, "it makes them purge a certain watery fluid and in three or four days they are healthy, clean, and disappear."[59] In this way, the animal facilitated the expulsion and circulation of fluids, while powders made from this animal could also dry and heal infected cuts.

Aside from these creatures, Delgar suggested that other animals were useful in mediating the circulation of fluids, but he did not generally classify them according to levels of humoral qualities. Nevertheless, such animals often provided cures that followed a Kallawaya notion of humors and their circulation. For example, a small animal named *ancocate tocotoco* possessed numerous curative properties when doctors made use of its hair and flesh in different forms. Described as a black and white creature the size of a pig, with a pouch for carrying its young, that is native to the Charcas region, the ancocate tocotoco was most likely a species of opossum. Its hair could be employed as a dry substance to end the loss of fluid through nosebleeds. Its meat, however, could do something quite the opposite, facilitating the expulsion of menstrual blood in women when consumed in dried or powdered form.[60] Delgar, however, offered no description of its intrinsic qualities of heat and cold.

Other animals proved useful in expelling kidney stones and resolving illnesses affecting the bladder and urination, restoring ideal, proper Kallawaya forms of fluid circulation. These included an animal named *quispicancho*, whose tailbone could be drunk as a powder mixed with lukewarm water and lemon juice. It also included the *achocalla*, a kind of Andean weasel known for covering itself with mud and then attacking snakes. Delgar suggested that patients, in order to break down kidney stones, grind the animal's bones into a powder and drink them mixed with wine—a blending of Andean Spanish ingredients.[61] A species of South American armadillo named *quirquincho* could also dissolve kidney stones when its tailbones were converted into a powder, mixed with lemon juice, and consumed as a beverage.[62]

Animal parts could also be manipulated according to their intrinsic properties to facilitate or limit the release of other fluids from the body, thus corresponding once again to broader Kallawaya humoral notions of the circulation of fluids. Delgar recommended using burnt llama wool to stop bleeding when removing teeth, to end nosebleeds, and to dry out cuts.[63] The feathers from the *guacamayo* (macaw) could not only be used in dry, powdered form to end nosebleeds, but could be mixed with egg whites and rosewater to form a plaster that would relieve sinus pressure in the face. Burning the feathers of the *tungui tungui* bird and mixing them with vinegar and other powders created a remedy for the overproduction of "salty phlegm." Likewise, condor skin could be used to prevent *el estómago flojo* (dysentery), and its fat could be used topically to treat not only joints swollen by gout, but also throat pain, hard tumors, and pinched nerves. Fat taken from the kidneys of a species in the chinchilla family named *viscacha*, moreover, could eliminate the broader problems of purging within the body, especially if administered in combination with other liquids. These practices returned the body to its normal cycle of circulating and processing fluids and semifluids, which the Kallawaya saw as fundamental to health.[64]

Fats from animals generally could also be used for softening tumors and resolving the hardening of certain abdominal organs, and they could be applied to the ear canal to relieve ear pain. For example, Delgar argued the guinea pig, which is known in his work as *coyo guanca* and in Cobo's as *cuy*, had obvious medicinal properties of this sort. Noting that it was bad for those suffering certain diseases because its meat increases their levels of pain, he suggested that its fat could relieve pinched nerves and tumors caused by poor digestion, among other things. It could also solve ear pain and glandular problems, as well as the hardening of the stomach and liver.[65] Fat taken from the quirquincho and a bird known as *yuris pájaro* could also serve these purposes. Delgar added claims about ear pain for the yuris pájaro that are missing from Cobo's work. Cobo, however, differed from Delgar in claiming that the bird, which he believed was a relative of the ostrich and common in the region near Buenos Aires, could be used to relieve pain caused by bleedings.[66]

Finally, making smoke out of animal parts could also have medicinal effects. One animal not found in Cobo's work is the *papa pahspa*, a nocturnal bird Delgar described as having a variety of potential uses. Delgar suggested that the smoke created by burning its feathers could provide relief from malaria fevers, while a stew made with its carcass could work against more general kinds of fevers. Spreading a paste made with the bird's brains on scars, moreover, could cause them to flatten, soften, and eventually disappear without a trace.[67] With respect to the llama, Delgar added to Cobo's work by arguing that its body could be used in the medicinal treatment of other animals. He suggested that "the smoke of its fat eliminates their struggle, and mixed with sulfur and a little mercury it heals irritation on horses' hooves and mange on rams."[68] In this way, the manipulation of animal parts for medicinal uses in Delgar's work reflects the influence of a rich indigenous Andean medical culture, one that he aimed to disseminate among other sectors of society for popular use. In many cases, it also reflects Kallawaya, and broader indigenous Andean, beliefs about humoral systems and the circulation of fluids within the human body.

Conclusion

Delgar's recetario suggests that beliefs about the ability of animal-based remedies to manipulate the circulation and processing of fluids in humans were widespread among indigenous peoples in the eighteenth-century southern Andean highlands, and were in fact similar to beliefs about the workings of plant-based medicinal remedies. His failure to categorize animals consistently and systematically according to degrees of heat, cold, moisture, and dryness most likely resulted from two factors: a European tradition of scientific writing that constrained how he cataloged animals, and the complicated nature of animals' bodies, in which different parts possessed different qualities. It likely did not stem from an absence of these categories among indigenous people, or from a lack of ideas about how animal-based substances and treatments affected the humoral cycle in colonial Kallawaya and broader indigenous Andean medical thought. Delgar's own work, moreover, attempted to move these beliefs from indigenous Andean medical systems into the popular medical pharmacopoeia and system of beliefs about the body in eighteenth-century urban Peru. There, he hoped such beliefs would form part of a complex medical system based on both indigenous Andean and Spanish ideas about humors.

Delgar's volume points to a long and complex practice of drawing on animals and manipulating animal substances as part of a broader set of palliative and curative treatments in the natural world. In this way, analysis of his medical recetarios suggests that Andean peoples of various cultural backgrounds understood the relationship between animals and themselves at least in part through a colonial, hybrid medical framework that integrated Spanish and indigenous medical knowledge. At the same time, Delgar's work sheds light on how much remains to be learned about the histories of indigenous Andean medicine, popular medicine, and medicinal uses of animals in the colonial period. His recetarios raise the question as to what degree indigenous Andean and other highland populations in Peru considered animals to be a separate category from plants with respect to their medicinal properties. Even less is known about how the Kallawaya and other groups hunted, captured, and exchanged such animals as the condor or the ancocate tocotoco for medicinal use.

Further research may be required to answer these questions with any reasonable certainty, but it does appear that Delgar saw animal-based treatments as central to indigenous Andean healing repertoires. For indigenous and nonindigenous poor residents of the colony, moreover, his work suggested, knowledge of the natural world could ultimately serve to extend human life. Such outcomes, however, ironically depended on the extinguishing of nonhuman animal life and the conversion of animal parts into remedies for human consumption. In this way, various Andean peoples appear to have approached healing and longevity as frequently necessitating the destruction of nonhuman animals for human medical benefit. Animal death and the process of "preparing" animal-based remedies thus likely proved central to how many members of colonial society understood their own bodies and health in relation to the natural world around them, fitting Erica Fudge's call to study animals as a window into "the wider world of human relationships with nature" and to think about animals as "a necessary part of our reconceptualization of ourselves as human."[69] In this way, the analysis of colonial medicine at the popular level requires adopting an approach to the past that recognizes the importance of the animal kingdom.

Notes

1. Some of the main works on this topic are Antonio Barrera-Osorio, *Experiencing Nature: The Spanish American Empire and the Early Scientific Revolution* (Austin: University of Texas Press, 2006); Daniela Bleichmar, "Visual Culture in Eighteenth-Century Natural History: Botanical Illustrations and Expeditions in the Spanish Atlantic" (PhD diss., Princeton, 2005); Jorge Cañizares-Esguerra, *Nature, Empire, and Nation: Explorations in the History of Science in the Iberian World* (Stanford: Stanford University Press, 2006); Paula De Vos, "An Herbal Eldorado: The Quest for Botanical Wealth in the Spanish Empire," *Endeavor* 27.3 (2003): 117–21; Paula De Vos, "The Science of Spices: Empiricism and Economic Botany in the Early Spanish

Empire," *Journal of World History* 17.4 (2006): 399–427; Londa L. Schiebinger, *Plants and Empire: Colonial Bioprospecting in the Atlantic World* (Cambridge: Harvard University Press, 2004).

2. A notable exception to this first trend is Miguel Asúa and Roger French, *A New World of Animals: Early Modern Europeans on the Creatures of the Iberian World* (Burlington: Ashgate, 2005).

3. Spain's Archive of the Indies (AGI) houses numerous examples of such animal shipments. For example, in 1789 the viceroy of New Granada, Francisco Gil y Lemos, sent a shipment of live animals and skeletons on behalf of the archbishop of Santa Fé de Bogotá, which landed in La Coruña. In another case, the bishop of Trujillo provided the viceroy of Peru with twenty-four boxes of samples to be sent to Spain, including one box of actual quadrupeds and another containing an extensive collection of birds. AGI, Audiencia de Lima 798, "Cartas y expedientes: Curiosidades para el Jardín Botánico."

4. Such practices in Lima culminated in the development of the *Examen de anatomía, fisiología, y zoología, que presentan en la Real Universidad de San Marcos de Lima a mañana y tarde, y consagran al Excmo. Sr. Virrey su fundador los alumnos del Colegio de San Fernando* . . . (Lima: Imprenta de Huérfanos, 1812). For surgery in Spain, see Michael E. Burke, *The Royal College of San Carlos: Surgery and Spanish Medical Reform in the Late Eighteenth Century* (Durham: Duke University Press, 1977).

5. The historical importance of animals in eighteenth-century southern Andean medical pharmacopoeia differs from what medical anthropologists have emphasized among southern Andean populations in the twentieth century: a heavy reliance on plant-based cures, the relative absence of curative substances made from animals, the dependence on animal-based substances only as vehicles for treatment, and the use of animals' bodies only for ritual healing purposes. Two studies that exemplify this trend are Joseph Bastien, *Healers of the Andes: Kallawaya Herbalists and Their Medicinal Plants* (Salt Lake City: University of Utah Press, 1987); Mario Polia Meconi, *"Despierta, remedio, cuenta . . .": Adivinos y médicos del Ande* (Lima: Pontificia Universidad Católica del Perú, 1996).

6. The epigraph quotation is given in Hermilio Valdizán and Angel Maldonado, eds., *La medicina popular peruana* (Lima: Imprenta Torres Aguirre, 1922), 3:141–42.

7. Other Spanish language recetarios focused on animal remedies included the anonymous work entitled "El médico verdadero: Prontuario singular de varios selectísimos remedios, para los diversos males a que está expuesto el Cuerpo humano desde el instante que nace: Compuesto por un curioso, para el alivio de todos los que se quieran" (Lima: unpublished manuscript, 1771). A version of this manuscript is included in Valdizán and Maldonado, *La medicina popular peruana*.

8. My work here examines Delgar's unpublished manuscript "Libro de medicina y cirugía para el uso de los pobres" (Biblioteca Nacional del Perú [BNP] Manuscritos D59 C), his *Nuevo thesoro de pobres* (BNP Manuscritos D12936), and a related work entitled "Terapéutica indígena boliviana," reproduced in the *Archivos*

de medicina boliviana 1.1–2 (1943): 187–245. I also base my analysis on the text from Arequipa entitled *Recetario eficáz para las familias*, republished by Valdizán and Maldonado. It is a verbatim match to Delgar's other works, and thus I conclude that Delgar is the author.

9. Much of this effort was centered in Lima, where a group of mostly creole Spanish doctors and natural philosophers sought to refashion themselves as the champions of medical reform in the colony. See Adam Warren, *Medicine and Politics in Colonial Peru: Population Growth and the Bourbon Reforms* (Pittsburgh: University of Pittsburgh Press, 2010); Adam Warren, "Piety and Danger: Popular Ritual, Epidemics, and Medical Reforms in Lima, Peru, 1750–1860" (PhD diss., University of California, San Diego, 2004).

10. In making this assertion, however, I do not mean to suggest that indigenous Andean medicine remains timeless, unchanged, or marked only by continuities connecting modern practices back to the colonial period. Rather, I argue that modern practices, while complex and ever changing, were built on a foundation of earlier beliefs about humors, the body, and healing that becomes evident in the work of ethnographers.

11. Erica Fudge, "A Left-Handed Blow: Writing the History of Animals," in *Representing Animals*, ed. Nigel Rothfels (Bloomington: Indiana University Press, 2002), 6.

12. Fudge writes that rather than merely look at animals as symbols, it is the historian's duty "to understand and analyze the uses to which animals were put. If we ignore the very real impact of human dominion—whether in meat-eating, sport, work, or any other form—we are ignoring the fundamental role animals have played in the past" ("A Left-Handed Blow," 7).

13. This section summarizes a more extensive analysis of the relationship between Delgar's recetario and Cobo's natural history in Adam Warren, "*Recetarios*: Sus autores y lectores en el Perú colonial," *Histórica* 33.1 (2009): 11–41.

14. Philippe Descola and Gísli Pálsson, "Introduction," in *Nature and Society: Anthropological Perspectives*, ed. Philippe Descola and Gísli Pálsson (New York: Routledge, 1996), 4. In this quote Descola and Pálsson paraphrase the argument of Roy F. Ellen in the same volume.

15. Francisco Mateos, "Introducción: Personalidad y escritos del P. Bernabé Cobo," in *Obras del P. Bernabé Cobo de la Compañia de Jesús*, ed. Francisco Mateos (Madrid: Atlas, 1964), v–xlvii.

16. Mateos, "Introducción."

17. Cited in Joseph Bastien, *Mountain of the Condor: Metaphor and Ritual in an Andean Ayllu* (St. Paul: West, 1978), 23.

18. Cobo's thinking about the medicinal applications of animals, however, was also shaped by his time in Lima. While residing in Lima, between 1613 and 1615, he attributed medicinal properties to the turkey buzzard (*gallinazo*), a bird commonly found in the city's squares. Cobo wrote that consuming its meat provided relief from syphilis, and he argued one could heal wounds by applying its skin in burnt form. Furthermore, he noted, when a city resident lost his mind in 1614,

"they cured him over fifteen days by giving him turkey buzzard broth drawn through a filter, and with this he recovered his judgment and healed so completely that, afterward, he became a religious and carried out mass" (Bernabé Cobo, *Historia del Nuevo Mundo*, book 7, chap. 14, in *Obras del P. Bernabé Cobo de la Compañia de Jesús*, ed. Francisco Mateos [Madrid: Atlas, 1964], 319).

19. The Jesuits administered the region as a *doctrina de indios* starting in 1576, and they attempted to convert its indigenous population. Francisco Mateos described the mission as going through its golden age while Cobo was there. It became a site of experimentation where priests developed tools for further evangelization projects, which were later applied "with such happy results for the evangelization of Paraguay, of Mojos, and other parts of the American interior" (Mateos, "Introducción," xxi).

20. Mateos, "Introducción," xiii.

21. Mateos, "Introducción," xiii.

22. It is worth noting that the Kallawaya do speak their own secret language within their communities. However, as an Aymara ethnic group who traveled extensively, it is very likely that they would have communicated with figures like Cobo and Delgar in Quechua and Aymara.

23. Few scholars have ever written about Martín Delgar. In fact, I have found only two relevant scholarly discussions. One is the brief introduction to "Terapeútica indígena boliviana." Elsewhere, Marcos Cueto described Delgar as "a surgeon who arrived in Peru from Europe in the mid-eighteenth century and distinguished himself with his 'miraculous' cures, especially in the provinces" ("Guía para la historia de la ciencia: Archivos y bibliotecas en Lima," in *Saberes andinos: Ciencia y tecnología en Bolivia, Ecuador y Perú*, ed. Marcos Cueto [Lima, Instituto de Estudios Peruanos, 1995], 163).

24. Cobo, *Historia del Nuevo Mundo*, book 9, chapter 61, 369–70.

25. Delgar, *Recetario eficáz para las familias*, 124. Cobo also mentioned that eating the animal's meat could assist those with mercury poisoning, but he provided no concrete examples.

26. Hipólito Unanue, "Decadencia y restauración del Perú: Oración inaugural que para la estrena y apertura del Anfiteatro Anatómico, dijo en la Real Universidad de San Marcos el D. D. José Hipólito Unanue el día 21 de noviembre de 1792," *El Mercurio Peruano*, 10 February 1793, 106–7.

27. See, for example, José Pezet's bachelor's thesis in medicine, in which he refers to the surgical practices of "Martinus Delgar" ("Conspectus Disputationis Medicae, Quam Pro Gradu Baccalaureatus Obtinendo, Auspice Deo, et Praeside D. D. Josepho Hippolito Unanue, Anatomes Professore, sustinebti Josephus Pezet, Baccalaureus Physicus, Regii Anatomes, Amphiteatri Alumnus" [Lima: Imp. de los Niños Huérfanos, 1798]).

28. Delgar, "Libro de medicina y cirugía para el uso de los pobres."

29. Scholars charted a shift in the colonial period from the idea of plants and animals as "wonders" of the New World to the notion that they could also provide empiri-

cal knowledge about the place. Antonio Barrera-Osorio charts several of these shifts in *Experiencing Nature*.

30. Other authors reproduced passages of Cobo's work in the early nineteenth century, several decades after Delgar had finished his "Libro de medicina y cirugía para el uso de los pobres." The famous botanist Antonio José Cavanilles, for example, published ten chapters in the *Anales de Ciencias Naturales*, in 1804. Several other scholars published chapters from Cobo's *Historia de la Fundación de Lima* over the course of the nineteenth century. Manuel González de la Rosa published the first complete version in Peru in 1882.

31. For the complete list, which is far from an exhaustive catalog, see Carmen Beatriz Loza V., *Kallawaya: Reconocimiento mundial a una ciencia de los Andes* (La Paz: Viceministerio de Cultura / Fundación Cultural del Banco Central de Bolivia / UNESCO, 2004).

32. Bastien, *Healers of the Andes*.

33. George M. Foster, *Hippocrates' Latin American Legacy: Humoral Medicine in the New World* (Langhorne, Penn.: Gordon and Breach, 1994), 155.

34. Foster, *Hippocrates' Latin American Legacy*, 155. Other Mexican recetarios included earlier work such as Agustín Farfán's 1579 text *Tractado breve de anothomía y chirugía, y de algunas enfermedades, que más comunmente suelen haber en esta Nueva España*, which was republished as the *Tractado breve de medicina* in 1592.

35. See, for example, "Viaje hecho al partido de Larecaja por el Dr. Dn. José María Boso el 2 de septiembre de 1821 en que se han descripto varias plantas particulares botánicamente," in *La medicina popular peruana*, ed. Hermilio Valdizán and Angel Maldonado (Lima: Imprenta Torres Aguirre, 1922), 3:317–415.

36. See George M. Foster, "On the Origin of Humoral Medicine in Latin America," *Medical Anthropology Quarterly, New Series*, 1.4 (December 1987): 355–93.

37. Foster, "On the Origin of Humoral Medicine in Latin America." Also see Foster, *Hippocrates' Latin American Legacy*.

38. See, for example, Bernard Ortiz de Montellano, *Aztec Medicine, Health and Nutrition* (New Brunswick: Rutgers, 1990); Alfredo López Austin, *The Human Body and Ideology: Concepts of the Ancient Nahuas* (Salt Lake City: University of Utah Press, 1988).

39. Foster, *Hippocrates' Latin American Legacy*.

40. Foster, *Hippocrates' Latin American Legacy*, 150–58.

41. Foster, *Hippocrates' Latin American Legacy*, 157.

42. See, among others, Joseph Bastien, "Qollahuaya-Andean Body Concepts: A Topographical Hydraulic Model of Physiology," *American Anthropologist, New Series* 87.3 (September 1985): 595–611; Bastien, *Healers of the Andes*.

43. Foster disagreed with Bastien but did not directly attack Bastien's work in his main monograph. Instead, he simply noted that Bastien acknowledged the possible influence of Greek humoral pathology and "feels it is quite possible that the humoral variant he finds among the people he studies—the Qollahuaya Quechua of Bolivia—is indigenous in origin" (*Hippocrates' Latin American Legacy*, 148).

44. Bastien, *Healers of the Andes*, 46.
45. Bastien, *Healers of the Andes*, 47, 55.
46. Bastien, *Healers of the Andes*, 47.
47. Delgar, "Libro de medicina y cirugía para el uso de los pobres."
48. Delgar, "Libro de medicina y cirugía para el uso de los pobres."
49. The most common animal products used as mediums for the application of me-
    dicinal treatments were eggs and honey. Delgar proposed they could be used to
    make plasters and ointments. Other animal products employed in this fashion
    were animal brains, especially bird brains, mud from bird's nests, and animal feces,
    a common ingredient for making poisons. Hair and fur could also be included in
    this list (Delgar, "Libro de medicina y cirugía para el uso de los pobres").
50. Cobo, *Historia del Nuevo Mundo*, book 9, chap. 22, 246.
51. Delgar, "Libro de medicina y cirugía para el uso de los pobres."
52. Bastien, *Healers of the Andes*, 46.
53. Delgar, "Libro de medicina y cirugía para el uso de los pobres."
54. Delgar, *Recetario eficáz para las familias*, 188.
55. Delgar, *Recetario eficáz para las familias*, 141.
56. Delgar, "Libro de medicina y cirugía para el uso de los pobres."
57. Delgar, *Recetario eficáz para las familias*, 196.
58. Delgar, *Recetario eficáz para las familias*, 195. I have been unable to find a reliable
    definition for *vinco vinco*.
59. Delgar, *Recetario eficáz para las familias*.
60. Delgar, "Libro de medicina y cirugía para el uso de los pobres."
61. Delgar, "Libro de medicina y cirugía para el uso de los pobres."
62. Delgar, "Terapeútica indígena boliviana," 216. Cobo added that the armadillo's
    shell could be ground into a powder that was useful in the care and perfuming of
    babies.
63. Delgar, "Terapeútica indígena boliviana," 188.
64. Delgar, "Libro de medicina y cirugía para el uso de los pobres."
65. Delgar, *Recetario eficáz para las familias*, 136.
66. Delgar, *Recetario eficáz para las familias*, 154.
67. Delgar, *Recetario eficáz para las familias*, 169.
68. Delgar, *Recetario eficáz para las familias*, 155.
69. Fudge, "A Left-Handed Blow," 5.

5

Pest to Vector

Disease, Public Health, and the Challenges of State-Building in Yucatán, Mexico, 1833–1922

HEATHER McCREA

In 1889 two young boys, Frank and Fred, otherwise known as "The Boy Travellers," headed for Mexico with their guide and mentor, Doctor Bronson, to chronicle "the land of the Aztecs, its history and resources, the manners and customs of its people, and the many curious things to be seen."[1] Such was the premise of Thomas W. Knox's novel, *The Boy Travellers in Mexico*. Knox wrote a series of juvenile travel stories that combined the textual tropes of the children's encyclopedia (maps, dates, histories, descriptions of manufacturing and manners, etc.) and the textual trope of the adventure story (exploration and danger), to teach his readers about the geography and culture of such exotic parts of the world as the Far East, Arabia, and the Americas. Knox claimed that the only fiction in his books was the names of the boy narrators.[2]

While traveling through Mexico's southeastern Yucatán Peninsula to view the ancient wonders of the Maya civilization, the well-heeled youths note the commingling of livestock and pets with humans in the markets and homes, and they continually grouse about the unrelenting heat and pesky insects.[3] They hear dramatic tales about venomous snakes, noisy iguanas, and gigantic spiders, and experience the infamous, relentlessly menacing *garrapatas* (ticks) (fig. 5.1).[4] In fact, one of the worst annoyances of their visit to the ancient Maya site of Uxmal was "that whenever they moved about they became covered with garrapatas . . . capable of making a bite or sting like that of a red ant or a hot needle."[5] One of the boy's guides to

5.1 "An Unwelcome Visitor." SOURCE: THOMAS W. KNOX, *THE BOY TRAVELLERS IN MEXICO* (NEW YORK: HARPER AND BROTHERS, 1890), 510.

Uxmal, an American living in Yucatán by the name of Mr. Benson, warns the boys that the garrapatas also attack dogs and other animals, "and the poor creatures are sometimes killed by them." For their readers, Frank and Fred recount a story, reported by the French archaeologist Claude-Joseph Déserié M. Charnay, about a pet dog belonging to the wife of the U.S. consul to Mérida, which was literally consumed by garrapatas at a hacienda in the Yucatecan countryside: "The little animal rolled on the grass and howled in agony, but the garrapatas kept on with their biting as though it was all fun to them."[6] The projection of human emotion onto the garrapatas as they "gleefully" destroy an innocent pet is typical of the book's pedagogic tone. The untamed and the tamed collide here in a murderous moment.

Themselves baffled by the "wildness" that surrounds them throughout southern Mexico, the boys Frank and Fred marvel at the way the locals endure such wild conditions. The perception held by travelers like Frank and Fred that Yucatán was infested with pests—both visible and invisible—merged into early twentieth-century public-health preventatives and discourse targeting insects and animals as agents of disease. Less than two decades after Frank and Fred surveyed Yucatán's countryside, the U.S.-based Rockefeller Foundation's International Health Board joined forces with the Mexican government and dispersed physicians throughout Mexico. Bent on eradicating yellow fever and malaria, the Rockefeller Foundation's campaigns focused on the elimination of the mosquito, the insect linked to the proliferation of these deadly tropical maladies. Herein, the relationship between humans, insects, and their shared environment is transformed. The insect is no longer perceived merely as a pest, as Frank and Fred viewed the garrapatas in their tale, but is recast within the context of modern medical science as a vector and a harbinger of disease.

In this chapter I explore the changing regimes of the human-animal interface beginning with Yucatán's experience, in the 1830s, of the first worldwide cholera pandemic. Cholera was brought into Yucatán on sea vessels traveling between the port cities of New Orleans, Tampico, Campeche, and Progreso in July 1833, and subsequently killed between 52,000 and 65,000 people, almost a tenth of the 574,496 residents of the peninsula.[7] The epidemic decimated the population and left public-health officials divided over key issues such as etiology and prevention. Professional bickering over the benefits of quarantine, purgatives, and prayer bled into an era of violent civil insurrection known as the Caste War, which began in 1847. In the second half of this chapter I highlight public-health challenges throughout the Caste War period, which endured until 1902, through the establishment of a revolutionary socialist state government in the 1920s. During this period, which included cholera epidemics, yellow fever and malaria infestation, war, and revolution, one scientific shift—from the notion of the pest to that of the vector—took place in the broader context of modernization in a frontier zone, where "civilization" confronted "barbarism." Taken together, the dual calamities of warfare and epidemic illnesses in Yucatán are conceptualized as part of a dynamic and constantly evolving landscape, a terrain wherein animals, insects, and humans consistently interfere with, irritate, and at times challenge state authority.[8]

In truth, the "wild" countryside that the boys Frank and Fred encountered in the 1880s and the obstacles that physicians working for the International Health Board faced during anti–yellow fever and antimalaria campaigns of the 1910s and 1920s possessed many of the same harsh characteristics denounced by sixteenth-century Spanish conquistadors, colonists, and missionaries. The Yucatán's physical environment and climate are quite severe. Geographically, coastal marshlands and jungles in the southeast buttress vast stretches of flat limestone terrain. A lack of water combined with extreme year-round heat and humidity create terrible conditions for raising crops. Although seasonal rains refresh the peninsula between May and mid-

September, they give rise to a stifling wet-heat that contemporary Yucatecans refer to as *bochorno*.

As Frank and Fred discover, ocean breezes in coastal cities and dips in ancient rain-filled *cenotes* (natural sinkholes that collect rainwater) provide reprieve from the harsh climate and are much appreciated by the natives and foreign visitors. The indigenous Maya of the peninsula revere the cenotes; many rural communities depend on aboveground cenotes to supply water for irrigation, cleaning, and cooking.[9] The cenotes also sustain wildlife—as well as providing ideal breeding grounds for insects. When cenotes are dry, they also serve as portals to an underworld of caves that crisscross the entire peninsula. In times of grave danger, the Maya used these caves to hide from conquistadors, slave traders, and threatening hurricanes.[10]

However, the boys—and through them, Knox—did remark on one of the characteristics of the Yucatán: its difference, in terms of ecology, climate, and culture, from the other territories within Mexico's modern borders. Rather, the peninsula seems to have more in common with its tropical Caribbean neighbors across the Gulf of Mexico and with the U.S. South. Yet, despite the severity of the climate and the chronic presence of tropical diseases, outsiders have continued to visit the peninsula since the Conquest and have therefore studied its ecology in order to make it safer. It is just this project—the conquest of the wild and seemingly untamed aspects of Yucatán's environment to make it safe not just for "human habitation," but for the particular kind of human habitation that is characteristic of modernity —that lends this work its methodological center. Humans, animals, and insects are situated within this chapter in "diseased moments" as constitutive of the environment they inhabit.[11] "Diseased moments" here serve as a methodological construct designed to offer insights into periods in which Mexicans struggled with the processes of state-building concurrent to waves of epidemic disease.[12] Specifically, I examine how politicians, inhabitants, and medical professionals used medical science as an instrument to construct a "healthy" Mexican state. The ideal of the healthy state required the regulation and control of animals, insects, and indigenous populations according to popularized notions associated with "good" and "bad" species and racial hierarchies.

The identification of a human, animal, or insect pest or vector provided statesmen, public-health officials, and educated elites with a powerful tool for recasting species status, which entailed reconfiguring the human-animal/insect boundary so as to call for pest or vector extermination in the name of protecting public health and safety. Nineteenth-century Yucatecan elites frequently expressed their irritation with the Maya and placed blame for violence, pestilence, and barbarity squarely on the shoulders of "vengeful" Maya. Print media frequently noted the contempt creole elites held for the Maya. Newspaper articles were loaded with characterizations of the Maya as "barbarians" who impeded economic growth by burning down productive export haciendas, stealing cattle, and living in squalor and filth with their animals. Medical circulars criticized the rural poor for their "ignorance" and "passivity," portraying them as members of society who either would not or could not comply with sanitation regulations.[13] Elites consistently placed nonhuman animals within the living spaces of the Maya to highlight a commingling of filth between nonhuman and human animals. Such associations merged nonhuman animal and human spaces together. Ironically, by drawing on these relational qualities of nonhuman animals and Indians, elites broke down—rather than reinforced—the human-nonhuman animal divide.

In Yucatán state power exerted through public-health campaigns strove to combat agents of disease proliferation by subordinating animals and insects to human power and laws. Debates generated by public-health campaigns articulated, in the vocabulary of medical and sanitary science, an ideology of hierarchies of both species and human races. Thus, a parallel emerged between taming and civilizing indigenous peoples and the project of controlling animals and insects to prevent the spread of disease.[14] Animals, insects, and frequently the indigenous Maya were all alike cast as purveyors of filth and illness. Frequently the indigenous Maya were classed with animals and insects as obstacles or impediments to progress, modernity, and salubrity. The reality, however, of Maya medical practices was far more complex than can be elucidated by a simple set of binaries along the lines of progress and modernity versus tradition and backwardness. According to the regional nineteenth-century historian Apolinar García y García, Maya shamans drew on a centuries-old therapeutics to treat the diarrheas and vomiting associated with cholera and yellow fever, using a combination of bitter *pozole* (a fermented corn beverage), boiled lemonade, and cooked *xkantumbú* (an indigenous herb). They did not generally employ Western methods of purgatives.[15] The Methodist missionary Richard Fletcher commented that during his twenty-five years serving in Corozal, beginning in 1857, he found the Maya preferred herbal remedies: "Of drugs they have a dread, and they trust themselves more readily in the hands of an Indian herbist than the most skillful doctor."[16]

The concept of diseased moments helps us move beyond binaries and see the nuances embedded in relationships between the state and its residents (human and nonhuman animals). Moreover, an examination of these diseased moments facilitates an understanding of instances wherein state and indigenous views of what constituted a clean and healthy environment collided outside of a framework that privileged the state viewpoint and labeled the indigenous as superstitious. Often these conflicts emanated from the unlimited power afforded city officials during an epidemic crisis wherein state-building efforts dovetailed neatly with public-health programs to eradicate disease by way of isolating alleged sites of infection like the slaughterhouse and the cemetery, as well as rural Maya communities, by representing them as antiquated relics filled with pests and disease.[17] We are thus engaged in an examination of biopolitics, as the role of humans, animals, and insects are shaped and reshaped in the civilizing and modernizing project.[18] The use of biopolitics in concert with diseased moments provides a unique analytical framework to explore the relationship between humans, animals (including insects), and the state.[19] From this perspective, state-building functions not just on the macrolevel, but also on the slaughter-house floor, the cemetery grounds, in the thoroughfares of busy urban suburbs, and in rural villages.[20] In essence, by understanding the field invested by governance in terms of dogs, lice, pigs, and so on, as well as people, we capture more than we could see by focusing exclusively on humankind.[21]

Cholera and the Maya Rebel

The first worldwide cholera epidemic, in the early 1830s, challenged the resources of the nascent Mexican Republic. Throughout Mexico, the most affected were the productive segments of society, those between sixteen and forty-five years of age. Mortality among this cohort crippled the labor force and slowed international trade.[22] Cholera is difficult to stop once it has been introduced into a community, as *Vibreo cholerae*, the cholera pathogen, spreads quickly through water supplies that have been contaminated with the fecal matter from any infected mammal. Thus, the disease can propagate in wells, in stagnant cisterns and cenotes, and even in muddied puddles on the streets, avenues, and park pathways. Once the victim is infected, he or she suffers from vomiting and diarrhea, leading to acute dehydration and—in untreated cases—a high chance of death (50–60 percent).

Despite cholera's demonstrated ability to infect and kill indiscriminately across race, class, and ethnic boundaries, to the eyes of nineteenth-century policymakers, statesmen, and public-health officials, the disease seemed to prey most on those with certain common weaknesses, such as a poor spiritual and physical constitution and an intemperate lifestyle, defined by the consumption of liquor, the use of tobacco, the enjoyment of spicy foods, and promiscuity.[23] Morality and cleanliness of body and spirit figured prominently in the framework of disease prevention. Medical discourse as early as the sixteenth century equated intemperance with illness. Over-indulgence, according to the dominant moralistic discourse, led humans down a path of immorality, thus stripping them of their human virtues and propelling their descent to the status of an animal.[24] Medical discourses of the nineteenth century absorbed this paradigm and medicalized its vocabulary, viewing intemperance as a conduit to illness.[25] In particular, this allowed the state to enfold the habits of the indigenous Maya into unsanitary and unhealthy categories.

One expression of the politics of public health was the increasing control over "unclean" or untended animals on both the community and state level. As Phillipe Descola and Gísli Pálsson eloquently argue about such nineteenth-century measures in the developed countries: "Though technically a space of domestication, this forsaken industrial forest retains the wild attributes of the natural forest it replaced. . . . In certain cases, 'wild' environments may be more satisfactorily controlled, socially, technologically and ideologically, than domesticated ones."[26] The presence of uncontrolled or loose animals was a constant target of state intervention. Often accompanied by swarms of insects, untended beasts in human-occupied zones (cities, houses, villages) inspired legislation designed to limit their contact with human populations, especially in response to disease epidemics like cholera.[27] *Bandos* (proclamations) stipulated that animals being raised for sustenance must remain contained in stables or corrals. Throughout Mexico, public-health regulations prohibited with "fuerza y vigor" (force and vigor) swine from wandering about urban locales, assigning fines to the owners of "vagrant" swine and allowing them to be picked up and disposed of by municipal officials.[28] Residents were forbidden from bringing their pets and livestock to public places during cholera outbreaks, with public-health officials often reinforcing the rules by decree during times of maximum panic, as for instance in the summer of 1833.[29] Public figures also began to voice concern about other parasites, such as *piojos* ("louses," a general term used for small parasites on animals), spreading from animals to humans. Rather hefty fines (starting at five pesos) were decreed in cases of animals being found in homes or wandering the streets, with the penalty increasing if the animals were infected with parasites.[30]

It was under the shadow of the epidemic threat that policymakers began to enforce a new regime segregating animals and humans, eradicating selected species of noxious animals and insects, and subjecting humans they deemed threats to public health to a number of measures seeking to control them, change their cultural habits, and, if necessary, exterminate or incarcerate them. Exercising authority over animals was a vector by which the state could monitor humans, insofar as the regulation of the human-animal interaction struck at the heart of the practices of various subcultures, particularly those rural residents who owned herds of livestock and domesticated animals. In general, urban residents supported the fight to exterminate mosquitoes or clear the streets of filthy swine: these policies garnered popular support for the administrators.

To laypersons and public-sanitation policymakers alike, insects and animals in public spaces were tangible evidence of the filth and disease that lurked in the community.[31] Guidelines for the prevention of cholera from 1833 consistently advised the need to contain or kill roaming animals, collect garbage, and ban the slaughter of livestock at public markets. Livestock roaming about *zócalos* (city squares), cemeteries, streets, and *rastros* (slaughterhouses), rummaging through garbage in the streets, mating, and defecating, were marked as public hazards in the community landscape. The correspondence Yucatán's Supreme Health Council received throughout the first cholera pandemic brimmed with complaints about disorder, filth, excrement, and invisible "miasmas" (unwholesome atmospheric conditions).[32]

The 1830s global cholera pandemic was followed, less than a decade later, by a second one, this time moving from Europe to Canada, where it claimed thousands of lives, before it entered New York and worked its way through the Americas via port cities. Cholera likely entered Yucatán once again because of trade with New Orleans, which penetrated the principal Mexican trade ports of Veracruz, Campeche, and Progreso.[33] This second wave of cholera coincided with warfare related to the eruption of the civil Caste War, in 1847, which lasted, with periods of greater and lesser intensity, until 1902.[34] The Caste War has been considered the largest and most successful peasant rebellion in Latin American history.[35] The governing elite saw the twin menace of cholera and the Caste War within the framework of "civilizing" Yucatán, putting efforts to monitor the movements of Maya rebels and to curb disease on the same level of surveillance. At its height, from about

1850 to 1855, the war was waged with extreme violence, both physically and rhetorically, in indigenous discourses of local power and creole fears of indigenous rebels known as *los bárbaros* (the barbarians). Demographically, the results of the Caste War were catastrophic. Estimates place the loss of lives, through disease, battle, or starvation, at over 24 percent of the peninsula's population.[36] Yucatán's local and regional public-health authorities were overwhelmed by the cholera epidemics, smallpox episodes, and warfare, which tended to merge in stunning cycles of destruction. Within this context, public-health workers saw free-roaming animals as intolerable threats as they hurried to isolate entire neighborhoods and initiate mass fumigations. In this context, the owner's inability or unwillingness to pen his or her animals transformed mere livestock or pets into pests.

The notion that nonhuman animals facilitated the spread of disease, particularly if living in close quarters with humans, seemed clear to Westerners and Yucatecan elites. For instance, Thomas Gann, a British archaeologist, noted in *The Maya Indians of Southern Yucatan and Northern British Honduras* (1918) that Mayan children tended to suffer from internal parasites such as hookworm because of their "earth-eating habits." Gann claimed the earth consumed by Maya children was "usually from the immediate vicinity of the house, where pigs and other domestic animals have their quarters. This disgusting habit no doubt accounts in part for the swollen bellies and earthy color of many of the children."[37] Gann and other Western travelers likely did not consider the practical benefits of keeping food sources close, as investments that needed protection and care. Creating physical boundaries between human and nonhuman animals only served to distance valued food sources from their owners and could thus compromise subsistence. In her work on native systems of thought and the environment, Virginia D. Nazarea argues that the scholarship of Harold Conklin, Brent Berlin, D. E. Breedlove, and P. H. Raven has led to a "radical shift in mindset" in how contemporaries view relationships between natives and the environment. Nazarea claims the new perspectives move away "from viewing native systems of thought as naïve and rudimentary, even savage," and that there is now "a recognition that local cultures know their plant, animal and physical resources intimately and are expert at juggling their options for day-to-day requirements."[38] Beyond practical concerns, another possible rationale for keeping livestock and pets within living quarters extends into a supernatural realm. Contemporary anthropological fieldwork points to a Maya belief attributing some illnesses to nonhuman animals. In particular, birds, horses, and dogs can cause illness if they have a "strong gaze" or evil eye. In this scenario, humans can succumb to the evil eye, or *ojoy balche'*, perpetrated by an animal, particularly if the animal is allowed to reach a state of excess. One tactic employed to avoid such excesses is to maintain a close relationship or watchful eye over these animals. An animal left to wander cannot be monitored, but one kept in close proximity to its owners can be kept in proper balance.[39]

Western notions about the link between disease and the proximity of livestock to humans did not resonate within the Maya etiological lexicon. For nineteenth-century Maya, the origins of disease derived from a complex web of Western and "traditional" notions about illness. For instance, the Maya believed that willful forces in nature caused diseases. Illnesses were associated with different types of winds, or *ik*. These winds functioned as semipersonified forces spreading malfeasance and disease. Very little documentation from the nineteenth century details Maya methods of healing. In fact, most of the sources available date to the late colonial period. Nonetheless, it is reasonable to assume the Maya of the nineteenth century combined notions such as *mal de aire* (bad air) articulated in sacred books such as the *Chilam Balam de Nah* with Western notions of the body's humors.[40] Ultimately, the Maya selected some elements from Western medicine and rejected others (e.g., smallpox vaccination), therein constructing a plural understanding about disease proliferation and therapeutics in the nineteenth century more in line with their worldview.

The categorization of an animal—or a human—as a pest is context-specific. Certainly, many of Yucatán's inhabitants viewed the uncared for and seemingly wild animals that roamed the streets during cholera epidemics as pests. But was this the same as believing that they were vectors of deadly infection? A few key distinctions between pest and vector are essential to understanding the human-animal/insect relationship during the late nineteenth century and early twentieth. Then as now, a pest could be almost any animal or insect that caused damage to crops, food, humans, or even domesticated pets—as in the Boy Travellers' story of the dog tortured by garrapatas. The categorization of an animal or insect as a pest is also determined by unique social, cultural, and environmental conditions. A pest in the southern tropics of Mexico may not be so considered in the western mountains of Oaxaca or Chiapas. A pest such as a domesticated dog can become a vector if the dog bites a human or another animal and spreads rabies. But it was not until the arrival of germ theory in the 1880s and 1890s that the transition from pest (a destructive entity) to vector (a channel of infection) became possible. Germ theory proposed that pathogens could travel undetected from one biological entity to another, and resided in specific niches. Older, miasma theories of disease gave way to these vanguard theories, although the latter were often presented to the public as nothing more than new possible explanations.[41] From today's perspective, the difference between a pest and a vector lies in the fact that the pest does not itself bring about disease, but only carries the microorganisms that cause it. Anything that allows the spread of organisms that transmit disease is a vector.[42]

During the height of the 1850s cholera epidemic, civilian complaints about garbage-strewn streets, fetid slaughterhouses, and crowded cemeteries overlapped with a palpable fear among residents about stray animals feeding on garbage, slaughterhouse scraps, and cadavers, especially as cholera's victims swelled the cemeteries past the point of maximum capacity.[43] Enforcement of legislation regulating loose animals, the collection of garbage, and the inspection of incoming ships lapsed as communities were overwhelmed by the epidemic.[44] The geology of the Yucatán peninsula made grave-digging a particularly frustrating task, as teams of workers had to chip through and remove huge slabs of stone to properly bury the dead. As a result, thousands of shallow graves were dug, which served as excellent feeding bins for animal and insect populations.

For instance, in October 1853, Don Esteban Herrera, the *juez de paz* (town magistrate) of Cacalchén took charge of burials during the cholera epidemic. Not only was the volume of the dead overwhelming, but Herrera also found himself dealing with other issues, such as the Church's requirement that there be a twenty-four-hour waiting period before the interment of the deceased (about which he received a notice from the local priest). Then there were the popular practices, especially overnight vigils and processions to the gravesite. To ban these, Herrera knew, would prompt fierce protests. Yet state public-health laws required immediate burial for victims of disease, and as a civil representative, he had to find a way to implement the law. Meanwhile, Herrera confronted shortfalls in the capacity of the community to even guard the dead, with sometimes horrifying consequences; one morning, for example, he arrived to bury victims who had remained exposed overnight and found a herd of wild pigs feeding on the cadavers. Frustrated, Herrera decided to build a cemetery farther outside town limits. Community members complained about having to travel so far from their homes to visit their dead; moreover, they claimed that the time they might spend constructing a new cemetery would have to be taken from the time they spent tending their *milpas* (small agricultural subsistence plots) and feeding their families.[45]

The archives reveal the level of complaints kept rising as a result of lack of cemetery space, garbage in the streets, stagnant and offensive smells, and phobias about infectious animals (even pets). Newspapers printed letters filled with exasperation, and occasional broadsides or announcements specified the everyday toll of horrors on Yucatan's citizens as they watched the municipal health councils struggle with limited funds, supplies, and manpower.[46] Letter-writers complained to Mérida's Superior Health Committee about the inconsistent public services: one week the garbage would remain uncollected in heaping piles throughout the community, and the next week officials would fine neighborhood residents for sanitary violations.[47]

The letters reveal a fearful and worried community. Residents suddenly focused on occurrences that had regularly transpired in their neighborhoods. In this moment of crisis, they moved to do something about the stray animals in their neighborhoods and foul smells rising from fetid garbage. To make matters worse, animals who regularly scavenged among the street's usual detritus expanded their foraging grounds to include the corpses of cholera victims. In particular, complainants were concerned that deadly miasmas were spreading the plague.[48] In the words of one irritated correspondent, such conditions held the potential to "rot our bodies and our communities to the core."[49] Citizens were quick to invoke the language of the civilizing mission to compel the regional health council to address their needs as part of their personal efforts to triumph over barbarism.[50]

One major focus of complaint was the possibility of consuming contaminated animal byproducts, which pointed toward the need to control animal slaughter. Legislation forged during Mexico's first two cholera epidemics moved toward tighter regulation of livestock rendering at city slaughterhouses and of the sale of animal byproducts. The public slaughterhouse in Mérida was poorly administered and unhygienic, as was pointed out in complaints from consumers and ranchers. Some ranchers expressed irritation at the fact that officials in charge of sanitation for the Mérida slaughterhouse were uninformed about veterinary science and unable to judge how animals should be butchered.

In December 1849 the state of Yucatán issued a decree requiring a census of all vaccinated cattle and setting a slaughtering fee of eight reales per head. Furthermore, it stipulated that any animal slaughtered for the sale of its meat had to be butchered in a public slaughterhouse, with violators subject to heavy fines. Similarly, ranchers were required to demonstrate that they were licensed to raise cattle or pigs and to present up-to-date vaccination records for their animals; violators would be fined and forced to pay the slaughterhouse fee. To ensure compliance, livestock "police" were stationed at slaughterhouses, collecting for themselves a standard 4 percent cut for each slaughter.[51] The inconsistencies in slaughterhouse inspections and the problem of corrupt inspectors underscored the unevenness (or absence) of state power at the community level. Likely cognizant of regional legislation affecting the operation of slaughterhouses, inspectors often simply took matters into their own hands by massaging laws to better suit their own circumstances and compelling consumers to seek other clandestine options.[52]

Often the fears aroused by the onset of a disease epidemic motivated the regulation of animal byproducts (such as meat, milk, and eggs), as consumers became more wary about the potential of these products to cause illness. Thus, public-health authorities extended their authority into the marketplace, butcher shop, and mill, as part of an overarching campaign to curb contagion and protect the public.[53] Projects designed to control the consumption of animal products inevitably came to include legislation regulating the animals themselves.[54] However, inspections were often not thorough, and inspectors were eminently bribable. Even with strict regulations regarding the health of the animal before slaughter, after it was rendered the meat was often left hanging in the open air all night, until the market opened the following morning. By the early 1900s, public-health authorities in each municipality were required to report their findings from regular inspections of dairy and meat cattle and pigs to regional health branches.[55]

The rendering of animals for human consumption has always constituted the ultimate edge of the human-animal boundary. After death, transformed from an object of nurture to an object of consumption, an animal, in the social imaginary, still had the "ability" to infect humans with deadly pathogens, or at least to cause massive digestive discomfort. Animals could still inhabit the role of pest, although at this point they were pure carriers of other, invisible pests. They became the object of transformation in the semiotic regime of the pest to the regime of the vector. With animals parceled out and strung up on hooks, the final conquest of humans over these beasts thus lies in human willingness to repeatedly internalize potential harbingers of disease.[56]

It is at this point that a familiar social metaphoric comes into play, in which the animal serves as a negative comparative term for the "inferior" human. As Philippe Descola has pointed out, a mere two years after Columbus's first voyage, a Spanish doctor was already writing of the natives that "their bestiality exceeds that of any beast."[57] By the terms of that metaphoric, if the consumption of an animal constituted the signal form of the human conquest of beasts, then the objectification process that made such an act feasible could also logically extend to inhabitants viewed by educated elites as inferior.[58] The press often depicted the indigenous Maya as "barbarous savages" on the same level as the beasts of the jungle.[59] Just like those beasts, the indigenous Maya presented an obstacle to progress that had to be removed—culturally exterminated, at best, physically, if necessary.[60] The rebel Maya were often singled out in the print media as *bárbaros*.[61] In military reports the rebel Maya were generally identified as chief impediments to the progress of the region, standing in the way of good relations with Guatemala and Belize.[62]

Broadly speaking, the prevailing ideology of late nineteenth-century elites mixed together social Darwinism, positivism, and Enlightenment ideals. Men of science who were bent on providing rational answers for the differences between themselves and "others" recorded the phenotypic, linguistic, and sociological tendencies of the origins of the world's "races." Public-health campaigns conceptualized animals and indigenous residents in terms of the positivistic narrative of antithetical couples: the civilized versus the barbarous, the tamed versus the wild, and the healthy versus the diseased. Given this narrative, it was easy to subsume animals, insects, and the indigenous Maya into negative categories all associated with filth, pestilence, and backwardness.[63]

The animal-barbarian metaphoric was widely distributed in the popular vernacular. In a diatribe penned to the justice of the peace in the neighboring peninsular state of Campeche, a complainant railed against the acts of the rebel Maya in his area, declaring, "We are not property of them [the rebels] and even less so their slaves, we have suffered enough of the unjust cruelty and tyranny of these brute animals."[64] In March 1855, at the height of Caste War violence, the prefect of Tekax wrote to Yucatán's governor about the "unhappy families in his city who remained in hiding because of the war of extermination led by the barbarous Indians." The prefect of Tekax pleaded for relief from the war while also extending his advice for dealing with the barbarous rebels: if they could not submit and obey laws, if they continued to threaten families, then the barbarous rebels were "animals," he concluded.[65] Frequently the print media pointed to the unhealthy conditions of the rebel Maya involved in Yucatán's civil insurrection. In the elite view the Maya rebels had to be "tamed"—another metaphor connecting the animal and the native—or they would pose a significant threat to state-building agendas. The notion of taming, whether of the jungle, of animals, or of the Maya, justified the harshest measures, such as the export of captured Maya rebels to Cuba, which occurred between 1848 and 1861.[66] Within the negative animal-native metaphoric, the Maya could be easily blamed for spreading illness throughout the peninsula and thus for threatening the health of the entire region. Moreover, Mayan traditions of consulting with local *curanderos* (healers), *yerberos* (herbalists), and midwives for healthcare did little to engender confidence among the elite policymakers who sought to impose a modern medical regime. In fact, they saw the continued use of non-state-sanctioned systems of healthcare as subversive of state authority and as cultural support of the rebel side in the Caste War. Finally, human-animal interaction in Mayan culture was established through different markers and boundaries than in the urban Mexican paradigm. As the anthropologist E. N. Anderson observes, "For the Maya, human society is intimately involved with plants and animals. The landscape is a human creation and is constantly being modified."[67]

Consequently, many state public-health and sanitation campaigns came to focus on regular fumigation and cleansing routines in Maya communities.[68] This was, however, easier to plan in theory than to implement in fact, given difficulties in communication and transport, limited access to funding, and the sheer expense of instituting regular inspections, fumigations, and scourings. The difficulties, in turn, were interpreted as further evidence that the Maya were resisting modern medicinal practices, including vaccination, sanitation, and animal control.[69]

Outside observers reflected elite public opinion. The U.S. foreign consul in Progreso, Edward Thompson, noted in December 1900 that most of the peninsula's unsanitary conditions could be found in Maya mud-and-thatch houses, where a steady stream of livestock and other animals streamed freely between outside and inside environs. Thompson stated that even "under the most favorable of circumstances sanitary decrees are carried into effect among them with the greatest difficulty."[70] Like Thompson, many foreigners regularly connected the filth they saw in the countryside to the Maya's allegedly unhealthy habits.[71] Most Euro-Western travelers saw Maya communities as dirty, an impression conveyed by the Maya blurring of the boundaries between animal and human realms.[72] The discoveries of medical science, particularly those concerning mosquito-transmitted diseases like yellow fever and malaria, gave a scientific basis to fears of uncontrolled populations of insects, as public health transformed from a miasmatic to a germ-centric model.

Yellow Fever Eradication and the Mosquito Menace

Louis Pasteur was the founder and most famous advocate of the germ theory of disease, which posited the link between microorganisms and various kinds of disorders in animals, man, and plants. Pasteur's discoveries revolutionized the paradigm of disease causation. Building on Pasteur and other bacteriological studies, the scientists Walter Reed, Carlos Finlay, and Jesse Lazear studied a number of diseases, looking for the microorganic link. Finlay's conclusion, in 1881, that the *Aedes aegypti* mosquito was the principal vector of yellow fever set the stage for further mosquito studies, and by 1897 the causative factors for malaria and yellow fever were well understood.[73] The discovery that mosquitoes, louses, ticks, and other pesky insects transmitted disease transformed the insect pest, whose bite or sting was evil in itself, into a vector, an unconscious carrier of disease, placing these creatures squarely at the center of the three stages of disease transmission: pathogen, insect host, human host.[74]

Public-health officials responded to the new science by seeking out insect breeding grounds and destroying them wherever they occurred: in private patios, cisterns, cenotes, and gardens. Public-health officials, municipal leaders, residents, and foreigners all viewed the elimination of mosquitoes as a means toward achieving a modern and healthy state that would, among other things, open up the peninsula to foreign investment. Within this context, residents renegotiated access to and control over their bodies, their animals, and their homes, thereby redefining the meaning of hygiene and sanitation. In this new paradigm, those who failed to comply with the modern tenets of sanitary science were labeled "ignorant," "ambivalent," or "resistant."[75]

To reduce the lethality of the tropical Yucatán environment, it seemed, all that was needed was compliance with regulations for keeping chicken coops clean, delousing children and animals, and fitting screens tightly across water cisterns to control mosquito reproduction. Through this process of modernization, the rough outlines of a Mexican citizenry—one with a duty to the common health—emerged. Conversely, those whose homes remained filthy, thus allowing insects to breed, or who refused to properly segregate human and animal spaces were considered rebels even if they did not take up arms.

From the vantage point of some animals (or alleged vectors), the world must have morphed into a frighteningly small place, as urban sprawl, booming trade ports, and intensified human landscape use pushed development deeper and deeper into all ecological niches. Increasingly, as humans sought to extend control over the environment down to the microbial level, a new, vigilant rationality would encompass animal pens, corrals, and slaughterhouses, and a new distance sprang up between humans and their animal food sources. But as urban areas grew, some insects flourished in the resulting cornucopia of water and blood, which sustained their short lifecycles.

One of the most powerful enemies the insect vector faced came about as work crews were employed in the task of mosquito eradication. The Rockefeller Foundation became a notable sponsor of such projects. Collaborating with regional public-health councils throughout Mexico, the philanthropic organization began an ambitious project aimed at improving the health of populations in developing regions throughout the world. By the 1920s, the Rockefeller Foundation's International Health Board spearheaded anti–yellow fever and antimalaria campaigns throughout Latin America, the Caribbean, and the Far East.[76] They built on important lessons they had learned from their hookworm campaigns in British Guiana (1914) and in the U.S. South (1915, in Arkansas and Mississippi). The foundation began negotiations with the Mexican government in 1915, but in the wake of the strong anti-*yanqui* sentiment—stemming in large part from the U.S.-led invasion of Mexico, in 1914, through the port of Veracruz—full-scale implementation of activities was delayed until late 1919. Outside of their own fieldwork, most Rockefeller physicians and administrators had learned about indigenous populations of the Americas, the environment, and non-human animals of the region from myriad travel logs, geographical surveys, and archeological publications. In these studies Western descriptions of nonhuman animals and Indians were often painted with the same broad-brush strokes. Depicted as "savage" and "wild" yet vibrant and exotic, the Indians and nonhuman animals of Yucatán and Latin America's tropics oozed with contradictions in nineteenth- and early twentieth-century Western literature and scholarship. Harlan Page Beach noted in *A Geography and Atlas of Protestant Missions* (1901) that "Central American plant and animal life most abounds in the moist, warm regions near the coast. Its human inhabitants, on the contrary, 'flourish in the drier parts, where agriculture presents fewest difficulties and the conditions of health are favorable.' "[77] From the vantage point of the Westerner, Latin America's tropics teemed with vital and abundant forms of exotic flora and fauna that called to mind Edenic lore. Other travelers to Latin America's tropics concluded, as did the Boy

Travellers, that while undoubtedly menacing at times, the annoying nonhuman animals and insects that irritated and allegedly spread illness nonetheless remained biologically fascinating to the outsider.[78]

The Yucatán in which the Rockefeller Foundation doctors began to work in late 1919 was, at least rhetorically, a much different place than that encountered by the Boy Travellers in 1880. A socialist governor, General Salvador Alvarado, had put in place a socialist framework that included the addition of more rural health clinics, *boticos* (pharmacies), and obligatory vaccinations for school-age children.[79] When Alvarado took office in 1915, he completely overhauled public works and began ambitiously modernizing, hoping to reverse Yucatán's international reputation as a "backwards" Mexican periphery. By now, part of that reputation was due to the oppression of the Mayan peasantry, whom he hoped to liberate.[80] Alvarado saw the elite group of *henequeñeros* (owners of large haciendas specializing in the export of henequen fiber used to make rope) as obstacles to his revolutionary agenda, blocking the gains of the revolution in Mexico's southeastern periphery. The national revolution had arrived late in Yucatán, beginning almost four years after the dictator Porfirio Díaz (r. 1876–1910) had fled Mexico and revolutionary fighting had erupted in central and northern Mexico. When the revolution came to Yucatán, in 1915, it assumed a distinctly socialist form, in contrast to the rest of Mexico.

Yucatán's socialist government disassociated modernization from the elite project of subduing the Maya, which qualitatively contributed to the success of the Rockefeller Foundation's anti–yellow fever and antimalaria projects. International Health Board physicians appropriated the connections and energies of socialist youth brigades and medical student collectives, and used local health leagues (*juntas*) in the mosquito-eradication campaign.[81] Initially, the foundation focused on both vaccination and mosquito eradication, but it became evident that the anti–yellow fever vaccine developed by the Japanese scientist Hideyo Noguchi posed too many dangers, which forced the International Health Board to abandon vaccination as a viable preventative.[82]

By 1921, Rockefeller physicians were concentrating primarily on eliminating mosquito larvae with house-to-house inspections. The *ligas de resistencia* (resistance leagues) facilitated the Rockefeller Foundation's entry into communities to vaccinate, inspect homes, kill mosquitoes and larvae, and educate residents about the benefits of acquiring screens and tight-fitting lids for water tanks, of placing mosquito-larvae-eating fish in water reservoirs, and of "bombing" cisterns, tanks, and cenotes to kill mosquito larvae.[83]

Once again, spatial boundaries forged between human, animal, and insect were the specific target of a transformation that created a new set of tasks (screens had to be regularly refitted and on occasion repaired to fit properly onto water cisterns; larvae-eating fish placed in cisterns had to be discarded when they perished) and new zones (mosquito nets that were hung around hammocks or beds, for example, divided the home life into smaller "safe" zones). By way of modern medical science and of state public-health committees, the home-dweller who did not abide by these new arrangements became the new "primitive."[84]

Ultimately, how residents responded to vector-eradication efforts helped define their relationship with municipal, regional, and even national officials. With the focus on insect-vectors removal, Yucatecan policymakers possessed a formidable tool to penetrate the private sphere in the name of public health, science, and the revolution. For instance, when public-health workers identified a boarding house for foreigners as a focus of infection in Mérida, the owner, not the guests, was heavily fined. The report generated from the inspection and subsequent fumigation of the premises (at the owner's expense) presented a character sketch of the owner as a disloyal citizen, ignorant, careless, and unscrupulous in claiming that his clientele, being mostly foreign, was inherently more susceptible to yellow fever anyway.[85]

The focus on mosquito elimination and house-to-house inspections ultimately yielded a reduction in Yucatán's *Stegomyia* mosquito indexes.[86] Rockefeller doctors stationed in Yucatán attributed their success to the willingness of residents to comply with home inspections.[87] Yucatán's Junta Superior de Sanidad (Superior Health Council) issued a bando in 1921 that made cooperation mandatory, with violators labeled as unpatriotic.[88] The Rockefeller Foundation started winding down the program in 1922, using it as a model to follow in other tropical regions plagued with persistent malaria and yellow fever cycles. Although not explicitly laid out in the International Health Board accounts, field methodology noticeably shifted to prioritizing door-to-door inspections and the training of local eradication teams over experimentation with Noguchi's anti–yellow fever serum.[89] Similarly, local public-health brigades began to prioritize cleaning up the environment—again within the limits of filth and cleanliness prescribed by the new norms of the human-animal interface—and educating residents about sterilization and hygiene.[90]

International Health Board physicians reiterated an old pattern by emphasizing control over the environment as the key to mosquito-borne-disease prevention. At the community level, however, such concerns translated into a heightened anxiety about the cleanliness of one's own household and one's neighbors that seeped into the fabric of daily life, ultimately contributing to fears that the very soil the people tilled and the air they breathed could promote the favorable conditions for the propagation of disease.[91]

The Rockefeller Foundation's International Health Board led an effective campaign against yellow fever by transferring the site of prevention from the human body to the mosquito, an insect with no economic importance (unlike the bee) and no defenders—an annoying, useless pest that carried acute illness and sought human blood. The mosquito was the perfect emblem of all the foes of modernity. The decision to highlight mosquito eradication over vaccination brought the Rockefeller Foundation and the International Health Board international acclaim for their victory over vectors. (They also worked to combat the louse in the transmission of typhus and hookworms, which perpetuated malnourishment.) In the eyes of the international community, the cooperative efforts between Mexico and the United States, funneled through regional Mexican health councils and the Rockefeller Foundation's International Health Board, facilitated the emergence of a healthful republic by the mid-1920s.[92] Overall, the Rockefeller Foundation's work in Mexico and throughout the globe totaled over nine million dollars for projects in eighteen different countries.[93] A *New York Times* article from November 1929 declared that the Rockefeller Foundation had spent $21,690,738 in 1928 to improve public health throughout the globe. Much of the work throughout the Caribbean and Latin America included "intensive work against malaria."[94]

With the help of the Rockefeller Foundation's International Health Board, Yucatán became a yellow-fever-free peninsula by the early 1920s. The Rockefeller Foundation's success was made possible because the socialist revolution provided a highly motivated, organized cadre of community servants ready to combat disease and advance revolutionary doctrine. On the level of the history of human-animal interactions, the Rockefeller Foundation's conquest of the mosquito was a long-sought victory over the "wild" and untamable insect vectors of the tropics, in a sense making the world safe for the Boy Travellers, just as President Wilson had entered the First World War to make the world "safe for democracy." In actuality, Mexico's countryside remained an unpredictable zone, but at least peripheral urban centers like Mérida exhibited modernity's triumph over nature as public-health services enforced controls over pets, livestock, and human activity.

Conclusion

The creation of sanitary and disease-free environments in Yucatán required the imposition of a system of new limits, relations, and values on the population. I have tried here to explore how these limits were interpreted by the elite ruling class, the policymakers, and the population at large. For the first two, animals wandering in public spaces and swarms of hovering insects signaled disorder and unhygienic conditions. On a local level, citizens complained about the miasmas arising from overcrowded cemeteries, unclean slaughterhouses, and animals in parks, streets, and other spaces frequented by humans. These reactions, under the sign of the pest, helped construct new boundaries of human-animal interaction and new public-health codes. As a regime of selective segregation began to regulate the urban environment, it allayed fears of disease due to close contact with animals.

Controlling animals and insects in the name of health and modernity served the Yucatecan state as both a metaphor and model for marginalizing, penetrating, and radically transfiguring indigenous Maya culture. The indigenous Maya were conceptualized as "barbarous beasts" and "savages," whose status as citizens or even as humans depended on changing their old ways completely. The precepts of public health served the elite and state institutions as tools for social change—which speaks to the powerful role medical science and public health held in processes of state formation.

In general, urban citizens viewed the presence of animals and insects in their public and private spaces as pests spreading filth and disease according to the unpredictable whims of animal and insect species. It is difficult to say what the public ideal was vis-à-vis animals. Perhaps, like the Spanish developed colonial caste systems and social hierarchies, postcolonial Mexicans envisioned a world where species difference was so clearly demarcated within their material world that extermination could easily target the noxious species. And yet, in reality, disease could be carried by a family pet or baby. Within these parameters, goals to beautify urban centers and construct a modern, clean, and healthy state and citizenry demanded the containment of all nonhumans in the name of public safety. Fears aroused by the worldwide cholera epidemics made unsupervised human-animal interactions suspect as new diseases, spread through human-animal/insect contact, could place humans at risk time and time again.

Many of the factors outlined above in the fight against dirt, wildness, and insects have reemerged in the fight against a new disease, dengue fever, in contemporary Mexico. This is, again, a disease transmitted by the *Aedes aegypti* mosquito that has become a chronic menace defining a tropical area in its appearance in the peninsular states of Yucatán, Quintana Roo, and Campeche. Although rarely is the original form of the disease fatal, it has mutated into a deadly form of hemorrhagic dengue, or dengue shock syndrome.[95] Public-health campaigns to combat dengue remain virtually unaltered in their basic approach to disease prevention, including a focus on endemic zones for treatment, the use of environmental and vector controls, and targeting at-risk populations for public-health and hygiene education.[96]

One common feature tying together early campaigns against mosquito-transmitted diseases of the early twentieth century and contemporary programs targeting dengue in Yucatán is the quandary of how to safely eliminate mosquito vectors. Fumigations, bombing of cisterns, and the application of chemicals all alter human–animal/insect relationships. On the microbial level, we now know, eradication campaigns can lead to resistant strains in the mosquito, while antibiotics can lead to resistance in the microbe. Selection pressures exerted by technological fixes have shown us just how unpredictable and how quickly diseases, and their vectors, can adapt to survive.[97]

Notes

1. Thomas W. Knox, *The Boy Travellers in Mexico* (New York: Harper and Brothers, 1890), n.p., preface. Thomas Wallace Knox founded a school in Kingston, New Hampshire, for young boys. After the Civil War, he headed west to prospect for gold, but ended up becoming a reporter for the Denver *Daily News*, which is where he found his métier. His Civil War–era letters were published as *Camp-Fire and Cotton Field: Southern Adventure in Time of War. Life with the Union Armies, and Residence on a Louisiana Plantation* (New York: Da Capo, 1865). In 1866 Knox began a world tour as a journalist, covering such events as the Paris universal exposition of 1878 while authoring his *Boy Travellers* series.

2. Review of the "Boy Travellers in the Far East Part 2: Adventures of Two Youths in a Journey to Siam and Java," *New York Times*, 19 November 1880.

3. Knox, *The Boy Travellers*, 446, 462.

4. Knox, *The Boy Travellers*, 510.

5. Knox, *The Boy Travellers*, 508–9.

6. Knox, *The Boy Travellers*, 509–10.

7. For information on Mexico's first cholera pandemic see Lilia V. Oliver, *Un verano mortal: Análisis demográfico y social de una epidémica de cólera: Guadalajara, 1833* (Guadalajara: Gobierno de Jalisco Secretaria General Unidad Editorial, 1986); and Rafael Valdez Aguilar, *El cólera: Enfermedad de la pobreza* (Culiacán, Sinaloa:

Universidad Autónoma de Sinaloa, 1993). A census taken in 1832 estimated the population of the peninsula at 574,496. The deaths from cholera are somewhat difficult to confirm given that this figure reflects only registered deaths. See Estado de Yucatán, "Introducción de Dr. Luis A. Solis Alpuche," *Monografía sobre la salud pública de Yucatán* (50 Aniversario de la Creación de la Secretaria de Salud), Servicios Coordinados de Salud Pública (Mérida, Yucatán: Talleres Gráficos del Sudeste, S.A. DE C.V., June 1993), 63.

8. For more on the relationship between environments and organisms, see Philippe Descola and Gísli Pálsson, "Introduction," in *Nature and Society: Anthropological Perspectives* (New York: Routledge Press, 1996), 5.

9. For a description of cenotes, see John Lloyd Stephens, *Incidents of Travel in Yucatan*, 2 vols. (New York: Dover, 1963 [1843]), 1:71.

10. For a description of one cave system, see Stephens, *Incidents of Travel in Yucatan*, 1:71.

11. I borrow the term *disease moments* from Charles Rosenberg, *The Cholera Years* (Chicago: University of Chicago Press, 1962), 113. I modify his notion in two ways: first, by expanding the idea to a frontier zone, Yucatán, between the West and the non-West (thus bringing into play non-Western patterns of healing); and second, by examining the disease moment in terms of the public policy attempt to capture and change relationships between humans, animals, and insects.

12. Heather McCrea, *Diseased Relations: Epidemics, Public Health, and State-Building in Yucatán, Mexico, 1847–1924* (Albuquerque: University of New Mexico Press, 2011), 7.

13. See, for example, *Siglo XIX*, no. 261, 31 May 1850, 1; *El Regenerador Periódico Oficial*, no. 1 Mérida and no. 124, 9 December 1853, 3.

14. Matthew Freye Jacobson, *Barbarian Virtues: The United States Encounters Foreign Peoples at Home and Abroad, 1876–1917* (New York: Hill and Wang, 2000), 137.

15. Apolinar García y García, *Historia de la guerra de castas* (Mérida, 1865), 130.

16. Excerpts from Richard Fletcher's writings can be found in "The Methodist and the Mayas," in *Maya Wars: Ethnographic Accounts from Nineteenth-Century Yucatán*, ed. Terry Rugeley (Norman: University of Oklahoma Press, 2001), 103–16, 109.

17. Archivo General del Estado de Yucatán (AGEY), Poder Ejecutivo (PE), Gobernación, box 139, Mexican Consulate in Belize, 1864.

18. For more on biopolitics and its application to state-building, see Marius Turda, "The Nation as Object: Race, Blood, and Biopolitics in Interwar Romania," *Slavic Review* 66.3 (fall 2007): 413–41, 413 and n3; Edward Ross Dickinson, "Biopolitics, Fascism, Democracy: Some Reflections on Our Discourse about 'Modernity,'" *Central European History* 37.1 (2004): 1–48, 3–4. In his definition of biopolitics Dickinson includes "medical practices from individual therapy and regimes of personal hygiene to the great public health campaigns and institutions; social welfare programs . . . ; the whole complex of racial science from physical anthropology to the various racial theories; eugenics and the science of human heredity; demography, scientific management and occupational health."

19. My use of *biopolitics* emulates Michel Foucault who coined the word *biopolitique* in the late 1970s. See Michel Foucault, *The Birth of Biopolitics: Lectures at the College of France, 1978–1979* (New York: Palgrave Macmillan, 2008).

20. Social histories that examine state-building and imperialism through the lens of medicine, disease, or public health include Alan M. Kraut, *Silent Travelers: Germs, Genes, and the "Immigrant Menace"* (Baltimore: Johns Hopkins University Press, 1994); Shirley Lindenbaum and Margaret Lock, eds., *Knowledge, Power, and Practice: The Anthropology of Medicine and Everyday Life* (Los Angeles: University of California Press, 1993); Libbet Crandon-Malamud, *From the Fat of Our Souls: Social Change, Political Process, and Medical Pluralism in Bolivia* (Berkeley: University of California Press, 1991); Warwick Anderson, *Colonial Pathologies: American Tropical Medicine, Race, and Hygiene in the Philippines* (Durham: Duke University Press, 2006); Ronn Pineo and James A. Baer, eds., *Cities of Hope: People, Protests, and Progress in Urbanizing Latin America, 1870–1930* (Boulder: Westview, 1998) (in particular David S. Parker, "Civilizing the City of Kings: Hygiene and Housing in Lima, Peru," 153–78; Ronn Pineo, "Public Health Care in Valparaíso, Chile," 179–217; Sam Adamo, "The Sick and the Dead: Epidemic and Contagious Disease in Rio de Janeiro, Brazil," 218–39); Enrique Florescano and Elsa Malvido, eds., *Ensayos sobre la historia de las epidemias en México*, 2 vols. (Mexico City: Instituto Mexicano del Seguro Social, 1982); Julyan Peard, "Tropical Disorders and the Forging of a Brazilian Medical Identity, 1860–1890," *Hispanic American Historical Review* 77.1 (February 1997): 1–44; Marcos Cueto, ed., *Missionaries of Science: The Rockefeller Foundation and Latin America* (Bloomington: Indiana University Press, 1994) (especially Steven C. Williams, "Nationalism and Public Health," 23–51; Armando Solorzano, "The Rockefeller Foundation in Revolutionary Mexico," 52–71); Nancy Leys Stepan, *The Hour of Eugenics: Race, Gender, and Nation in Latin America* (Ithaca: Cornell University Press, 1991); David Sowell, *The Tale of Healer Miguel Perdomo Neira: Medicine, Ideologies, and Power in the Nineteenth-Century Andes* (Wilmington, Del.: Scholarly Resources, 2001); Donald Cooper, *Epidemic Disease in Mexico City 1761–1813: An Administrative, Social, and Medical Study* (Austin: University of Texas Press, 1965); Oliver, *Un verano mortal*; Lilia Oliver Sánchez, "Una nueva forma de morir en Guadalajara: El cólera de 1833," in *El cólera de 1833: Una nueva patología en México causas y efectos*, Colección Divulgación (Mexico City: Instituto Nacional de Antropología e Historia, 1992), 89–104; Andrew Lewis Knaut, "Disease and the Late Colonial Public Health Initiative in the Atlantic Ports of New Spain" (PhD diss., Duke University, 1994); Anne-Emanuelle Birn, *Marriage of Convenience: Rockefeller International Health and Revolutionary Mexico* (Rochester: University of Rochester Press, 2006); and Ann Zulawski, *Unequal Cures: Public Health and Political Change in Bolivia, 1900–1950* (Durham: Duke University Press, 2007).

21. On different approaches to the treatment of animals in historical analysis, see Erica Fudge, "A Left-Handed Blow: Writing the History of Animals," in *Represent-*

ing Animals, ed. Nigel Rothfels (Bloomington: University of Indiana Press, 2002), 3–18, 8–9.

22. Lynda Sanderford Morrison, "The Life and Times of José Canúto Vela, Yucatecan Priest and Patriot (1802–1859)" (PhD diss., University of Alabama at Tuscaloosa, 1993), 117–18.

23. Archivo General de la Nación de México (AGN), Gobernación, sin sección, 20, box 365, 1849, Decretos y leyes, relativo al cobro y préstamo del Ayuntamiento, por la Epidemia de cólera morbus. Archivo Histórico de la Secretaría de Salubridad y Asistencía (AHSSA), Salud Pública (SP), Epidemias, box 1, exp. 24, 1 June 1850; AHSSA, SP, Epidemias, box 1, 25, 2 September 1850, Bandos de la Epidemia de cólera.

24. Erica Fudge, *Brutal Reasoning: Animals, Rationality, and Humanity in Early Modern England* (Ithaca: Cornell University Press, 2006), 61–62.

25. Centro de Apoyo de Investigaciones Históricos de Yucatán (CAIHY), Folleto, "Método profiláctico y curativo del Cólera Morbo, por el Doctor F. Pedrera" (Mérida: Imprenta de *La Revista de Mérida*, 1892), 3; National Archives Washington, D.C. (NAWDC), Department of State (DOS), Consular Dispatches (CD), Mérida and Progreso, RG 59, vol. 3., U.S. Consul Louis H. Aymé in Mérida to Honorable W. Hunter, Acting Secretary of State, 27 November 1882; Molly M. Mullin, "Mirrors and Windows: Sociocultural Studies of Human-Animal Relationships," *Annual Review of Anthropology* 28 (1999): 201–24, see 204–5.

26. Descola and Pálsson, *Nature and Society*, 10.

27. Estado de Yucatán, *Monografía sobre la salud pública de Yucatán*, 72.

28. AHSSA, SP, Epidemias, box 1, exp. 4, 20 July 1833, Summary: Bando sobre las precauciones y medidas de asistencia a los entermaos de cólera morbus.

29. AHSSA, SP, Epidemias, box 1, exp. 4, 20 July 1833.

30. AHSSA, SP, Epidemias, box 1, exp. 4, 20 July 1833; Archivo General del Estado de Campeche (AGEC), Gobernación "Periodo Yucateco 1820–1857," box 9, exp. 689.

31. *Boletín Oficial* (Mérida), no. 4, 4 August 1849, 2; Dr. D. Francisco Javier Santero, *Elementos de Higiene Privada y Pública*, vol. 2 (Madrid: El Cosmos Editorial, 1885); Dr. L. A. Calvi de Turin, *Instrucción sobre el cólera-morbo asiático, y método curativo que le conviene*, trans. El Conde de la Cortina (Mexico City: Impreso de F. Escalante y C., 1854).

32. See for instance Archivo Histórico de Archediócesis de Yucatán (AHAY), Box: Defunción cólera morbuz 1833–1853. Recommendations for burials—which included digging deeper graves in limestone and regulating the burial of bodies under church floorboards—were formulated so as to prevent the spread of cholera. On the 1830s cholera epidemic in Guadalajara, Mexico, see Oliver, *Un verano mortal*.

33. Mitchel Roth, "Cholera, Community, and Public Health in Gold Rush Sacramento and San Francisco," *Pacific Historical Review* 66.4 (November 1997): 527–51; and Ramon Powers and James N. Leiker, "Cholera among the Plains Indians:

Perceptions, Causes, Consequences," *Western Historical Quarterly* 29.3 (autumn 1998): 317–40.

34. For more on the Caste War, see Nelson Reed, *The Caste War of Yucatán* (Stanford: Stanford University Press, 1964); Victoria Bricker, *Indian Christ, Indian King: The Historical Substrate of Maya Myth and Ritual* (Austin: University of Texas Press, 1981); Moisés González Navarro, *Raza y tierra: La Guerra de Castas y el Henequen* (Mexico City: El Colegio de Mexico, 1970); Howard F. Cline, "Civil Congregations of the Indians in New Spain, 1598–1606," *Hispanic American History Review* 29 (1949): 349–69; Howard F. Cline's Appendix C, "Remarks on a Selected Bibliography of the Caste War and Allied Topics," in *The Maya of East Central Quintana Roo*, by Alfonso Villa Rojas (Washington: Carnegie Institute, no. 559, 1944), 165–78; Lawrence Remmers, "Henequén, the Caste War and Economy of Yucatan, 1846–1883: The Roots of Dependence in Mexican Region" (PhD diss., University of California, Los Angeles, Department of History, 1981); Robert W. Patch, "Decolonization, the Agrarian Problem, and the Origins of the Caste War, 1812–1847," in *Land, Labor, and Capital in Modern Yucatán: Essays in Regional History and Political Economy*, ed. Gilbert Joseph and Jeffery T. Brannon (Tuscaloosa: University of Alabama Press, 1991), 51–82; and Terry Rugeley, "The Maya Elites of Nineteenth-Century Yucatán," *Ethnohistory* 42.3 (1995): 477–93. For a reexamination of the Caste War, see Terry Rugeley, *Yucatán's Maya Peasantry and the Origins of the Caste War*, Symposia on Latin America Series (Austin: University of Texas Press, 1996).

35. Rugeley, "The Maya Elites of Nineteenth-Century Yucatán," 477.

36. Remmers, "Henequen, the Caste War and Economy of Yucatán," 312. This population figure is for 1850, and the percentage is from the population figure for 1846. The population of the capital city of Mérida fell almost 25 percent between 1846 and 1850. See Colección de Doctors Laviada (private family archive), Mérida, Yucatán, collected in 1998–99, various ephemera, newspaper clippings, and personal writings on the topic of cholera.

37. Thomas W. F. Gann, *The Maya of Southern Yucatan and Northern British Honduras*, Smithsonian Institution, Bureau of Indian Ethnology Bulletin no. 64 (Washington: U.S. Government Printing Office, 1918), 37. Excerpts from Gann's publication can also be found in Rugeley, *Maya Wars*, especially "The Thin Veneer of Civilization and Restraint," 170–86, 178.

38. Virginia D. Nazarea, "A View from a Point: Ethnoecology as Situated Knowledge," in *The Environment in Anthropology: A Reader in Ecology, Culture, and Sustainable Living*, ed. Nora Haenn and Richard Wilk (New York: New York University Press, 2006), 34–39, see 35. Also see Harold Conklin, "The Relation of Hanunuo Culture to the Plant World" (PhD diss., Yale University, 1954); Brent Berlin, D. E. Breedlove, and P. H. Raven, *Principles of Tzeltal Plant Classification* (New York: Academic Press, 1974); Brent Berlin, *Ethnobiological Classification: Principles of Categorization of Plants and Animals in Traditional Societies* (Princeton: Princeton University Press, 1992).

39. Hernán García, Antonio Sierra, and Gilberto Balám, *Wind in the Blood: Mayan Healing and Chinese Medicine*, trans. Jeff Conant (Berkeley: North Atlantic Books, 1999), 50.

40. Ralph L. Roys, ed., *The Book of Chilam Balam of Chumayel* (Norman: University of Oklahoma Press, 1967). Also see McCrea, *Diseased Relations*, 30.

41. Linda Nash, *Inescapable Ecologies: A History of Environment, Disease, and Knowledge* (Berkeley: University of California Press, 2006), 51.

42. H. F. van Emden and M. W. Service, *Pest and Vector Control* (New York: Cambridge University Press, 2004), 8–9.

43. Pamela Voekel, *Alone before God: The Religious Origins of Modernity in Mexico* (Durham: Duke University Press, 2002), 70. Voekel notes occasions when pigs and dogs rooted in Mexico City's San Lázaro cemetery. For fumigation procedures in cemeteries, see AHSSA, SP, Epidemias, box 1, 23, 12 July 1850. On nineteenth-century smallpox episodes, see Estado de Yucatán, *Monografía sobre la salud pública de Yucatán*, 64.

44. AGEY, Justicia, Tribunal Superior de Justicia Penal Abuso de Autoridad, 24 October 1854–March 1855, Citizens of Cacalchén against Don Esteban Herrera; Raúl E. Casares G. Cantón, ed., *Yucatán en el Tiempo: Enciclopedia Alfabética*, 6 vols. (Mérida: Inversiones Cares, S.A. de C.V. Col. México, 1998–2001), 2:162.

45. AGEY, Justicia, Tribunal Superior de Justicia Penal, Abuso de Autoridad, 24 October 1854–March 1855, Citizens of Cacalchén against Don Esteban Herrera; Heather McCrea, "On Sacred Ground: The Church and Burial Rites in Nineteenth-Century Yucatán, Mexico," *Mexican Studies/Estudios Mexicanos* 23.1 (winter 2007): 33–62.

46. AGEY, PE, Gobierno, Junta Superior de Sanidad, box 95, 1853.

47. *El Repertorio Pintoresco, ó miscelanea instructiva y amena consagrada a la religion, la historia del pais, la filosofia, la industria y las bellas letras*, ed. and graphics D. José D. Espinosa Rendon, ed. D. Crescencio Carrillo (Mérida: Imprenta de José D. Espinosa, 1861), 11.

48. AGEY, PE, Gobierno, Junta Superior de Sanidad, Mérida and Maxcanu, 1853, box 95, folder 11.

49. AGEY, PE, Gobernación, box 254, 1889. Also see Carlos Contreras Cruz, "Urbanización y modernidad en el Porfiriato: El caso de la ciudad de Puebla," in *Limpiar y obedecer: La basura, el agua y la muerte en la Puebla de Los Ángeles (1650–1925)*, ed. Rosalva Loreto López and Francisco Javier Cervantes Bello (Puebla, Mexico: Centro de Estudios Mexicanos y Centroamericanos / Universidad de Puebla, 1994), 187–219; David S. Parker, "Civilizing the City of Kings: Hygiene and Housing in Lima, Peru," in *Cities of Hope: People, Protests, and Progress in Urbanizing Latin America, 1870–1930*, ed. Ronn Pineo and James A. Baer (Boulder: Westview, 1998), 153–78.

50. AGEY, PE, Gobernación, box 139, Consulado de México en Belice, 1864; Biblioteca Cresencio Carillo y Ancona (BCCA), *Observación crítico-histórico ó defensa del clero yucateco* (Mérida: Imprenta de José Dolores Espinosa é hijos, 1866).

51. Alonso Aznar Pérez, ed., *Colección de leyes, decretos y órdenes ó acuerdos de tenden-*

cia general: Que comprende todas las disposiciones legislativas, desde 1 de enero de 1846, hasta fin de diciembre de 1850, 3 vols. (Mérida, México: Government of Yucatán, 1851).

52. On clandestine butchering in Mexico City, see Jeffrey M. Pilcher, *The Sausage Rebellion: Public Health, Private Enterprise, and Meat in Mexico City, 1890–1917* (Albuquerque: University of New Mexico Press, 2006), 59.

53. *Boletín de Instrucción Pública: Órgano de la Secretaría del Ramo* (Mexico: Tipografía Económica, 1905), 5:64. For more on the regulation of meat and public health in nineteenth-century Mexico City, see Pilcher, *The Sausage Rebellion*.

54. CAIHY, folleto 7, 1850 no. 11, Mérida, Medidas sanitarias adoptadas por el Ayuntamiento de Mérida para el caso de que el Cólera-Morbus invada esta capital, de José Cristóbal Hernández, 12 pages; and AHSSA, SP, Higiéne Pública (HP), Serie: Inspección de alimentos y bebidas (IAB), box 1, exp. 2, 20 October 1849.

55. AHSSA, SP, HP, Higíene Industrial (HI), box 1, folder 22, 1907–1908.

56. For more on the role that meat consumption plays in understanding human-animal relationships, see Carol J. Adams, *The Sexual Politics of Meat: A Feminist-Vegetarian Critical Theory* (New York: Continuum, 1990), 92–93.

57. Philippe Descola and Nora Scott, *In the Society of Nature: A Native Ecology in Amazonia*, Cambridge Studies in Social and Cultural Anthropology (Boston: Cambridge University Press, 1996), 2.

58. For a summary of Aristotle's discussion of the differences between animal and human, see Gary Steiner, *Anthropocentrism and Its Discontents: The Moral Status of Animals in the History of Western Philosophy* (Pittsburgh: University of Pittsburgh Press, 2005), 78–79.

59. AGEY, PE, Justicia, Juzgado de Paz de Campeche, Campeche, box 99, folder 14, 1854.

60. For accounts of foreigners stricken with illness in Yucatán, see Karl Bartolomeus Heller, *Alone in Mexico: The Astonishing Travels of Karl Heller, 1845–1848*, trans. and ed. Terry Rugeley (Tuscaloosa: University of Alabama Press, 2007), 159; Gustavus Ferdinand von Tempsky, *Mitla* (n.p., 1857), 274; Alice Dixon Le Plongeon, *Here and There in Yucatan: Miscellanies* (New York: John W. Lovell, 1889).

61. *Siglo XIX: Boletín Oficial*, no. 264, 4 June 1850, 1–3.

62. AGEY, PE, Gobernación, Correspondencia de los estados, box 76, various places, 1850.

63. Michael Adas, *Machines as the Measure of Men: Science, Technology, and Ideologies of Western Dominance* (Ithaca: Cornell University Press, 1989), 23–24.

64. AGEY, PE, Justicia, Juzgado de Paz de Campeche, Campeche, 1854.

65. AGEY, Milicia, box 100, PE, Varios Lugares, 1855.

66. AGEY, PE, Gobernacíon, Lista de Pasaportes (incluye tráfico de indios a Cuba), 1848–1851, box 71; Carlos R. Menéndez, *Historia del infame y vergonzoso comercio de Indios vendidos a los esclavistas de Cuba por los políticos Yucatecos desde 1848 hasta 1861: Justificacion de la revolucion indigena de 1847, documentos irrefutables que lo comproueban* (Mérida: Talleres Graficos de La Revista de Yucatán, 1923).

67. E. N. Anderson with Aurora Dzib Xihum de Cen, Felix Tzuc, and Pastor Valdez Chale, *Political Ecology in a Yucatec Maya Community* (Tucson: University of Arizona Press, 2005), 167.

68. AGEY, PE, Beneficencia, Salubridad, Hospitales, box 394, 1903, Mérida.

69. Dr. D. Francisco Javier Santero, *Elementos de Hygiene Privada y Pública*, vol. 2 (Madrid: El Cosmos Editorial, 1885).

70. NAWDC, DOS, CD, Mérida and Progreso, RG 59, Consul Edward Thompson to the Honorable U.S. Secretary of State, Progreso, 20 December 1900.

71. Pastor Solis, "Higiene Pública por el socio titular," *La Emulación* (Mérida) 1.10 (October 1873): 119; "Estado consecutivo a los frios y calenturas," *El Estudiante de Medicina: Revista Mensula de Medicina, Ciencias Accesorias y Variedades* (Mérida de Yucatán) 1.2 (5 June 1890): 23–24.

72. For more on the views of travelers in foreign lands and illness, see Kraut, *Silent Travelers*; Mary Louise Pratt, *Imperial Eyes: Travel Writings and Transculturation* (New York: Routledge, 1992).

73. François Delaporte, *The History of Yellow Fever: An Essay on the Birth of Tropical Medicine* (Cambridge: Massachusetts Institute of Technology Press, 1991), 23–24; also see George Cheever Shattuck, M.D., *The Peninsula of Yucatán: Medical, Biological, Meteorological and Sociological Studies* (Washington: Carnegie Institution of Washington, 1933), 356. At the time of the publication, Dr. Shattuck was assistant professor of tropical medicine at Harvard University Medical School. The study was conducted in collaboration with thirteen other professionals. Also see Nash, *Inescapable Ecologies*, 106–7.

74. Creatures identified as "vermin" in early modern England share many characteristics with those defined as "pests" and "vectors" in nineteenth- and twentieth-century Mexico. Mary E. Fissell, "Imagining Vermin in Early Modern England," in *The Animal/Human Boundary: Historical Perspectives*, ed. Angela N. H. Creager and William Chester Jordan (Rochester: University of Rochester Press, 2003), 77–114, 77–78.

75. AGEY, PE, Gobernación, Secretaria de estado y del despacho de gobernación, box 100, México, 1855; AGEY, PE, Beneficencia, Sanidad, Salubridad y Hospitales, box 784, 14 February 1912, Mérida.

76. Anne-Emanuelle Birn, *Marriage of Convenience: Rockefeller International Health and Revolutionary Mexico* (Rochester: University of Rochester Press, 2006), 24, 29.

77. Harlan P. Beach, *A Geography and Atlas of Protestant Missions: Their Environment, Forces, Distribution, Methods, Problems, Results and Prospects at the Opening of the Twentieth Century* (New York: Student Volunteer Movement for Foreign Missions, 1901), 1:72.

78. A few nineteenth-century travel writings focused on Yucatán are B. M. Norman, *Rambles in Yucatan: Including a Visit to the Remarkable Ruins of Chi-chen, Kabah, Zayi, Uxmal etc.* (New York: J and H Langley, 1844); von Tempsky, *Mitla*; José Fernando Ramírez, *Viaje a Yucatán: 1865* (Guadalajara: Ediciones Et Cetera, 1971); William Parish Robertson, *A Visit to Mexico, the West India Islands, Yucatan*

and the United States, 2 vols. (London: Simpkin, Marshall, 1859); Stephen Salisbury Jr., *The Mayas, the Sources of Their History: Dr. Le Plongeon in Yucatan, His Account of Discoveries* (Worcester: Press of Charles Hamilton, 1877); Le Plongeon, *Here and There in Yucatan*.

79. Rockefeller Archive Center (RAC), Rockefeller Foundation (RF), Record Group (RG) 5, series, 1.2, Various Private Sector/International Health Division (IHD) Correspondence: Typhus—1916, box 29, folder 461. After a conference with Señor Cabrera of Mexico, the head of the Mexican delegation on the American and Mexican Joint Commission, on 23 October 1916, the Rockefeller Foundation's Dr. Rose reported that the two had discussed the possibility of the foundation's International Health Board (IHB) aiding in the control of typhus fever in Mexico. Rose stated that the IHB would need a public statement from Mexico that verified the country's desire for cooperation, emphasizing that the program would not be politically connected and that the work would not be exploited by the press. Also see RAC, RF, RG 1.1 (Projects) IHB, series 3, box 13, sub-series 323 A (Mexico) (Medical Sciences), 1920, 1922, 1925, folder 93, letter to James R. Sheffield from George Vincent, 5 May 1925.

80. See, for instance, John Kenneth Turner's *Barbarous Mexico* (Austin: University of Texas Press, 1990 [1911]); and John Reed, *Insurgent Mexico* (New York: International Publishers, 1969 [1914]).

81. AGEY, PE, Junta de Salud y Salubridad (JSS), Copiador de oficios (asunto), 1920: 1 January 1920 through 18 February 1920, vol. 8.

82. Details of adverse human reactions to the Noguchi vaccine can be found in RAC, RFA, RG 5, Correspondence, series 1.2, sub-series 323, box 96, folder 1329, letter to Pruñeda from Noguchi regarding yellow-fever-vaccine development, 23 November 1920. See also Hideyo Noguchi, "Leptospira Icteroides and Yellow Fever," *Proceedings of the National Academy of Sciences USA* [Communicated by S. Flexner. Read before the Academy, 10 November 1919], no. 3 (New York: Rockefeller Institute for Medical Research, 1920), 110–11.

83. RAC, RF, RG 5, International Health Board, series 2, box 33, sub-series 323 (Mexico), folder 194, 1922.

84. AGEY, PE, Beneficencia, Sanidad, Salubridad, box 679, 1919, Mérida; AGEY, PE, Junta Superior de Sanidad, Copiador de oficios (asunto), 1920: 3 January 1920–28 February 1920, vol. 85, Mérida. For more on spatial separations and disease prevention, see Alexandra Minna Stern, "Buildings, Boundaries and Blood: Medicalization and Nation-Building on the U.S.-Mexico Border, 1910–1930," *Hispanic American Historical Review* 79.1 (February 1999): 41–81.

85. AGEY, PE, Beneficencia, Salubridad, Hospitales, box 394, 1903, Mérida.

86. *Stegomyia* a genus of mosquitoes that includes approximately six hundred species, some of which are vectors of disease, while others are pests. *Stegomyia* includes *A. aegypti*, a vector of yellow fever and dengue. See RAC, RF, RG 5, International Health Board, series 3, box 147, sub-series 323 (Mexico), folder 1757, letter from Dr. Russell to Dr. Vasconcelos, 20 January 1923.

87. RAC, RF, RG 12.1 (Diaries, Officer's Logs), F. Russell 1922–1924 (vol. 1 of 4), box 53, interviews from Thursday, 4 April 1924 with Dr. M. E. Connor, 137; RAC, RG 5, series 3, box 147, M. E. Connor, "Yellow Fever Reports, 1922." Armando Solorzano argues that Noguchi's vaccine was given too much credit for protecting Mexican soldiers from yellow fever, but, in fact, the vaccine was rendered useless by the 1920s ("The Rockefeller Foundation in Revolutionary Mexico: Yellow Fever in Yucatan and Veracruz," in *Missionaries of Science: The Rockefeller Foundation and Latin America*, ed. Marcos Cuerto [Bloomington: Indiana University Press, 1994], 52–71, 64–65).

88. AGEY, PE, Junta Superior de Sanidad, 1921, Mérida, vol. 91: 3 February 1921–30 April 1921.

89. Hideyo Noguchi's anti–yellow fever serum was derived from fieldwork he completed in Guayaquil, Ecuador. While in Guayaquil, Noguchi isolated a spirochete microorganism in guinea pigs that "produced lesions suggestive of yellow fever." The microorganism, *Leptospira icteroides*, had also been isolated in human patients suffering from yellow fever. Thus Noguchi concluded that *Leptospira icteroides* was the disease-causing agent, and he subsequently synthesized it into an anti–yellow fever serum and vaccine. See Marcos Cueto, "Sanitation from Above: Yellow Fever and Foreign Intervention in Peru, 1919–1922," *Hispanic American Historical Review* 72.1 (February 2002): 1–22, 5 and n15.

90. RAC, RF, RG 5, International Health Board, series 2, box 33, sub-series 323 (Mexico), folder 194, 1922.

91. AGEY, PE, Junta Superior de Sanidad, Mérida, 1920, vol. 87: 1 July 1920–31 July 1920; also see "Reglamento Interior de la Junta Superior de Sanidad."

92. "Control of Yellow Fever," *New York Times*, 5 March 1922, 84.

93. "Rockefeller Foundation Leads World Fight on Disease," *Science News-Letter* 8.272 (26 June 1926): 2–3.

94. "Rockefeller Help to World Reviewed," *New York Times*, 10 November 1929, 14.

95. María A. Loroño-Pino, José A. Farfán-Ale, Alicia L. Zapata-Peraza, et al., "Introduction of the American/Asian Genotype of Dengue 2 Virus into the Yucatan State of Mexico," *American Society of Tropical Medicine and Hygiene* 71.4 (2004): 485–92; Carl Kendall, Patricia Hudelson, Elli Leontsini, Peter Winch, Linda Lloyd, and Fernando Cruz, "Urbanization, Dengue, and the Health Transition: Anthropological Contributions to International Health," Contemporary Issues of Anthropology in International Health, *Medical Anthropology Quarterly, New Series* 5.3 (September 1991): 257–68.

96. "Mexico, Venezuela Implement Counter-Dengue Measures," Xinhua General News Service, 29 August 2007, http://news.xinhuanet.com/english/2007-08/29/content_6623616.htm.

97. R. Danis-Lozano, M. H. Rodriguez, L. Gonzalez-Ceron, and M. Hernandez-Avila, "Risk Factors for Plasmodium Vivax Infection in the Lacandon Forest, Southern Mexico," *Epidemiology and Infection* 122.3 (June 1999): 461–69, 466.

6

Notes on Medicine, Culture, and the History of Imported Monkeys in Puerto Rico

NEEL AHUJA

In 2007, nearly seventy years after the first large-scale importations of "old world" monkeys to Puerto Rico, the Commonwealth's Department of Agriculture proposed a regulation prohibiting the importation, trade, and possession of rhesus, patas, and squirrel monkeys—three species designated as "detrimental to agricultural interests and a threat or risk to the life and security of humans."[1] If, as Roberto Esposito has argued, modernity's politics of life is established through a paradigm of immunization, safeguarding against the proliferating risks of public, communal life, then the public discourse on monkeys in Puerto Rico has undergone a reversal since the 1930s, when monkeys were established as technologies for the pharmaceutical engineering of U.S. American immunity.[2] At that time, the U.S. biosecurity apparatus figured nonhuman primate spinal material as vital to polio-vaccine development and other areas of biomedical research, and the Puerto Rican archipelago was seen as a necessary breeding ground for monkey bodies. In 1939 the comparative psychologist Clarence Ray Carpenter, with the backing of the Columbia University School of Tropical Medicine at San Juan, personally transported over four hundred rhesus macaques from northern India to the islet of Cayo Santiago in the Puerto Rican archipelago. Although primate species such as the African vervet had arrived via slave ships on the nearby Caribbean islands of St. Kitts, Nevis, and Barbados as early as the 1560s, neither "new world" nor "old world" monkeys are indigenous to the Caribbean.[3] The twentieth-century research station founded by Carpenter, which inspired similar efforts across the United States and one in Brazil, was the first free-ranging colony of Asian-origin primates in the Americas. In U.S. American print media, officials justified Cayo Santiago and called for expanded U.S. imports of Indian rhesus macaques based on their value as pharmaceutical raw materials that could protect the nation against an increasingly wide array of diseases.

What are the terms on which monkeys, as "raw materials" for the production of scientific knowledges and pharmaceutical commodities, become rhetorically aligned with the imperial nation? And in a multiply colonized, globally interconnected space like twentieth-century Puerto Rico, how does the concomitant alignment of monkeys with biomedical progress collide with the ascendant nationalisms of imperially dominated populations? Or even with the life-practices and ecologies of the animals themselves? In this chapter, I sketch the history of imported rhesus and patas monkeys in Puerto Rico with a focus on their political significance and the divergent institutional sites, discourses, and landscapes in which they appear as imperial migrants. I document monkeys' contradictory status as, on the one hand, figures of progress aligned with modern biomedical technology, and, on the other, as "invasive species" that symbolize the multiple violences of U.S. imperialism and neoliberal development policy. Drawing in part on scientists' official histories of primate institutions, I trace the historical conjunctions of nonhuman primate bodies, medical research institutions, Puerto Rican nationalism, cultural fears of government secrecy, and the transnational and transcolonial politics of the global primate trade. Monkeys' fate in Puerto Rico—as well as their relationships with humans and Puerto Rican ecosystems—has been intimately linked to the dynamics of U.S. imperial power on the archipelago, especially as it has been expressed economically and militarily. The history I outline is marked by four key transitions in primate biopolitics in Puerto Rico: the establishment of U.S. biomedical institutions in the 1920s and 1930s; the rise of Puerto Rican nationalism and the subsequent takeover of Cayo Santiago by the Puerto Rican government in the 1940s; the establishment of federally funded spin-off colonies in the 1960s and 1970s; and finally, new research and trapping agendas run by Puerto Rican scientists and conservation authorities.

I conclude by reflecting on how a history of imported monkeys in Puerto Rico might help scholars theorize knowledge production in the humanities, as well as in emerging interdisciplinary research sites including biopolitical theory, critical species studies, and science studies. I focus in particular on the politics of the archive, issues of representation and agency, and the difficulties in writing history against imperial discourses that silence the histories of colonized subjects and yet produce extensive, if conceptually limited, archives of animal representation in the form of behavioral studies and official institutional histories. Such archives register the paradoxes of the segregation of humans and animals in scientific writing, which at times privileges animal objects over human ones. At the same time, these representations often allow us to think against the grain of historical knowledge to address what Martha Few and Zeb Tortorici in the introduction to this volume call "the types of looking and interaction that take place among and between different species." Tracing histories of interspecies living and representational practices contributes to a critique of the discursive and material practices of imperialism in the Americas as well as to understanding the complex politics of difference produced by modern, imperial circuits of biopower.

Biomedical Imperialism and Primate Universality

Like the other new United States possessions annexed in the years 1893–1904, Puerto Rico became a target of a variety of biopolitical interventions that served both immediate public-health agendas and the interests of experimental researchers from the continental United States. Much of the scholarly attention on this point has focused on two particular forms of intervention: first, the initial public-health colonialism, in the decades immediately following the takeover of Puerto Rico in 1898, emblematized by the successful antihookworm campaign; and second, regulations of female sexuality and eugenic campaigns that culminated in the infamous sterilizations of the 1930s through the 1970s.[4] As Laura Briggs suggests, such dramatic health and eugenic campaigns have been deeply imbricated in an imperial discourse that pathologizes Puerto Ricans, establishing both a racial ordering and a disavowal of the U.S. role on the archipelago.[5] If, as Jorge Duany argues, post-1898 Puerto Rico was envisioned as a sort of laboratory for processes of modernization, it was also a site where the controlled reproduction of primates made it a key material link in the emerging apparatuses of experimental medical research at midcentury.[6]

Rhesus macaques of Indian origin had, following the path of polio research established in the 1920s, become central to national disease-research efforts. Key European and U.S. researchers insisted on using rhesus models for polio research. Although differences between rhesus and human development of polio stalled polio research for decades, the eventual realization of a successful vaccine in the 1950s was directly based on rhesus modeling.

By this point, rhesus had become a favored subject for physiological research across a variety of ailments and medical specializations, yet the vast majority of rhesus—an estimated 120,000 of 150,000 imported annually by the mid-1950s—were actually killed for use in vaccine production.[7] The process of polio vaccine testing as practiced in Europe and the United States worked as follows: researchers would inject a population of monkeys with poliomyelitis, wait for signs of illness, and then kill the animal, grinding its spinal matter to extract the fluid in which the virus had reproduced. Once this fluid had been processed as a serum, researchers would test the effectiveness of the experimental vaccine on another set of monkeys.[8] When a workable vaccine was finally available for mass production in the 1950s, pharmaceutical corporations paid Indian harvesters and imported large populations of animals, at times in cramped cages holding up to a hundred animals each.

Such demands for animals in research and pharmaceutical production by midcentury—increasingly supported by strategies to establish national health entities like the National Institutes of Health (NIH)—moved public-health investments away from "premodern" strategies like quarantine and toward biomedical alternatives that penetrated the human body, making body systems rather than populations the privileged objects of health intervention. Yet there were a variety of limitations on importing key laboratory subjects like the Indian rhesus. Clarence Ray Carpenter's idea in the 1920s of establishing a rhesus colony in Puerto Rico to shore up the number of research subjects inspired the development of a national policy on animal "research resources," eventually leading to the establishment of today's system of National Primate Centers run by the NIH. Puerto Rico's role in the biomedical primate trade and in the production of breeding facilities thus helped make American medicine "modern."

Carpenter's colony at Cayo Santiago was only possible because of its close association to a broader initiative in tropical medicine controlled by Columbia University's College of Surgeons and Physicians. In 1926 a joint effort by Columbia University and the University of Puerto Rico transformed the former U.S. Army Institute of Tropical Medicine into Columbia's School of Tropical Medicine (fig. 6.1). The school's history indicates an unequal exchange of scientific knowledge and labor between the mainland and the islands. As Annette B. Ramírez de Arellano notes, the partnership was initially hailed as a sign of progress for U.S.-occupied Puerto Rico. The *Herald Tribune* claimed it would produce "a very strong bond of fellowship and sympathy between the United States and one of the most interesting of

6.1 School of Tropical Medicine, San Juan. SOURCE: *ILLUSTRATED LONDON NEWS*, 13 AUGUST 1938. COURTESY OF THE UNIVERSITY OF PUERTO RICO, MEDICAL SCIENCES CAMPUS, CONRADO F. ASENJO LIBRARY, SPECIAL COLLECTIONS.

its territorial possessions" while addressing "the complaints made by Puerto Ricans that their interests are being neglected."[9] But as the school's establishment of Cayo Santiago demonstrates, its priorities were often guided by the mercurial and self-interested agendas of mainland researchers.

At the time of the founding of Cayo Santiago, the Columbia comparative psychologist Clarence Ray Carpenter was navigating a number of divergent interests in Puerto Rico. Carpenter's immediate interest was not in the sort of biomedical practices that depended on rhesus for physiological modeling and vaccine production. Instead, his work was in the behavioral sciences, specifically field studies of primate sexuality, social organization, and behavior, with a particular interest in the "way a population organizes itself."[10] The appeal of the rhesus colony, however, had to go beyond behavioral sciences to include pharmaceutical research and production that could be intimately tied to the security of the imperial nation. George Bachman, director of the School of Tropical Medicine, latched on to the idea because of an interest in securing experimental research subjects. Other interested parties, including Columbia experts in endocrinology and anatomy, found use in either free-ranging or captive populations of rhesus and gibbons. In public media Carpenter's justifications for the colony raised the specter of vital "medical" research first, both drawing on fear of disease and aligning monkeys' roles in "laboratory" work with ideals of security and progress.[11] Rhesus of Indian origin were in high demand, were inaccessible

6.2 Aerial photograph of Cayo Santiago, 1938. PENN STATE UNIVERSITY ARCHIVES, PENNSYLVANIA STATE UNIVERSITY LIBRARIES.

during wartime shipping disruptions, were subject to British-Indian export restrictions, and were ecologically threatened in Northern India. Carpenter was likely the first U.S. scientist to call for a national study of nonhuman primate research resource needs and coordination of imports from India, thus marshaling federal diplomatic authority.[12]

Carpenter's colleagues at the school were able to secure the islet of Cayo Santiago from the sugar baron Antonio Roig just after the Great Depression brought an end to the U.S. capital-driven sugar boom in Puerto Rico. The Roig sugar empire—born during Spanish rule and spanning 12,500 acres of cane plantations at Humaçao in the 1930s—used Cayo Santiago for goat pasture near its coastal shipping station.[13] Running 600 meters from north to south and 400 meters from east to west, Cayo Santiago lies a half mile southeast of Punta Santiago, Humaçao (fig. 6.2). When it was initially leased to the school, it was covered primarily with brush, grasses, and coconut groves.[14]

Yet the islet was reshaped for Carpenter's project, revealing some of the subtly colonialist and racialist assumptions that became clear when scientists designed new habitats for monkeys. Primate colonies, which proliferated globally in the 1920s and 1930s, advertised what Donna Haraway has called a "simian orientalism," situating nonhuman bodies in difference marked by geography and behavior.[15] Haraway references Edward Said's critique of colonial knowledge production, which understands the production of Western self-images and institutional hegemony over colonized Asia as a matter of positioning the "Orient" as geographically, culturally, and temporally distant, and further as a scene of origins for language and civilization. As orientalism displays the Western imagination of the origins of civilization and city, primatology displays an imaginary of the origins of society itself as found within a "natural" animal order. Racial and sexual signifiers become key indicators of the natural order in this mirroring of human and animal.

In the history of Cayo Santiago, simian orientalism involved the attempt to design an imagined "natural" landscape for the rhesus, and thus for the colony to produce properly "natural" rhesus bodies and forms of social organization. In the main historical description of the colony construction offered by Richard Rawlins and Matt J. Kessler, the landscaping of the island in advance of the monkeys' settlement demonstrates how scientists tropicalized Cayo Santiago in advance of the monkeys' arrival. The islet was landscaped with forests of coconut and mahogany, as well as with tropical tubers; media such as the *Illustrated London News* could print postcard pictures of Cayo Santiago's beaches and palms (fig. 6.3).[16] Presuming that tropical tubers and imported produce would nourish animals and that the sea would enclose them, the scientists were surprised when the monkeys destroyed most of the imported vegetation and, in small numbers, swam to the main island. In escaping and literally tearing down the imperialist fantasy of "tropical nature," the animals forced scientists to establish feeding stations.

At the same time that simian orientalism advertised the tropical "scene of origins" of human sociality via the bodies of nonhuman primates, an emerging public discourse embraced the universality of the primate and saw rhesus as appropriate models of humanity (not just representatives of a sort of primitive psychology underlying social relations). Hansel Mieth's early *Life* photographs from Cayo Santiago—captioned, against Mieth's wishes, to joke that a rhesus "misogynist" was escaping female "chatter" by swimming out to sea—figure the monkey as a symbol of wartime alienation, stranded after the dislocations brought by the violence and dislocations of modernity (fig. 6.4).[17] When the Cayo Santiago station was unveiled to the media in the opening days of the Second World War, it was presented as a new line of defense in the nation's biosecurity arsenal. This line of reasoning followed a particular racialized geography of difference, with Indian mon-

6.3 View of Cayo Santiago from Puerto Rico. SOURCE: *ILLUSTRATED LONDON NEWS*, 13 AUGUST 1938. COURTESY OF THE UNIVERSITY OF PUERTO RICO, MEDICAL SCIENCES CAMPUS, CONRADO F. ASENJO LIBRARY, SPECIAL COLLECTIONS.

keys representing both the failures of British imperialism and the possibilities of a new U.S. American empire founded on scientific rationality. Carpenter repeatedly denounced both the supposed filth of "disease-carrying humans" in India and the British Indian animal-welfare regulations protecting monkeys in transport.[18] In 1939 a *Life* article introducing the Cayo Santiago station initially focused on familiar colonial tropes exoticizing India: "Because he is considered sacred in India," claimed the unnamed writer, "the rhesus is domineering, undisciplined and bad tempered."[19] Reinforcing British colonial discourse that saw the backwardness of Indian religions as justification for external rule, the article also naturalized Puerto Rico as a site for the demonstration of the technical-scientific rationality underpinning biomedical progress. Nearby residents in Punta Santiago were "alarmed" by the *Life* article's mention of plans to use the colony of free-ranging monkeys to research cures for polio, tuberculosis, and leprosy (now called Hansen's disease).[20] A community group met with colony scientists to voice their concerns about risk of disease transmission to humans in the area.[21] Although this initial public resistance to the colony apparently died down after a forum attended by scientists, the history of Cayo Santiago's founding testifies to the imbrication of biomedicine within histories of unequal ex-

6.4 *A Misogynist Monkey Seeks Solitude in the Caribbean off Puerto Rico*, 1939. PHOTOGRAPH BY HANSEL MIETH. COLLECTION CENTER FOR CREATIVE PHOTOGRAPHY, UNIVERSITY OF ARIZONA. © 1988 THE UNIVERSITY OF ARIZONA FOUNDATION.

change between colonized spaces and the imperial mainland. Utilizing the institutional resources of tropical medicine, the labor of primate traders in British India, and the infrastructure of the sugar trade, Cayo Santiago represents an odd trajectory of transcolonial relations that enabled the transition to "modern" biomedicine.

Renationalizing the Puerto Rican Monkey

Cayo Santiago's rise and fall were meteoric, and the early years of the colony were characterized both by disruptions to the rhesus population and by turnover in colony administration. By December 1938, after Carpenter had sold a number of his animals to pay for travel expenses, caretakers began releasing the remaining 406 rhesus macaques, 14 gibbons, and 3 pigtail macaques onto the islet. By 1940, field research was under way and the monkeys were reproducing rapidly (194 infants were born by 1942). Well over 500 animals were also sold through the end of the Second World War for use in mainland U.S. research labs. During the major studies of the early 1940s, 350 rhesus remained on Cayo Santiago; by 1944, there were only an estimated 200. Social reorganization after resettlement in Puerto Rico was a major threat to young rhesus, as many were killed by other monkeys in the first year.

In the early 1940s a number of behavioral field studies were conducted on the sexual and social organization of the monkeys, and colony administrators had to both provision monkeys with commercial monkey chow and take serious steps to contain disease. For both scientific and health purposes, monkeys were tagged and trapped for census-taking and medical inspections, especially given outbreaks of diarrhea and shigella (fig. 6.5). (Scientists were successful at this point in eradicating tuberculosis among the population.) There is little accounting of the numbers of escaped monkeys during this period, but there were reports of monkeys swimming to the Puerto Rican mainland. One rhesus escapee was reported captured and returned by the scientist in charge.[22]

The main historical work on Cayo Santiago notes that both the world war and the end of major funding for the colony in 1941 triggered a period of neglect. Yet it is important to also situate the abrupt decline of the colony in this period within the changing relationship between Puerto Rico and the United States. In the 1940s a variety of relationships between universities, industry, the island government, the federal government, and foreign governments were transformed. Operation Bootstrap, which brought tax breaks to encourage new factories in Puerto Rico, came along with the liberal policies of Rex Tugwell, including land reform and the eventual approval of Puerto Rico's "Free Associated Status" as proposed by the Popular Democratic Party leader Luis Muñoz Marín. The federal acceptance of both the funding and political strategies associated with this nationalism was integrated within a larger program to use Puerto Rico in Cold War battles over decolonizing nations. The U.S. State Department established its Point Four Program, intended to showcase Puerto Rico as a capitalist development model to Third World elites, while disavowing Puerto Rico's colonized status. More than 30,000 Third World officials stayed in Puerto Rico and were shown "the industrialization projects, the health system, and other construction work."[23] During the era of what Laura Briggs calls "modern-

6.5 Michael Tomilin and assistants tagging a rhesus macaque at Cayo Santiago, n.d. PENN STATE UNIVERSITY ARCHIVES. PENNSYLVANIA STATE UNIVERSITY LIBRARIES.

ization nationalism," "Puerto Rico was explicitly a 'laboratory' in which development—foreign aid, industrialization . . . , import substitution, and population control—was being tested as a global policy."[24] The establishment of Cayo Santiago occurred just at the moment when, with both an ascendant nationalism in Puerto Rican politics and an infusion of New Deal funding and federal interest in Puerto Rican institutions, officials on the archipelago made a sustained push for an expansion of basic public health on the island. Aggressive demands for the basic training of doctors and nurses transformed the relationship between Puerto Rico and the Columbia officials. In 1938, with nationalist leaders stressing the development and independence of the school, Puerto Rico demanded an increased voice in decision making, leading to the replacement of George Bachman as the school's head. In 1942 the chancellor set up a committee to launch a new medical school devoted primarily to medical and nursing training. In 1946 Columbia shifted the focus of its tropical medicine research to immigrants in Washington Heights, New York, announcing that its funding for the Puerto Rico institution would end in two years.[25]

The late 1940s were a turning point that eventually brought decentralization, reimportation, and a new player—the National Institutes of Health—to the management of Puerto Rican monkeys. After Carpenter's initial studies at Cayo Santiago in the early 1940s, many of the caretakers left, and funding was scarce, directed toward more immediate health needs. The dean of the School of Medicine put Cayo Santiago's monkeys up for sale. Yet the colony was saved by an NIH grant, largely engineered by José Guillermo Frontera, a Puerto Rican biologist studying in Michigan, who convinced the school to delay closure of the colony.[26] This grant, which funded the colony in 1949, was the springboard for a new era of federal involvement in primate institutions in Puerto Rico and for a much more diverse array of research activity.

The grant invigorated interest in the establishment of new, highly controlled colonies by the government in order to ensure a steady supply of monkeys for both behavioral research and biodefense. Although the NIH explored setting these colonies outside of U.S. continental borders, it eventually established a system of Regional Primate Research Centers (now the National Primate Research Centers) at major continental universities and research facilities. This institutional history has been discussed elsewhere by several scientists involved with Cayo Santiago and its spinoff institutions across Puerto Rico.[27] These spinoff institutions, which offered research monkeys for work on blindness, neurological disease, and heart disease, eventually included laboratories and colonies of monkeys in San Juan, Sabana Seca, and three small islets in the northwest and southwest of the archipelago: Cueva and Guayacán—which together formed the La Parguera colony, funded by the NIH and the FDA, in the southwest—and Desecheo Island, where ecological studies of free-ranging rhesus took place in the northwest. The La Parguera monkeys were eventually transferred to populate new colonies at Morgan Island, South Carolina, and the German Primate Center in Göttingen.

A cartoon that appeared in *El Mundo* in December 1949 represents a rhesus strutting confidently, head cocked back, after learning of the initial NIH grant; if federal funding ensured the continued ability of medical institutions to support imported monkeys, it also initiated a set of changes that would remake the image of monkeys as an invasive presence, draining resources and land. Federal funding was thus a mixed blessing: while it sustained a type of research program not usually available to U.S. possessions, it made funding conditional on the increased access of mainland researchers to Puerto Rican monkeys and land. In particular, FDA grants that attempted to increase the number of monkeys used for polio vaccine production led to a new problem: the monkey as an "invasive species," responsible for millions in damage to agriculture and threatening to indigenous species. In the postwar era the decentralization of monkey settlements in Puerto Rico resulted in the establishment of large free-ranging populations of introduced monkeys in Puerto Rico. On Desecheo Island, rhesus decimated the population of brown boobies that remained after decades of U.S. military bombing and training exercises. In the agricultural southwest escapees from Cueva and Guayacán caused sustained damage to crops and contributed to the tense politics surrounding the maintenance of agriculture in the region given expanded development, decades of import-substitution policies that favored industry over agriculture, and the proposed establishment of U.S. military installations. These developments are thus directly related to the lengthy history of U.S.-led development schemes on the archipelago, along with policies that brought new waves of foreign investment, expansion of heavy industry (including pharmaceuticals), and export-oriented tax policies.[28]

Millennial Monkeys: From Universal Primate to El Chupacabras

In 1962–63 the School of Medicine's Laboratory of Perinatal Psychology, which operated as a captive colony of Cayo Santiago-derived rhesus in San Juan, established a free-ranging colony on the islets of Cueva and Guayacán, just off the coast from La Parguera. La Parguera is a fishing village in southwest Puerto Rico with a number of mangrove islets among the phosphorescent waters surrounding its bay. The objective of the colony was to test whether breeding cycles and other behaviors of rhesus were ecologically adapted. While some animals were transferred from Cayo Santiago, the ecologist Carl Koford also traveled to India to acquire new stocks of rhesus. By 1963, 278 monkeys had been released on the islets, whose combined area amounts to approximately 190 acres. Because of the extreme proximity of these islets to the mainland—shores were as close as fifty meters—monkey escapes occurred from the beginning of the program. The researcher John Vandenbergh claims that "local people did not object to the presence of these renegades" at the time, and that the facilities supervisor Carlos Nagel acted as an effective diplomat to the villagers.[29] His mention that the colony paid off frustrated landowners, however, betrays the tension that occurred from the outset over escaped monkeys in the southwest.

Escapes of rhesus and African patas monkeys, which were introduced at La Parguera in 1971, steadily increased free-ranging populations in the Lajas Valley of southwestern Puerto Rico. When the Food and Drug Administration became interested in La Parguera to increase the supply of monkeys for polio vaccine production, things got worse. The FDA's 1974 grant helped maintain Puerto Rican primate facilities facing continuing funding problems. Yet with a grant to expand La Parguera, the FDA attempted to increase the colony size to 2,000 animals, many times larger than either the original population or the population at Cayo Santiago. Escapes increased, with entire troops of animals swimming freely across the channels of the bay. Scientists at La Parguera did not keep estimates of the number of escapes, but by the end of the 1980s, they had trapped over 250 monkeys in the southwest of the main island and reported other monkeys living free in cattle-grazing areas on the Sierra Bermeja mountain range.[30] Two population studies have been carried out in the area. The first, completed in the early 1990s, claims that the number of monkeys in the southwest was in the low hundreds for both rhesus and patas species.[31] An unpublished 2006 study finds a total of well over 500 patas alone in the large troops in this area.[32] The total population of free-ranging monkeys is now likely over 1,000 and growing.

Multiple factors related to farm activism and public-health concerns converged in the late 1990s to bring more attention to the free-ranging monkey population in the southwest. First, sporadic human encounters with feral monkeys in increasingly urban areas brought evidence of possible zoonotic health risks, especially with the discovery of B-virus in a monkey that died in San Juan.[33] Second, an agricultural group in the southwest, El Frente Unido Pro-Defensa del Valle de Lajas (United Front for the Defense of the Lajas Valley), took up the issue of crop damage as part of its push to establish the Lajas Valley as a protected agricultural area. Finally, with mounting public pressure, dramatic events related to the Puerto Rican government's capture and culling of monkeys gained international media attention.

Government officials attempted to address animal-welfare concerns, but as farmers' protests grew louder and the threat to health and environment became more apparent, the government took increasingly strong measures to control and eliminate monkey populations. Farmers from Lajas had reported losses due to monkeys for years, but it was only in the late 1990s that Puerto Rico took significant action. In 1999 a new wildlife plan established the authority to manage invasive species through a variety of non-lethal and lethal means, including proscribed hunting. While trapping and removal of animals was ongoing in Lajas, there was no coordinated study of it until 2006, when the government also first proposed significant funding ($1.8 million) for monkey removal. In 2008 Puerto Rico gained international media attention when it initiated a trap-for-export program, beginning with the transfer of fifteen rhesus to a private safari park in Florida; after the animals escaped by swimming across a moat, they led county officials on a six-month chase. No other institutions were willing to take more trapped animals until early 2009, when Iraq's National Zoo in Baghdad agreed to take a shipment of monkeys for public display. People for the Ethical Treatment of Animals and other international animal-rights groups denounced the measure for placing rhesus in a war zone.

By December 2008, the government was openly shooting trapped monkeys to prevent their spread across Puerto Rico, attracting the attention of media and activist organizations worldwide. The move risked stoking stereotypes regarding animal cruelty in Puerto Rico, especially given the widely publicized slaughter of feral dogs earlier in the year.[34] The final solution is still being negotiated, with the Caribbean Primate Research Center pressuring the Puerto Rican government to take more steps to ensure that monkeys are captured and utilized for scientific purposes, and with several animal-welfare organizations calling for humane population-control measures.[35]

The new monkey-control initiatives of the DRNA—Puerto Rico's Department of Natural Resources and Environment, whose animal-control efforts have been criticized in Puerto Rican newspapers—signal a reversal in the image of the monkey as an indicator of universal scientific progress. Patas and rhesus monkeys began to serve as figures of invasion as early as the 1990s. In 1998 news reports claimed that farmers faced 20 percent losses and were switching from profitable fruit and vegetable exports to less-profitable crops and, in some instances, leaving the business altogether.[36] The new media attention was not without its exaggerations. The numbers of monkeys were regularly reported to be significantly larger than the population surveys indicate. Yet the monkey problem was occurring within a larger context of economic decline, transformations in people's livelihoods, and a tense situation regarding the presence of the U.S. military. It was within this context that the most sensational stories of Puerto Rico in the 1990s were disseminated internationally. The legend of the cryptid *chupacabras*—first reported in 1994 in Canóvanas, northeastern Puerto Rico—quickly spread to southwestern Puerto Rico, a region which in the 1990s became the epicenter for reports of paranormal activity on the archipelago: UFOs, alien landings, abductions, and the death of thousands of farmed animals associated with precision bloodletting. Monkeys were thus one of several signs of U.S. imperial presence that fomented anxiety in the figure of el chupacabras; while monkeys helped form the visual impression of this cryptid, they were also consistently presented as the "real" beings behind the scare, mobilized to dismiss the supposed superstition of rural Puerto Ricans. This was common practice even as drought and other economic-environmental factors threatened farmed animals. For example, in response to a report of two sheep deaths in Lajas in 1996, officials quickly claimed that the animals had been attacked by monkeys in the area.[37]

Imagined as an extraterrestrial vampire that blends characteristics of reptiles, dogs, and monkeys, the chupacabras legend spread first to Mexico, then throughout the Spanish-speaking communities of the Americas; it became a staple of U.S. televisual representations of the paranormal during this time. According to Lauren Derby, the chupacabras legend must be understood as part of a broad "culture of suspicion" regarding the U.S. presence in Puerto Rico.[38] Derby mentions in particular the pervasive and secretive U.S. military installations as promoting a particular "state effect": "The state in Puerto Rico is . . . pervasive yet remote; commanding yet invisible, since much of the actual muscle of U.S. imperial power resides on the island because the U.S. armed forces have enormous holdings on Puerto Rican soil."[39] Economic and environmental concerns went hand in hand with suspicion over the military presence. Chupacabras sightings clustered in areas associated with U.S. government or industrial presence, including new pharmaceutical plants that sprang up in the 1990s after the North Atlantic Free Trade Agreement (NAFTA) went into effect. In the agricultural southwest, which was facing a prolonged drought, the phenomenon was linked to paranormal activity at the site of a new U.S. military radar project. The United Front for the Defense of the Lajas Valley formed in part to oppose the siting of a U.S. military radar project in Lajas. The navy proposed the project—consisting of a transmitter on Vieques Island, a receiver in Lajas, and a control center in Virginia—in the face of protests against the military installation on Vieques, a large island in the east of the archipelago. Ostensibly proposed to monitor Caribbean drug trafficking, the project was seen by antimilitary activists as an excuse to justify continuing presence of the military in Vieques, where bombing runs were denounced as damaging the environment and for causing unusually high cancer rates. In 1994 the United Front linked with other groups, including El Comite Pro-Rescate y Desarrollo de Vieques (Committee for the Rescue and Development of Vieques), to promote a cultural nationalist discourse against the military that linked the presence of U.S. installations and technologies to environmental, health, and economic destruction.[40] Thus economic, environmental, and military forces combined to promote a discourse of skepticism whereby certain "open secrets" of the government formed a spectacle. Thus, chupacabras sightings in the southwest cannot simply be dismissed as "superstition." As Robert Michael Jordan notes, key socioeconomic forces gave rise to the chupacabras legend, first in Puerto Rico, then elsewhere in the Americas: "perceptions of U.S. economic, cultural, and political imperialism," "pollution," and "fragmentation of rural society" wrought by the post-NAFTA spread of industry to rural areas.[41] Although within a decade interest in the chupacabras in Puerto Rico had mainly become limited to paranormal and cryptozoological communities, the legend's emergence demonstrates the ways in which nonhuman animals become incorporated into complex cultural negotiations over colonized space.

As the free-ranging monkeys of the southwest were being discussed as a "problem" for Puerto Rico, new changes were taking place at Cayo Santiago. Adaris Mas became Cayo Santiago's first Puerto Rican director, and, along with other facilities now linked as the Caribbean Primate Research Center (CPRC), Cayo Santiago received funding ensured via an NIH grant aimed at supporting a program of AIDS research at the University of Puerto Rico (UPR). Scientists at UPR have been at the center of a pan-American AIDS-research agenda, grounded in studies on local disease transmission and possibilities for vaccines targeted at particular HIV strains. Monkeys remain part of this research agenda, both as experimental animals and as essential links in the history of the disease. (Populations of captive monkeys in Puerto Rico faced early outbreaks of AIDS, beginning in the 1960s, and were important in the isolation of simian retrovirus.)[42] At the same time, monkey institutions often remain hidden from public view, linking them to the economies of secrecy associated with the U.S. state effect on the archipelago. Officials at UPR have become acutely aware of the possible distance between the scientific venture and the community. Edmundo Kaiselbard, the director of the Caribbean Primate Research Center, has argued for the need to "give back" to the local fishing communities and to establish a closer relationship between the scientific institution and the local community, particularly highlighting the necessity of tourism and the potential damage that government trapping programs may cause to the tourist industry. Fishing boats and kayaks increasingly bring tourists to view Cayo Santiago from the sea, supplementing declining fishing revenues. Kaiselbard has also proposed an onshore museum and library to attract visitors.[43]

As scientists in Puerto Rico navigate the complex Puerto Rican national politics regarding health, development, and scientific research, they continue to be confronted with international attention that pits science against race- and class-biased depictions of Puerto Ricans. A 2006 article in *Science*, which associates Puerto Ricans with poverty and drug use, portrays the NIH funding for the CPRC as offering unrecognized public-health benefits: "Good HIV / AIDS care and strong research in this U.S. commonwealth often mean little to the island's many heroin addicts."[44] Still, with the development of a robust indigenous research agenda, Puerto Rican scientists are attempting to straddle the at times conflicting interests of medical research, conservation, and economic development.

The most recent development in the history of monkey importation in Puerto Rico was a failed attempt to bring Carpenter's initial vision of Puerto Rican primate biodefense full circle, as it would have established private breeding operations in Puerto Rico that were to be integrated into a global network of biomedical primate distribution. In May 2009 international animal-rights NGOs including Physicians Committee for Responsible Medicine and the British Union for the Abolition of Vivisection launched a campaign against Bioculture, a Mauritius-based breeder of standard pathogen-free research monkeys, which intends to establish a new breeding facility in Guayama, Puerto Rico. In order to ensure the standardization of research subjects (including their certification as pathogen-free), the proposed facility would have held an estimated 4,000 newly imported rhesus and cynomolgus monkeys, bypassing the ongoing government trapping operations. Bioculture, which exports research monkeys through several international distributors, including the Charles River Laboratories in Massachusetts, is one of the world's major suppliers of research monkeys. However, local officials in Guayama as well as Puerto Rican legislators took action to prohibit Bioculture's facility, and its permit was denied, effectively stopping any new large-scale importations of monkeys to the archipelago.[45]

Coda: Species Critique and the Archive

My overview of the cultural, political, and medical history of imported monkeys in Puerto Rico has benefited from traditional print sources including newspapers, scientific research publications, and photographs and other visual cultural sources. What is more unusual in my discussion is the use of self-produced histories written by researchers working at scientific institutions. Although scholars in the humanities and social sciences have for decades been producing rich work on the cultural and social study of science, it is unusual for scientists themselves to undertake a sustained project of historical research of their own institutions, research practices, and research subjects as have the scientists associated with Cayo Santiago. The uniqueness of Cayo Santiago—its status as the first free-ranging colony of Asian-origin monkeys in the Americas, its location far afield from mainland research institutions, its immersion in Second World War and Cold War history, and its status as one of a few remaining sites of free-ranging monkeys used in research—perhaps accounts for this fascination.

Thus two archives have been produced by scientists following the importation of monkeys to Puerto Rico. The first comprises the field studies that began with the work of Carpenter in the early 1940s. Such studies relied on the migration and settler-colonialism of colonial tropical medicine, and they established observation protocols through which to interpret the semiotic and social activities of colony-dwellers as indicators of population characteristics. Donna Haraway's excellent description of Carpenter's original research project at Cayo Santiago demonstrates that an understanding that monkeys share human abilities to communicate, as well as particular social forms that produce a population, played an important role in the interpretation of social and sexual behavior.[46] There is therefore an investment in observing monkey life in order to extrapolate lessons regarding social organization understood to apply to humans. During the wartime era, this was seen as significant for defense priorities in engineering systems of command and control.

Of course, this history of primate sociality and sexuality was produced in the very act of divorcing the animal from the material contexts of imperial capitalist development and primate trading, as well as from the complex linkages that both monkeys and humans established across the lines of species difference. Because of this constraint, a second form of scientific writing—the institutional history—emerged to explain the significance of colonial primate institutions, situating monkeys in relation to the resolutely political dynamics of scientific "progress" and biosecurity. Key researchers and other officials from both the mainland and Puerto Rico—including C. R. Carpenter, William Windle, J. G. Frontera, J. G. Vandenbergh, and Jaime Benítez—have been involved in self-consciously producing Cayo

Santiago histories, from institutional and scientific priorities to economic contestations and ecological impacts. They have showed an interest in, and even an occasional identification with, imported monkeys and their life experiences. In 1980 William Windle wrote an important article outlining the history of monkeys and scientists at Cayo Santiago, their difficulties, and the colony's role in spurring new research in Puerto Rico and elsewhere. In 1986 Richard Rawlins and Matt Kessler published a volume detailing the history, behavior, and biology of the Cayo Santiago rhesus; in a poetic tribute to the monkeys, at the end of the volume Kessler laments the long transoceanic journey, lack of provisions, and sporadic institutional attention that the monkeys received. In 1989 the *Puerto Rico Health Sciences Journal* published the proceedings of Cayo Santiago's fiftieth anniversary celebration, with the notable inclusion of the historical perspectives of Puerto Rican researchers and one university official. In a humorous account of his visit to the colony in these proceedings, the former UPR chancellor Jaime Benítez describes how rhesus became the springboard for discussing the history of mistrust in scientific exchanges between Puerto Rico and the mainland United States at the time of the Puerto Rican nationalists' shooting incident at the U.S. capitol in 1954.

In each of these examples, archival practices and official histories frame the subjects and objects of history via particular epistemological frameworks, conventions of documentation, and affective investments that determine which topics count and which don't. It is through the performative nature of historical narrative and archiving that historical knowledge aligns with the political.[47] The scientific institutions were of course tied to modernization and U.S. imperialism, but they were also invested in an affective politics that made the ideal, universal monkey a much more central object of historical narrative than the human social fields surrounding the island laboratories. In contradistinction, other Puerto Rican perspectives—whether in cryptozoological theories or in the writing on the political history of scientific institutions by Jaime Benítez—clearly view the monkey as tied to the U.S. imperial institutions that brought them to the archipelago. Given this imperial dynamic, it is perhaps no wonder that both Benitez's narrative and the chupacabras stories more often than not describe *humans being watched by monkeys* rather than describing the monkeys from the gaze of the human narrator, as in behavioral and institutional histories; the monkey here is aligned with the surveillance technologies and secretiveness of the imperial state, rather than emerging simply as a victim or subaltern migrant within imperial economic flows.

Many aspects of monkeys' existence in Puerto Rico have escaped these historical accounts: monkeys' interactions with humans in the southwest, their transformation of the agricultural, ecological, and even visual cultural landscape of Puerto Rico, the new travel linkages they establish with the mainland and with other parts of Puerto Rico through tourist and biomedical traffic, their presence in responses to neoliberalism that travel across Latin America, and the violence enacted on their bodies in practices of vivisection. These examples indicate that, in large and small ways, the life-practices of monkeys can constitute forms of representation that fall out of the purview of biomedical institutions, the imperial modernization models, conservation discourse regarding invasive species, and nationalist discourses on development. A critical history of Cayo Santiago as a key space in the development of the biomedical primate trade therefore must, on the one hand, follow the prerogatives for telling particular "animal histories" that would work against what Few and Tortorici identify as the colonial archival "blurring" of specific microhistories into the categorical representation of delimited animal species, and, on the other hand, attempt to account for nonhuman animal representations as they appear in the margins of the histories emerging from largely human-derived archives.[48] Following works by Eduardo Kohn and Anna Tsing, we must recognize that nonhuman actors can engage in complex forms of representation that require us to read new presences in culture and relations between species in the landscape.[49] Imported rhesus and patas monkeys in Puerto Rico are co-travelers in the domains of culture and politics, and their performances speak to the alternative marks that they leave in memory, in the landscape, in the margins of the archive and the official history.

Scientific histories have at times ignored the question of the relationship between research institutions and local populations, save the cases of escaped monkeys and tourist revenues. In the extant histories, the roles of the many animal handlers and other facilities workers at Cayo Santiago—most of them Puerto Rican—are largely left out in order to focus on the monkeys and the prerogatives of federal funders and mainland researchers. These caretakers' long-term intimacy with monkeys and their habitats may make them at times more knowledgeable about rhesus behavior than are the official investigators from the mainland who visit for months-long research stints.

These exclusions allow for the telling of a certain type of history deeply impacted by the imperial dislocations that made Cayo Santiago and other monkey institutions possible. Monkeys are inextricably tied to the politics of development, the cultural memory regarding U.S. imperial control of Puerto Rican land and institutions, and the rhetorical linkage between modernity and science that framed the wartime and early Cold War eras in Puerto Rico. At the same time, the fact that monkeys can escape the boundaries of the experimental research site, transform built environments and economic linkages, and become symbols of both national progress and imperial domination, reflects the difficulty in figuring the animal as a sort of subaltern written outside of dominant histories.

Erica Fudge has eloquently written that "animal history" is first and foremost a human history of animals.[50] The case of imported monkeys in Puerto Rico demonstrates that even the microhistories that would highlight monkeys' labor, movement, and representational practices cannot help but perform the anthropocentric function of secular historical narrative that, as Dipesh Chakrabarty argues, violently translates diverse life-worlds and temporalities into a unitary conception of space and time.[51] There has been a persistent will to write the monkey as a historical figure in Puerto Rico, either to document its immersion in the modern systems of biodefense or to situate it as the alien invader to the nation. Given the deep socioeconomic divisions that this investment—forged out of histories of imperialism and neoliberal development—indexes, it is impossible to extricate monkeys from the contexts of social difference that produce a will to tell "monkey history." A critical monkey history must pay attention to the divisions of humanity through which monkeys emerge into humanist historical discourse (for the will to tell a monkey history is itself implicated in the circulations of biopower); recognizing the human differences through which monkey history emerges also forces us to think through the radical conjunctions and segmentations of human and monkey bodies in biological and social assemblages. This understanding is somewhat different than Fudge's solution to the dilemma of transspecies representation in her concept of a "holistic" history that understands how the concept of the human is formed through the animal, its other. While Fudge's work is attentive to the complex ways in which telling animal history is, in a sense, always telling a history of the human, this form of historical writing remains within a binaristic model that situates animal species in categorical difference as demeaned objects of modernity. Reifying "nonhumanness" risks postponing, first, a primary critique of human social subjection through which animals become historical objects, and second, the deep constitutive difference that conjoins life across bodies and species. Working in the binary mode threatens to universalize a Eurocentric understanding of animality as located within the realms of the secular and the phenomenal.[52] In the case of Puerto Rico, a multiply colonized space where monkeys are both visible signs of progress and the objects of suspicion, where the taxonomy of the monkey is caught up in narratives of the paranormal, and where the bodies of monkeys have been literally implanted into some humans to ensure their survival, such an understanding fails to account for the complexities of interspecies living—the ways in which animals perform historical relations with the many species, places, and institutions they encounter and with whom they share space, affect, communication, resources, and even bodies.[53] Such relations are central to the definition of imperial biopower, which materializes in geographically and historically delimited circulations of affect, investment, body parts, and knowledge.

Notes

Special thanks to Gabriel Troche, who shared his past experiences as a research assistant at Cayo Santiago.

1. Estado Libre Asociado de Puerto Rico, *Reglamento para designar como animales perjudiciales a ciertas especies detrimentales a los intereses de la agricultura y de la salud pública* (San Juan: Departamento de Agricultura, 2007).
2. Roberto Esposito, *Bíos: Biopolitics and Philosophy* (Minneapolis: University of Minnesota Press, 2008).
3. Michael T. McGuire, *The St. Kitts Vervet* (Los Angeles: S. Karger, 1974), 5–7; Michael A. Evans, "Ecology and Removal of Introduced Rhesus Monkeys: Desecheo Island National Wildlife Refuge, Puerto Rico," *Puerto Rico Health Sciences Journal* 8.1 (April 1989): 139–56, 140.
4. Bailey Ashford and Pedro Gutiérrez Igaravidez, *Uncinariasis (Hookworm Disease) in Porto Rico: A Medical and Economic Problem* (Washington: Government Printing Office, 1911); Annette B. Ramírez de Arellano and Conrad Seipp, *Colonialism, Catholicism, and Contraception: A History of Birth Control in Puerto Rico* (Chapel Hill: University of North Carolina Press, 1983); Laura Briggs, *Reproducing Empire: Race, Sex, Science, and U.S. Imperialism in Puerto Rico* (Berkeley: University of California Press, 2002).
5. Briggs, *Reproducing Empire*, 14.
6. Jorge Duany, *The Puerto Rican Nation on the Move* (Chapel Hill: University of North Carolina Press, 2002), 42–43.
7. Donna Haraway, *Primate Visions: Gender, Race, and Nature in the World of Modern Science* (London: Routledge, 1989), 413n34.
8. Debbie Bookchin and Jim Schumacher, *The Virus and the Vaccine* (New York: St. Martin's, 2004), 20.
9. Quoted in Annette B. Ramírez de Arellano, "Columbia's Overseas Venture: The School of Tropical Medicine at the University of Puerto Rico," *Medicine's Geographic Heritage* 5 (December 1989): 36. See also Francisco Joglar, "Cover Note:

The University of Puerto Rico School of Medicine," *Academic Medicine* 79.6 (June 2004): 596; Briggs, *Reproducing Empire*, 62.

10. C. R. Carpenter. "Sexual Behavior of Free-Ranging Rhesus Monkeys (*Macaca mulatta*)," *Journal of Comparative Psychology* 33 (1942): 113–62; Donna Haraway, *Primate Visions*, chap. 5.

11. C. R. Carpenter, "Rhesus Monkeys (*Macaca mulatta*) for American Laboratories," *Science* 92 (27 September 1940): 285–86.

12. Carpenter, "Rhesus Monkeys (*Macaca mulatta*) for American Laboratories," 286.

13. César Ayala, *American Sugar Kingdom: The Plantation Economy of the Spanish Caribbean, 1898–1934* (Chapel Hill: University of North Carolina Press, 1999), 144; Richard G. Rawlins and Matt J. Kessler, "The History of the Cayo Santiago Colony," in *The Cayo Santiago Macaques: History, Behavior and Biology*, ed. Richard G. Rawlins and Matt J. Kessler (Albany: State University of New York Press, 1986), 22; Matt J. Kessler, "Establishment of the Cayo Santiago Colony," *Puerto Rico Health Sciences Journal* 8.1 (April 1989): 15; José Guillermo Frontera, "Cayo Santiago and the Laboratory of Perinatal Physiology: Recollections," *Puerto Rico Health Sciences Journal* 8.1 (April 1989): 21.

14. Rawlins and Kessler, "The History of the Cayo Santiago Colony," 22. Cayo Santiago was eventually annexed by Puerto Rico.

15. Haraway, *Primate Visions*, 10–13, 19–25.

16. Constance M. Locke, "Peopling an Island with Gibbon Monkeys: An Ambitious West Indian Experiment in Biology," *Illustrated London News*, 13 August 1938, 290–91.

17. On Mieth's photograph, see Dolores Flamiano, "Meaning, Memory, and Misogyny: *Life* Photographer Hansel Mieth's Monkey Portrait," *afterimage* 33.2 (September–October 2005): 22–30.

18. Carpenter, "Rhesus Monkeys (*Macaca mulatta*) for American Laboratories," 285.

19. "First American Monkey Colony Starts on Puerto Rico Islet," *Life* 6.1 (2 January 1939): 26.

20. Rawlins and Kessler, "The History of the Cayo Santiago Colony," 24.

21. Rawlins and Kessler, "The History of the Cayo Santiago Colony," 25.

22. William Windle, "The Cayo Santiago Primate Colony," *Science* 209.4464 (26 September 1980): 1488; Richard G. Rawlins, "Perspectives on the History of Colony Management and the Study of Population Biology at Cayo Santiago," *Puerto Rico Health Sciences Journal* 8.1 (April 1989): 33.

23. Ramón Grosfoguel, *Colonial Subjects: Puerto Ricans in a Global Perspective* (Berkeley: University of California Press, 2003), 108–9.

24. Briggs, *Reproducing Empire*, 111–12. See also Duany, *The Puerto Rican Nation on the Move*, 122–23.

25. Ramírez de Arellano, "Columbia's Overseas Venture," 38–39.

26. José Guillermo Frontera, "Cayo Santiago and the Laboratory of Perinatal Physiology: Recollections," *Puerto Rico Health Sciences Journal* 8.1 (April 1989): 21–28.

27. W. Richard Dukelow and Leo A. Whitehair, "A Brief History of the Regional

Primate Research Centers," *Comparative Pathology Bulletin* 27.3 (1995): 1–2; Victoria A. Harden, "Interview with Dr. William I. Gay," 15 July 1992, Office of NIH History, http://history.nih.gov/NIHInOwnWords/docs/gay_01.html; Windle, "The Cayo Santiago Primate Colony."

28. For overviews of these transformations, see Laura Briggs and Palmira N. Ríos, "Export-Oriented Industrialization and the Demand for Female Labor: Puerto Rican Women in the Manufacturing Sector, 1952–1980," in *Colonial Dilemma: Critical Perspectives on Contemporary Puerto Rico*, ed. Edwin Meléndez and Edgardo Meléndez (Boston: South End, 1993), 89–101.

29. John G. Vandenbergh, "The La Parguera, Puerto Rico Colony: Establishment and Early Studies," *Puerto Rico Health Sciences Journal* 8.1 (April 1989): 118.

30. Eric Phoebus, Ana Roman, and John Herbert, "The FDA Rhesus Breeding Colony at La Parguera, Puerto Rico," *Puerto Rico Health Sciences Journal* 8.1 (April 1989): 157.

31. Janis Gonzáles-Martínez, "The Introduced Free-Ranging Rhesus and Patas Monkey Populations of Southwestern Puerto Rico," *Puerto Rico Health Sciences Journal* 23.1 (2004): 39–46.

32. M. Masanet and J. Chism, "The Abundance, Distribution, and Habitat Use of the Imported Patas Monkey Population in Puerto Rico," paper delivered at the American Society of Primatologists Annual Conference, West Palm Beach, Florida, 21 June 2008. The figure leaves aside small single-sex bands and considers only "heterosexual" troops.

33. Kristen Jensen et al., "B-Virus and Free-Ranging Macaques, Puerto Rico," *Emerging Infectious Diseases* 10.3 (March 2004), available at the Centers for Disease Control website, http://wwwnc.cdc.gov/eid/content/10/3/contents.htm.

34. "Puerto Rico Hunting, Killing Troublesome Monkeys," *Orlando Sentinel*, 21 December 2008; Kirk Semple, "Scrutiny for Puerto Rico over Dog Treatment," *New York Times*, 8 March 2009, http://www.nytimes.com/2008/03/09/us/09dogs.html.

35. "Advierten Peligro de Monos Rhesus," Associated Press, 19 December 2008, wapa.tv, http://www.wapa.tv/noticias/locales/advierten-peligro-de-monos-rhesus_20081219214708.html; "Nuevo Secretario del DRNA Revisa Plan de Captura de Monos," Associated Press, 21 January 2009.

36. Gladys Nieves Ramírez, "Urge Acción Oficial ante la Amenaza de los Monos," *El Nuevo Día*, 27 May 1998; Aura N. Alfaro, "Cultivando para los Monos," *El Nuevo Día*, 27 July 2008.

37. Scott Corrales, *Chupacabras and Other Mysteries* (Murfreesboro, Tenn.: Greenleaf, 1997), 115–16.

38. Lauren Derby, "Imperial Secrets: Vampires and Nationhood in Puerto Rico," *Past and Present* 199, supplement 3 (August 2008): 290–312, 310.

39. Derby, "Imperial Secrets," 294.

40. Katherine T. McCaffrey, *Military Power and Popular Protest: The U.S. Navy in Vieques, Puerto Rico* (New Brunswick: Rutgers University Press, 2002), 138–46. The U.S. military left its bases by 2004 following the protests.

41. Robert Michael Jordan, "El Chupacabra: Icon of Resistance to U.S. Imperialism" (master's thesis, Department of History, University of Texas, Dallas, May 2008), 4.

42. Jaap Goudsmit, *Viral Sex: The Nature of AIDS* (Oxford: Oxford University Press, 1998), 167.

43. June Carolyn Erlich, "A Look at Cayo Santiago," *ReVista: Harvard Review of Latin America* (spring 2008), available at the David Rockefeller Center for Latin American Studies website, http://www.drclas.harvard.edu/revista/articles/view/1084.

44. "Rich Port, Poor Port," *Science* 313 (28 July 2006): 475–76.

45. "PETA Honors Guayama Mayor for Saving Monkeys," *Puerto Rico Daily Sun*, 1 March 2011, http://www.prdailysun.com/news/PETA-honors-Guayama-mayor-for-saving-monkeys.

46. Haraway, *Primate Visions*, 97–101.

47. See, for example, Jacques Derrida, *Archive Fever*, trans. Eric Prenowitz (Chicago: University of Chicago Press, 1996), 4n1.

48. See Tortorici and Few's introduction to this volume.

49. Eduardo Kohn, "How Dogs Dream: Amazonian Natures and the Politics of Transspecies Engagement," *American Ethnologist* 34.1 (2007): 3–24; Anna Tsing, *Friction: An Ethnography of Global Connection* (Princeton: Princeton University Press, 2004), chap. 5.

50. Erica Fudge, "A Left-Handed Blow: Writing the History of Animals," in *Representing Animals*, ed. Nigel Rothfels (Bloomington: Indiana University Press, 2002), 3–18, 9.

51. Dipesh Chakrabarty, "Translating Life-Worlds into Labor and History," *Provincializing Europe* (Princeton: Princeton University Press, 2000), chap. 3.

52. I see this tendency in a number of field-defining texts that follow in the lineage of Peter Singer, who in the 1970s identified the relationship of human to animal as one defined by *speciesism*, analogous to sexism and racism. This analogy of social subjection works rather crudely in the case of animals because it reifies the singularity of the human and universalizes European taxonomic knowledge as the basis for turning the animal into an object of analysis. See, as examples, Cary Wolfe, *Animal Rites: American Culture, the Discourse of Species, and Posthumanist Theory* (Chicago: University of Chicago Press, 2003); and the Animal Studies Group, ed., *Killing Animals* (Champaign: University of Illinois Press, 2006). Jacques Derrida has been the most insistent on problematizing this object "animal," situating it as a sacred figure of otherness in European philosophical traditions. See Derrida, *The Animal That Therefore I Am*, trans. David Willis (New York: Fordham University Press, 2008).

53. Donna Haraway has been the most visible proponent of a critical species scholarship (rather than the alternative of "animal studies"). See especially her analysis of "companion species" that cannot be dissembled into the categories of "human" and "nonhuman" or even "human" and "animal" (*When Species Meet* [Minneapolis: University of Minnesota Press, 2008]).

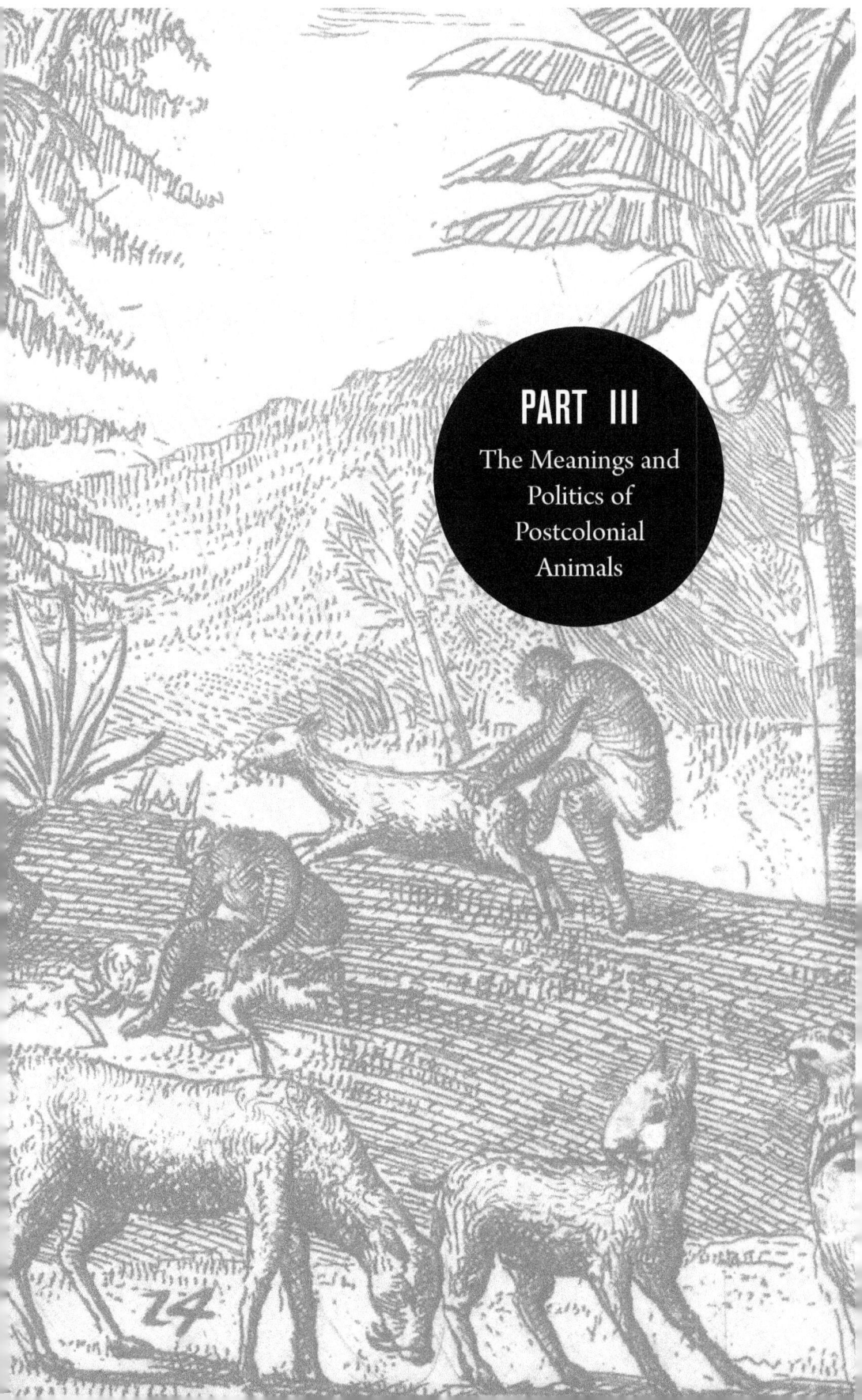

PART III

The Meanings and
Politics of
Postcolonial
Animals

7

Animal Labor and Protection in Cuba

Changes in Relationships with Animals in the Nineteenth Century

REINALDO FUNES MONZOTE

Translated by Alex Hidalgo
and Zeb Tortorici

During the 1990s, Cuba faced an acute economic crisis as a result of the collapse of the Eastern European socialist bloc and the disintegration of the Soviet Union, the nation's principal trade and political partners, which absorbed over 80 percent of the commercial relations of the only socialist country in the Americas. Among the most visible changes was significant growth in the use of animals for agricultural work and for the transportation of people and merchandise. Many Cubans were unaccustomed to this image, and the use of animals was interpreted by some as a sign of regression during the so-called "special period in times of peace." This critique was preceded by profound transformations in previous decades, such as the notable increase in agricultural mechanization. For example, the number of tractors in Cuba increased from 9,000 in 1960 to 85,000 in 1990. In contrast, during the same period, the number of oxen on the island decreased from 500,000 to 163,000, and the number of horses fell from 800,000 to 235,000. The rapid decline in imports such as fossil fuels, automotive vehicles, replacement parts, agrochemicals, and other external raw materials had a strong social and economic impact. However, this decline served as an impetus for many to seek alternatives to a highly industrialized agriculture, and consequently the use of animal labor was also reconsidered in the final decade of the twentieth century. In this economic context, the number of oxen increased radically from the aforementioned 163,000 in 1990 to nearly 400,000 as the year 2000 neared.[1]

At the same time, a rapid expansion of alternative modes of transportation—the rising use of bicycles and the return of wagons and animal-drawn carriages—could also be observed.[2] With this resurgence of animal labor and transport, an old preoccupation that had seemingly been forgotten, at least since draft animals had largely ceased to be the routine companions of humans in urban spaces, reappeared. For example, in the "Open Letters" section of the journal *Granma*, from 24 September 1996, one reader asked which institutions were responsible for monitoring and enforcing current legislation dealing with individuals who had committed acts of cruelty against animals, if indeed such laws existed. Frequent and commonplace scenes of the abuse of horses, often to the point of heart failure and death, motivated that particular inquiry and others like it. The journal's response to the reader showed that even after the Unión Nacional de Juristas de Cuba (National Union of Jurists of Cuba) had been consulted, it remained unclear whether or not there were specific laws that prohibited such acts. But, noted the journal, regulations limiting the number of passengers and the hours an animal could be made to work could be elaborated, alongside local rules and regulations to monitor this activity.[3]

The issues regarding the use of animals in agriculture and transportation here must be framed within a historical context that encompassed the gradual replacement of animals as the main driving force of the industrial era, with its new methods of transportation and the growing use of fossil fuels. At the same time, in various European nations and other parts of the world, legislative measures were enacted to reduce, regulate, or prohibit certain practices in relation to both domesticated and wild animals. In a similar vein, and partly as a result of the founding of the Society for the Prevention of Cruelty to Animals in England in 1824, the global development of societies dedicated to the protection of animals multiplied and spread, which significantly influenced changes in social perceptions and attitudes toward animals, as well as corresponding public policies.

In this chapter I focus on two interrelated themes that, though distant in time, remain highly relevant to Cuban history. First, in order to contextualize historically the relationships between animals, slavery, and technology, I examine the changing use of oxen in Cuban sugar plantations, from when steam engines and railroad transport first emerged to the beginning of the sugar centralization process in the late nineteenth century. During the nineteenth century, Cubans experienced major shifts regarding the use and conception of working animals that coincided in significant ways with antislavery movements and with slavery's eventual abolition. These shifts in attitude were marked by class and were simultaneously local in context and influenced by foreign technologies and ideologies. Keeping in mind that, especially in the production of sugar, humans used oxen (as well as other animals) as a labor force that was subject to overwork and other common abuses, I then explore the founding of the Sociedad Cubana Protectora de Animales y Plantas (Cuban Society for the Protection of Animals and Plants) in the 1880s. Influenced by the proliferation of other international animal-protection societies, the Sociedad Cubana Protectora de Animales y Plantas aimed, among other things, to improve the treatment of animals. In doing so, however, it also entered a wider discourse of reform proposed by local elites who sought to transform the particular type of economy and the society that had been branded by the slave plantation.

Works which have addressed the economic, environmental, and social history of Cuba have placed greater emphasis on the role of plants than on that of animals. One might contrast, for example, the large number of historical and anthropological works on Cuba's principal commercial crops —sugarcane, tobacco, and, to a lesser extent, coffee, which for centuries determined the island's participation in the world market—with the dearth of studies on Cuba's equally important cattle and ranching industries. Research on domesticated animals—whether the focus be on economic impact, roles in hunting and leisure, or pet-keeping practices—and on Cuba's undomesticated fauna has tended to focus on practical matters: descriptions of species, diet, illnesses and remedies, animals' productive potential, the possibilities of genetic modification, and so on. Unfortunately, the authors of such studies have rarely approached their topics of research from a historical perspective; however, the central role of animals in Cuba's economic, social, technological, cultural, and ecological processes, from the nineteenth century to the present, is undeniable.[4]

Oxen, Sugar Mills, and the Beginnings of Mechanization

During the nineteenth century, the island of Cuba became the principal producer and exporter of sugarcane, a result of the void created by the Haitian Revolution of 1791. The expansion of sugar plantations came to occupy a sizable portion of the western half of the island, while deforestation allowed for the establishment of smaller plantations on Cuba's eastern side. In addition to the massive introduction of slaves, other economic, political, and institutional factors—such as the relaxation of trade restrictions with foreign powers, the lack of restrictions on deforestation, and tax exemptions—contributed to the development of the sugar industry. Beginning in 1820, the industrial era ushered in a number of new technologies and practices that also directly impacted the sugar industry. Steam engines used to power sugar mills during that decade, multiple-effect vacuum evaporators in the boiling rooms during the 1840s, and the centrifuges of the 1850s revolutionized the sugar-making process. Around the same time, starting in 1837, the island became one of the first places in the world to develop an extensive railway system, used primarily for the transportation of sugar from plantations to the shipping ports. During the 1860s, Cuba not only had the highest concentration of slave plantations, but also enjoyed an extensive railroad network, with more kilometers of railroad per capita than any other nation. Both factors, in conjunction with the European and U.S. consumer markets becoming the principal destinations of Cuban sugar, brought about qualitative changes that fostered a substantial increase in the scale of sugar production.

Cuba's nineteenth-century sugar boom, with the profound transformations ushered in by the birth of the industrial era, did not, however, render animal labor unnecessary to the industry. Instead, a process of reorganization and a gradual substitution of some tasks for others took place, at least until the advances of agricultural mechanization that occurred during the twentieth century. In this sense, an analysis of the slave plantation system could contribute to the debate over the transition from the predominant use of endosomatic energy, based primarily on both human and animal strength, to the greater consumption of exosomatic energy sustained by the growing advances in fossil-fuel technology and the new technical tools of the industrial revolution.[5]

Animals played a decisive role in Cuban sugar production during the slave plantation era, which ended with the abolition of slave labor in 1886. Oxen, horses, and mules performed a variety of functions in the sugar-making process, although the first group was the most commonly used both on the plantation and to transport goods to markets throughout Cuba.[6] Cuba's experience is similar to that of Brazil, making Gilberto Freyre's insight particularly apt for both: "The most faithful ally that the African slave had in agricultural labor, in the daily routine of planting sugarcane, and in the very industry of sugar, was the ox; it was these two—the Black and the ox—that formed a living basis of the sugar economy."[7]

In the final decade of the eighteenth century, the landowner José Ricardo O'Farril noted that an average sugar mill, capable of producing 10,000 *arrobas* of sugar, consisted of 30 *caballerías de tierra* with 100 slaves, 40 yokes of oxen, and 30 mules.[8] Of those 40 yokes, 24 served to power the mill, while the rest transported cane from the fields.[9] The sugar boom that followed the revolution in Haiti in 1791 sparked the need for animal labor in sugar transportation, as the number of sugar plantations increased and so did their productive capacity. These circumstances as well as other factors contributed to a significant decline in cattle ranching in general, causing a shortage of oxen to power sugar processing. As Antonio Morejón y Gato wrote in 1797, "The scarcity of these [oxen] is, in my opinion, one of the factors that have motivated some sugar mill owners to procure machines that will allow them to save [money]; but if the species should be improved, doubtless there will be an abundance which will lower their cost. Because of [oxen's] poor quality, many cattle ranchers cannot get by on their own, but if they seek to improve their lot, they will achieve savings."[10]

In effect, the urgent necessity of draft animals for agricultural and processing work on sugar plantations stimulated the search for alternatives to substitute animal labor. It was not until 1797 that the first attempt to employ steam power to operate the sugar mills was undertaken; similar efforts were made to utilize hydraulic and wind energy. However, the latter two proved largely unsuccessful at that time, and the transition to the steam engine took place gradually, over two decades from its initial application to sugar production in Cuba.[11] Until that time, the consensus was that mills powered by oxen were the best suited for conditions on the island. Within this context, demand for cattle increased, especially for oxen to operate the mills and transport cargo. In the late 1840s, the count of Pozos Dulces, José Francisco de Frías y Jacott, wrote about this matter: "Oxen yokes and horses, then, experienced a sharp increase in cost . . . and the natural reproduction of herds and pastures could barely supply the demands of the agricultural industry . . . , [which] injected new life into cattle industry; one could say that it was then that the cattle economy reached its highest level of prosperity."[12]

Yet the increasing use of steam engines gradually brought about a steady decrease in the use of oxen for milling sugarcane, although some plantations combined steam-powered mills with mills powered by animal force. In 1827 only 26 sugar mills in Havana and Matanzas possessed steam engines. In the 1860 sugar-mill census of Rebello, however, the number of mills with steam engines in the Departamento Occidental was 829 (77.4 percent), compared to 231 operated by oxen. In contrast, the Departamento Oriental consisted of 120 of the former (40 percent) and 178 of the latter (59.4 percent).

Addressing these issues in 1832, the French technician Alejandro Dumont noted, "Perhaps some will object that a steam engine requires copious amounts of water and firewood. This is true, but it is also true that mills cannot exist without fertile water. And, aren't vast amounts [of water] needed to maintain 56 yokes of oxen to operate the mills?"[13]

Given the decline in the relative importance of oxen as beasts of burden for the sugar mills, the increasing substitution of animals (oxen, mules, and horses) for the transportation of mill products and supplies followed the expansion of the railroad in 1837. It was during the mid-nineteenth century that the new technologies of the burgeoning industrial era became partly responsible for the decline of the Cuban cattle industry. The count of Pozos Dulces made reference to the crisis in "the industry of carting and muleteering" as one of the "principal factors that paralyzed the breeding of livestock, which can be seen today," concluding that "much time would pass before the poorly illustrated declarations against the perfected railway would cease to be heard on a daily basis, as its construction coincided with the demise of towns that owed their prosperity to the old system of muleteering."[14] Around the same time, Antonio Bachiller y Morales proposed, "The introduction of train tracks [and] steam engines for sugar mills has greatly reduced the cost of cattle for draft labor. A yoke of oxen, which previously cost a considerable amount in pesos, has dropped literally to half its value. Cattle will be so reduced in number that they will soon be raised mainly for [human] consumption, as a source of food."[15]

In spite of these changes, animals continued to perform a fundamental role on the sugar mills, and on many ranches the tendency was to increase their numbers in order to simultaneously boost the scale of production. Some of their tasks included carrying sugarcane from the plantations (which were increasingly located in Cuba's interior) to the railroad switches or nearby shipping ports, returning with goods for the mills, and hauling firewood and sugarcane pulp (fig. 7.1). Not coincidentally, the management of oxen played an important role in the administration of plantations, as evidenced by the fact that the ox-driver was one of the most highly ranked salaried workers at the sugar mills, ranked, in fact, just below the foreman.

Statistical data on animals during the nineteenth century is highly problematic. Nevertheless, census records for 1827, 1846, and 1862 offer an approximation of the changes that took place as a result of the introduction of steam engine technology, providing a glimpse into the importance of oxen on the western plantation zone, especially in 1846 (table 7.1). Yet between 1846 and 1862 the number of cattle decreased in that same area by nearly

7.1 Oxen hauling a cart of sugarcane. SOURCE: *CUBA REVIEW AND BULLETIN* (JULY 1914).

30,000. The most significant factor that contributed to this change was the demolition of numerous sugar mills around Havana. In the Matanzas region, however, which up until then represented the principal nucleus of the slave-based sugar plantations, their numbers remained fairly stable between 1846 and 1862, nominally increasing from 26,242 to 26,526. Cattle in Cárdenas numbered 35,368 in 1846, and although the region was subdivided, giving rise to the jurisdiction of Colón, cattle in both regions totaled 37,339 in 1862.[16] In the meantime, the increase in oxen registered in Cuba's central zone resulted from the establishment of new sugar plantations, in the mid-1830s, in the jurisdictions of Cienfuegos and Sagua la Grande. In the latter, where sugar production began around 1835, the numbers increased dramatically from 6,734 in 1846 to 18,000 in 1862.

In spite of the relative decline in the numbers of oxen on slave plantations, oxen continued to receive preferential treatment. In 1836 Andrés de Zayas called for them to be given better care, since they were "as instrumental for plantations as [slave] field hands."[17] The comparison here between animal and slave labor is telling, and Zayas's sentiments highlight how an economic need to exploit human and animal labor sometimes went alongside calls for more "humane" treatment for both groups. Zayas recommended that foremen and ox-drivers strive for a more efficient way to care for the oxen in order to prevent the high mortality that accompanied each sugarcane harvest. On the one hand, he lamented how both white and black ox-drivers and muleteers struck the oxen when they paused en route due to

TABLE 7.1 Number of Oxen in Cuba according to Census Data, 1827, 1846, and 1862

Zone	1827	1846	1862
West	111,092	172,390	142,617
Central	20,487	53,502	84,544
East	8,960	31,128	19,495
Total	140,539	257,020	246,656

Sources: *Cuadro estadístico de la siempre fiel Isla de Cuba, correspondiente al año de 1827* (Havana: Oficina de las viudas de Arazoza y Soler, 1829); *Cuadro estadístico de la siempre fiel Isla de Cuba, correspondiente al año de 1846* (Havana: Imprenta del Gobierno y Capitanía General, 1847); *Noticias estadísticas de la Isla de Cuba en 1862* (Havana: Imprenta del Gobierno, Capitanía General y Real Hacienda, 1864).

At the time of the 1862 census, only two departments existed: east and west. The term *zone*, as used here, represents approximate spaces of the jurisdictional boundaries of each census in order to make comparisons.

lack of energy, insufficient water, or the poor quality of roads. On the other hand, he sought stricter rules against placing slaves in charge of feeding the oxen and providing them with water, since without the presence of a foreman or an ox-driver, he asserted, the slaves shirked their responsibilities, preferring instead to rest or to feed their pigs.

Such oversights, Zayas maintained, appeared at the close of each crop cycle, "when droves of oxen were undernourished [and overworked] to the point that 10 or 15 percent typically died, or, in better times, a mere 5 or 8 percent."[18] The animals that survived were generally thin and aged, worth only half of their original cost. Zayas considered fortunate those owners who did not incur further losses, since it was not uncommon for an entire herd to die as a result of "unknown illnesses, poor grass, drought, floods, or insufficient pastures." However, "it is never said that the oxen's demise is the result of insufficient care, of having been beaten, or of simply not having been fed."[19]

Two decades later, José Montalvo y Castillo, a third-generation sugar planter and a key figure in Cuba's late eighteenth-century sugar boom, in *Tratado general de escuela teórico-práctica para el gobierno de los ingenios de la Isla de Cuba en todos sus ramos* (1856), emphasized the need to care for and not to mistreat oxen: "Under the threat of severe punishment, no ox should be beaten."[20] Montalvo y Castillo's resolve signals the oxen's important role on the plantations: "Animal labor [*fuerza animal*] saves both time and human labor. Moreover, it allows for certain improvements in that all operations can be completed on time and without anything left outstanding."[21] After pointing out some specific methods for ensuring the well-being of oxen, such as treatment for ticks, proper nourishment, and acceptable means of use, Montalvo y Castillo concluded that "this branch, on which the sugar harvest depends entirely, is deserving of great care and assistance."[22] Such attitudes represent an important shift in elite public discourse on the proper care and treatment of working animals, and it is perhaps no surprise that these sentiments were increasingly publicized and disseminated at a time when the slave trade and the use of slave labor in Cuba were subject to mounting scrutiny and opposition.

The number of oxen on sugar plantations (*fincas azucareras*) certainly depended on the animals' productive capacity. In 1873, for example, two of the largest sugar mills in Cuba generated the following data: the plantation of Juan Poey, Las Cañas, was 102 caballerías in size (47 caballerías being dedicated to sugarcane production) and was worked by a crew of 450 black slaves, 230 Chinese indentured servants, 500 oxen, and 40 horses. The sugar mill of Juan de Zulueta, España, had a total of 91 caballerías (of which 65 were devoted to sugarcane) and was worked by a crew of 530 slaves, 86 Chinese servants, 500 oxen, 14 mules, and 30 horses.[23] Although both plantations surpassed the average in terms of scale of production, they represent a considerable increase in sugar output, due in large part to animal labor, at the end of the nineteenth century.

Oxen thus represented a significant contribution to the labor force of the sugar plantation. One manual, *Guía del administrador de ingenio*, written by Antonio de Landa in 1857, suggested a ratio of five workers per wagon: two *macheteros* (brush-cutters), two pickers, and the wagoner. If the crop, however, was located at a distance, Landa recommended lowering the ratio to twenty cutters and twenty pickers for every fourteen wagons, each with an individual driver. The number of oxen per wagon could consist of one, two, or three yokes, depending on both distance and the condition of the roads.[24] A clearer understanding of the importance of oxen in slave plantations can be gleaned if we agree with the estimate that each hour worked by an ox equaled some 3.8 hours of human labor.[25] With their help, Landa concluded, each worker should be able to cut approximately five wagonloads of sugarcane per day, and even more in places where "healthy sugarcane [*caña buena*] and dense plantations" existed.[26] Given fourteen wagons, each making six or seven trips, one could expect an average daily yield of between 95 to 105 wagonloads of sugarcane.

Not surprisingly, the ox-driver played an important role both in supervising the wagoners and slaves who were in charge of feeding the oxen and in providing care for the oxen maintained in pastures during the *tiempo muerto*, the "dead season," that followed the sugar harvest.[27] The Cuban historian Manuel Moreno Fraginals points to the fact that an increase in droves of oxen created the problem of needing more extensive land for grazing.[28] In practice, the expanse of pastures varied significantly by plantation. In 1873 the plantation of Las Cañas had twenty-seven caballerías devoted to pasture land, while España dedicated only five caballerías to such use. For many plantations like España, the solution consisted of leasing neighboring pasture lands.

Whatever the variant, the employment of oxen and other animals as natural suppliers of fertilizer to increase agricultural yield on Cuban plantations was rarely taken advantage of. Nor was much attention given to the development of manmade feed or other sources of nourishment that might have mitigated the effects of droughts on cattle.[29] In 1828 Ramón de Arozarena and Pedro Bauduy argued as much in an extensive report, commissioned by the Real Consulado, that evaluated Jamaica's developments in sugar production.[30] The report confirmed the importance of sowing *la yerba de guinea* (guinea hen weed), which thrived only when a sufficient quantity of fertilizer had been used for the annual replanting of sugarcane. They also described the Jamaican "flying-pens," a system of portable corrals where a great quantity of yerba de guinea was used to feed the animals, causing them to replenish the land with necessary nutrients through their manure.

As the nineteenth century advanced, the crisis in the slave labor system created a more favorable atmosphere for the use of agricultural science on sugar plantations, although without the same impetus for technological renovation that could be seen in the mills. For this reason, an interesting relationship ensued between the definitive abolition of slavery in 1886 and the increasingly widespread use of the so-called *hornos de bagazo verde* (green bagasse furnaces), a technological advancement that greatly reduced the amount of both human and animal labor on sugar plantations and mills.[31] Although earlier models of such furnaces had previously been introduced, the abolition of slavery served as an incentive to accelerate implementation.[32] This incentive is reflected in a promotional pamphlet from 1890 that praised the qualities of the Fiske brand of green bagasse furnaces and their ability to replace human and animal labor. The pamphlet included a testimonial by Edwin Atkins, proprietor of the Soledad plantation, who assured the public that a single Fiske furnace could replace the labor of sixty to seventy plantation hands and thirty yokes of oxen.[33]

During the same period, the first steps were taken to reduce the number of animals necessary for agricultural production. One example includes the so-called *carros ballena*, a low-lying model of car with four wheels, wide tires, and a front axle shorter than the rear one, which was promoted in a pamphlet in 1877. To demonstrate the car's potential, the pamphlet described how this innovation in transportation could carry five to six thousand more cases of sugar than could the traditional two-wheeled wagons. This meant that, according to the propaganda, a carros ballena could easily replace twenty-six human laborers and thirty-four yokes of oxen. Among its advantages were more efficient loading and the ability to make a greater number of trips. The car also allowed for cutting the cane at its root rather than into various pieces, for delaying the harvest in order to gain larger yields, and for working during the rainy season.[34]

Beginning in 1873, portable railcars, which could be set in motion by animal labor or by small locomotives, appeared in Cuba and were increasingly used to transport sugarcane to the mills of the large plantations.[35] During this period, railcars were frequently powered by animal labor and served to symbolize the transition toward mechanization of the agricultural sector. Finally, the proliferation of private railroads proved vital to the expansion of sugarcane plantations at much greater distances and served to guarantee that such prime resources would arrive in the least amount of time. Yet it was the twentieth century that witnessed the technological revolution brought about by the internal combustion engine, which almost completely replaced oxen with tractors and trucks in the Cuban sugar-producing industry.

Several decades earlier, in 1876, the Spanish naturalist Miguel Rodríguez Ferrer described the role of oxen in terms that synthesize the characteristic image of animal labor in the time of slave plantations:

> What a strange and incoherent thing! It is a fact in Cuba, a nation of slavery, in which only force prevails in the social order, that female animals [*las hembras de los animales*], like those [women] of men, are here most excluded in their miserable condition, yet they somehow appear privileged. The ox bears the full burden of carting and pulling without sharing it with the cow, unlike in Europe, particularly among the duties of the poor. . . . The male on this island [of Cuba] endures an entire life of

hard work. . . . Some accuse the Cuban ox of suffering some sort of degeneration since it is quite common there to see entire yokes of oxen pulling cargo that, in Europe, would only take one or two [oxen] at most. But this does not take into account, which we have seen, how quickly they are put in the service of men; or their diet, entirely herbaceous and lacking any grains; or the fact they carry triple the load that those in Europe do; or the difficult conditions of a region that lacks bridges, trails, and roads.[36]

The Animal and Plant Protection Society, 1882–1890

It was amid these far-reaching changes in Cuban society, technology, sugar production, slavery, and animal labor that concomitant elite attitudes on the proper treatment and "protection" of animals gained particular momentum. In 1882, within the context of a boom associated with the end of the Ten Years War (1868–78) and with the extension of the right of assembly afforded by the Spanish constitution, the Sociedad Cubana Protectora de Animales y Plantas was founded (fig. 7.2).[37] Especially interesting is how, in the midst of political, social, and economic changes during the final years of slavery, an organization emerged with the explicit goal of reforming the relationship between human beings and nature.[38] Although in actuality the society had far-reaching goals in a variety of areas, I limit the scope of this discussion to the subject of animals, since the topic has received relatively scant attention in the historiography of Cuba.

The appearance of the Sociedad Cubana Protectora de Animales y Plantas must be understood in terms of both local factors and a larger international movement that was initiated in England, in 1824, with the creation of the Society for the Prevention of Cruelty to Animals (which converted into the Royal Society for the Prevention of Cruelty to Animals in 1840).[39] Following the society's lead, similar societies appeared in the 1830s in British cities and in countries such as Holland, Germany, Prussia, Switzerland, and France, where the Parisian society was founded in 1845.[40] In the United States the first societies appeared in New York in 1866 and in Boston in 1868.[41] In Spain the Sociedad Protectora de Animales y Plantas de Cádiz was founded in 1872, and after several unsuccessful efforts starting in 1874, Madrid founded its own society in 1878. Similar societies were eventually established in Barcelona and Seville. By 1881, close to 270 such societies existed worldwide, including an International Society for the Protection of Animals.[42]

Among the explanations for the growth of these international societies is

SOCIEDAD CUBANA
PROTECTORA
DE
ANIMALES Y PLANTAS.

Presidencia.

7.2 Logo of the Sociedad Cubana Protectora de Animales y Plantas. JUAN SANTOS FERNÁNDEZ FOLDERS, FONDO DE ACADÉMICOS DEL MUSEO HISTÓRICO DE LAS CIENCIAS CARLOS J. FINLAY, HAVANA, CUBA. PHOTO BY REINALDO FUNES MONZOTE.

the historical development of a philosophical and juridical trend in favor of animal rights.[43] Also contributing to this dialogue were the new, modern sensibilities of bourgeois reformism, making ideals about animal protection a largely class-based endeavor. From this perspective, urbanization and a sense of urban alienation from rural life contributed greatly to the preoccupation, at least in certain sectors of society, with preventing animal cruelty. Such sentiments mixed also with a desire to defend economic interests, a moral and philanthropic stance, and a purported means of curbing violence among humans.[44] Moreover, these animal-protection societies and the regulations against the mistreatment of animals can be seen partly as a response to the increasing exploitation to which animals were subjected at the beginning of the new industrial era.[45]

Throughout the nineteenth century, various countries introduced provisions with the goal of prosecuting instances of cruelty against animals. On 22 June 1822, the British Parliament promulgated the Martin's Act, the first law that made it illegal to mistreat certain domesticated animals, and on 2 July 1850, in France, the Grammont Law, which punished those who abused animals in public, was approved. Other laws were more species-specific, protecting, for instance, the birds used in agriculture (Prussia in 1850 and Austria in 1868) or abolishing bear-baiting and bull-baiting (England in 1835) and cockfighting (England in 1849). In the fields of research and medicine, lawmakers passed regulations against vivisection for scientific experimentation, as in Britain in 1876.[46]

The goals of the societies for the protection of animals were broad, and they varied from one country to another, and even between individual cities. Similar to its counterparts in the Spanish metropole, the Sociedad Cubana Protectora de Animales y Plantas also included the protection of plants in their agenda. Nevertheless, proposals dealing specifically with "animal rights," antivivisection, and vegetarianism proved to be marginal. In their formation, the societies tended to emphasize the more narrow ties between social and moral questions. The fact that among the supporters of such societies in England and the United States one could find many abolitionists, also fighting to eliminate slavery and the slave trade, can be seen as part of the very same (and not entirely unproblematic) process of expanding civil rights to historically marginalized groups.[47] If in the late nineteenth century supporters of the Sociedad Cubana Protectora de Animales y Plantas regarded both nonhuman animals and African slaves as exploited labor forces, then we might actually see the Cuban animal-protection society conceptualizing nonhuman animals as part of the category of exploited laborers alongside humans.

The first calls in Cuba that articulated the need for a local society for the protection of animals date back, at least, to the middle of the nineteenth century. In *Memoria sobre la industria pecuaria en la Isla de Cuba* (1849), the count of Pozos Dulces gave the example of London's Royal Society for the Prevention of Cruelty to Animals as a means through which advances for livestock could be achieved. In his work the count of Pozos Dulces emphasized vigilance regarding the observance of ordinances that imposed fines and punishments on those who mistreated animals, the free distribution of related published material in schools, and contests and the awarding of prizes to the best works on the subject of animal protection.[48]

The recommendations of the count of Pozos Dulces decidedly had more to do with the practical side of the work laid out by the societies for the protection of animals. However, less utilitarian interpretations were often viewed with animosity, as seen in the commentaries about one particular article sympathetic to the notion of animal welfare, "Beneficencia para con los animales," reproduced in *Memorias de la Sociedad Económica de La Habana* in 1844. The magazine's editors described the essay as one of the "most fantastic, strange, and newfangled" writings offered to the public, made up of "exaggerated paintings" that could be "dangerous for moral sanity."[49] The editors accused the author of having a cunning tendency toward materialism, and of confusing animal instinct with human intelligence, which they ascribed to an ignorance of philosophy and natural history, in effect branding the author "an atheist."

The idea of forming a society for the protection of animals in Cuba resurfaced in 1865 under the auspices of Manuel Presas, a young botanist and malacologist who, in the periodical *Aurora del Yumurí*, referred to the example of France, where there was "a society of charitable men dedicated to compensating all that was predisposed to the protection of domesticated animals."[50] After highlighting their activities, he asked, "When will we have a similar society, or when will we be able to treat those animals that help us get by in our daily lives as they deserve?"[51] In a similar vein, in 1877, *Revista Económica* praised the existence of laws and provisions "in those self-fashioned civilized nations" that attempted to contain and punish cruelty against animals.[52] Cuba was no exception, yet such laws were rarely observed, and beasts of burden especially suffered "merciless and barbaric" (*despiadado y bárbaro*) mistreatment. Attention was directed toward animal owners, with an appeal to the "natural interest that every man should have for the conservation of his property."[53] Notwithstanding, they distanced themselves from what they termed "exaggerations of sentimentalism": "We will not be in favor of the formation of such societies amongst ourselves while we lack hospices, charitable schools and asylums endowed or budgeted with sufficient resources, and so well directed and administered that they succeed in eliminating from our streets and plazas so many unfortunate [human] beings, [such as] the infirm, the desolate, [and] the unemployed, finding a means of eliminating their bad fortune."[54]

In December 1878 the journal *El Ingenio* reported that an assembly of protection societies convened in Baltimore, where they discussed, among other topics, the abuse that beasts of burden regularly suffered.[55] One month later, the publication *Revista de Agricultura* pledged its full support for the establishment of such a protection society in Havana, under the criteria that all civilized nations protect their animals against the fury of "other beasts that usurp the title of 'man' and commit unjust cruelties."[56] For their part, *La Nueva Era* interpreted such protective societies as a positive sign for the moral progress of Cuba, stating that there existed "no other country in the world that needed it more."[57]

In the realm of medical science as well, the respected *Crónica Médico Quirúrgica de La Habana* lent its support, albeit with some reservations. An editorial note stressed that the Sociedad Cubana Protectora de Animales y Plantas could provide immense support in countering the mistreatment of animals, in favor of particular interests, humanity, public health, and hygiene. But it also admonished some of the "exaggerations" of its counterparts abroad: "The society for the protection of animals becomes ludicrous when moving away from its intended goals. . . . They propose [for example] challenging vivisections, which would be equivalent to prohibiting slaughterhouses or using animals and plants as sources of food."[58]

In spite of this opinion, many physicians actively supported the Sociedad Cubana Protectora de Animales y Plantas. One of them, Esteban Borrero Echevarría, questioned the general disposition in Cuba, where, according to him, it was still necessary to moralize the relations between men in order to guide "the most appropriate and humane approach in man's relations with animals and plants."[59] With exceptional discernment in his argument, Borrero Echevarría appealed to scientific advances in order to harmonize this relationship: "Our life finds itself physically and morally bound to animal life and vegetal life; to protect one another is also humane—it is to protect man."[60]

We should also take into consideration that municipal ordinances contained regulations dedicated to the protection of animals. In 1881, for example, municipal statutes in Havana specified the maximum weight allowed for wagonloads drawn by oxen or carts pulled by mules. If the animal suffered a weakened condition and the load was disproportionate to its weight, the load should be limited, consistent with the "spirit of the law [which is] to avoid the mistreatment of animals, prohibiting in a general manner, and with humanitarian awareness, any task that exceeds their capacity."[61]

On 14 October 1882, colonial authorities approved the bylaw of the Sociedad Cubana Protectora de Animales y Plantas.[62] Article 1 proclaimed the society's industrial and charitable nature, so as to encourage the livestock and agricultural industries and promote their development and perfection with support from public authorities. With that goal in mind, the society presented an agenda that included the following: the acquisition of land for cultivation, with the intention of improving native plants and acclimating exotic ones; the establishment of educational chairs for agricultural instruction; the founding of a library, a cabinet, and a museum; the exchange of seeds and products with corporations and institutes that shared the society's values; and the holding of exhibitions, competitions, and other functions of "honest recreation." Regarding animals, the society proposed forming depositories for both native and imported breeds, in order to promote their growth, improvement, and training for practical purposes. Agents of the society were to undertake the inspection of markets, stables, and slaughterhouses, with the express goal of administering fines for the following infractions: selling inferior products; disregarding the presence of sick animals within the general animal populations; improperly or insufficiently providing for the animals; abandoning animals; subjecting animals to excessive work; inflicting cruel punishments; and killing animals in prejudicial ways, or in ways that were harmful to public safety.

One of the originators of the idea for a Cuban society was Juan García Villarraza, a dentist from Málaga, who served as president of the society's first board of directors, alongside a number of other notable individuals. Among those first selected for the honorary presidency were the Cuban gobernador general, the affluent Marques de Campo of Valencia, and the president of the American Society for the Prevention of Cruelty to Animals (ASPCA), Peter Bergh from New York.[63] In honor of the latter's life, the Cuban poet and revolutionary José Martí (who lived in New York at the time) published a portrait of Bergh, a founding member of animal-protection societies in the United States, in the Buenos Aires newspaper La Nación on 29 April 1888. One section read, "The more they ridiculed him, the more Bergh preached; he did so with such success that there is scarcely one State in the Union that has not included in its legal code his ideas against the mistreatment of animals, since to mistreat animals, more than simply being wicked, is something that bestializes man."[64]

Bergh's writings—including a letter to Villarraza, dated 22 December 1882, in which the author expressed his most "ardent sympathies" for the "progress" of Cuban society—as well as the writings of many of his compatriots who were also lobbying for the protection of animals, appeared in translation in the Boletín of the Sociedad Cubana Protectora de Animales y Plantas.[65] In January 1883 the society decided to send a proposal to various animal-protection societies abroad, outlining an international exchange of relevant literature. Shortly thereafter, publications, regulations, bylaws, and reports from various animal-protection societies in New York, San Francisco, Chicago, Massachusetts, Pennsylvania, and elsewhere began to arrive in Cuba.[66] In effect, the most free-flowing external relations and ideas—such as the implicit links between freedom from slavery and animal protection—developed between Cuba and its neighboring countries, as the following example from 1883 illustrates: "There are, in the United States, forty-eight societies analogous to ours, with branches and agencies among less important towns . . . something made possible by the resolution and perseverance of men with high hopes who, seventeen years ago [during the Civil War], took on such a thorny task . . . in spite of the thousand difficulties of laying out the issues and in the face of scathing satires initially hurled at them by some fatuous individuals."[67]

In the case of the metropolis, the board of directors of Madrid's Sociedad Madrileña Protectora de los Animales y de las Plantas sent congratulations on hearing of the establishment of its counterpart in Havana, destined to defend in "those rich provinces, our sisters, the expedience and justice from which our ideas prevail."[68] Another epistolary exchange took place with J. F. Kuhtmann, founder and president of an animal-protection society in imperial Germany, who expressed his felicitations for the "energetic propaganda" against bullfights—an effort seconded by Cuba's autonomist newspaper *El Triunfo*—and who desired "with all his heart the support and blessings of the Almighty in such noble efforts to suppress that barbarous spectacle."[69] He also compared those efforts with the struggle against vivisection, "a malady whose roots run deep and extend throughout Germany."[70] Although there is little evidence of the sustained continuity of many such international ties, support for the Sociedad Cubana Protectora de Animales y Plantas by international animal-protection societies is evident. Years later, García Villarraza would establish direct contact with Barcelona's Sociedad Barcelonesa Protectora de Animales y Plantas, on the occasion of a trip to the Philippines, from late 1886 to early 1887, in search of textile plants to foster new industries in Cuba.

Confronting cruelty against animals in Cuba had a discursive side, through the society's *Boletín* and other periodicals, as well as a practical side, by means of actions that sought to enforce the penalties imposed by the municipal ordinances for abusive behavior or draft animals' excessive loads. Among those subjects that received the most attention were the disapproval of bullfighting and cockfighting, the abysmal state of slaughterhouses, and haphazard practices of animal feeding (figs. 7.3–7.4). In general, however, the arguments were unambiguously anthropocentric in their emphasis. The predominant criterion was that the Cuban animal-protection society did

7.3 "Corrida de toros." SOURCE: FEDERICO MIAHLE, *ÁLBUM PINTORESCO DE LA ISLA DE CUBA* (HAVANA: B. MAY Y CÍA. 1855), CHROMOLITHOGRAPH NO. 25.

7.4 "Valla de gallos." SOURCE: FEDERICO MIAHLE, *ÁLBUM PINTORESCO DE LA ISLA DE CUBA* (HAVANA: B. MAY Y CÍA. 1855), CHROMOLITHOGRAPH NO. 15.

not embrace, "in its aspirations and tendencies," the treatment of animals "with all due respect," but rather it aspired to "a higher objective still," which consisted of both improving man's habits and acquiring the practice of "reciprocal respect."[71]

An article from February 1883, "Nuestros Deseos" (Our Wishes), described the alarming situation that the Sociedad Cubana Protectora de Animales y Plantas would have to face. Draft animals commonly suffered beatings with whips and sticks. Horses spent day and night outdoors, often without even a bit of hay on which to rest, and were constantly obligated to work, often suffering from fatigue, hunger, and thirst, which simply led the carriage drivers to double their blows. Mules suffered equally poor or even worse treatment, forced to pull wagons loaded with excessive weight, reminiscent of the tasks of oxen on plantations and sugar mills. Domesticated fowl, also ubiquitous in farms' corrals, were taken to market tied by their feet "in such a manner that often . . . they remained immobile and paralyzed for days on end."[72]

In one piece of writing from 1883, "¡Moralidad!" (Morality!), also published in the society's *Boletín*, José Romero Cuyás lamented that the situation for domesticated animals was much the same in the domestic sphere, where families and individuals looked at the lives and sufferings of animals, primarily used for the amusement of children, with indifference. In addition to rejecting bullfighting and cockfighting, Romero Cuyás pointed to the slaughterhouse as a "center of perversion" (*centro de perversión*) and a "school of assassins" (*escuela de asesinos*).[73] In his sermon, he argued that animals, like human beings, had an equal right to live and not be mistreated, and that one day the time would come in which "man understands that to sustain his existence, it is not absolutely indispensable to sacrifice other beings, [which are] annihilated today without necessity."[74] Romero Cuyás, like others, deemed necessary the preparation of future generations of Cubans, since, according to him, there was nothing more powerful than education.

The renowned Cuban zoologist Felipe Poey, already in his eighties in the penultimate decade of the nineteenth century, gave the Sociedad Cubana Protectora de Animales y Plantas his unwavering support, partly by considering municipal ordinances dealing with animals completely ineffective if not actively enforced. He was especially partial to the French example, where agents monitored the streets, denounced infractions, and offered rewards (for bringing cases of abuse to light). Poey also suggested confronting more sensitive matters, such as "[taking] charity and morality to the next level by bringing their case before the Autoridad Superior, so that academics at the University [of Havana] abstain from practicing vivisections in the presence of students, under the pretext of instructing by visual means, and running the risk of infusing the dismal habit of cruelty on their minds."[75]

Juan Cristóbal Gundlach, a German naturalist and taxonomist who resided in Cuba, took a different approach. His piece, "Rehabilitación de algunos animales cubanos, perseguidos y maltratados por preocupación vulgar" (Rehabilitation of Some Cuban Animals, Persecuted and Mistreated by Vulgar Preoccupation), challenged the notion that animals, plants, and minerals were objects created "for the specific utility [of man] or for the exclusive entertainment of the human race," a common assumption among those who did not "contemplate the works of nature in their true worth."[76] He began with bats, creatures which the "vulgar" populace feared and misconstrued as a harmful species. Yet, he asserted, they fed on those insects detrimental to humans and agriculture, while their excrement served as valuable fertilizer, comparable to the best guano from Peru. Among birds, he mentioned the barn owl, which was popularly seen as a symbol of malevolence and bad omens, and was frequently charged with attacking domesticated fowl and pigeon houses, although this happened infrequently, and the damage inflicted "was of little importance compared with that caused by the hundreds of mice on which they feed."[77]

Cristóbal Gundlach continued with reptiles and amphibians, including chameleons, lizards, frogs, and toads, which served the common good by feeding on noxious insects. The *majá de Santa María*, a species of boa native to Cuba that, because of its size, occasionally wrought havoc in henhouses and pigeon sheds, overall produced a positive effect on warehouses and maize fields, as well as on squash, sweet potato, and yucca fields.[78] Gundlach sought to protect "those species that science has always deemed worthy of conserving in their generic form" or because they could someday be useful (to human industry), as happened with the silkworm or with bees. His hope was to eliminate the misconceptions engendered by a deficient education and a lack of observational spirit, so as not to falsely attribute faults to species that provided more utility than harm. In favor of those efforts, he stated, "Fortunately, in Havana a society for the protection of plants and animals has been founded, and there also exists a *ley de caza* [hunting law]; but, today the former is strictly local while the latter is not yet fully observed."[79]

The Sociedad Cubana Protectora de Animales y Plantas formally inaugurated its functions in Havana on 3 November, 1882, with a visit to a stable on Apodaca Street, after receiving news of "unprecedented cruelty" against nineteen sequestered horses, three of which died due to hunger and thirst. On site, the president, the secretary, and the veterinary inspector of the animal-protection society confirmed that the horses had died of starvation; they determined that the stable manager was obligated to feed the horses and, in his absence, put someone in charge of their care. In addition to the ten-peso fine established by municipal ordinances for animal cruelty and neglect, other expenses and fees would be billed to the Sociedad Martínez y Peña, which was held legally responsible for ownership of the sequestered horses.[80]

That same day, in a meeting of the Sociedad Cubana Protectora de Animales y Plantas, those in attendance unanimously approved the measures taken and agreed to inform the governador de la provincia of their actions. Likewise, they accepted a proposal to create a special police force composed of the members of the animal-protection society.[81] By that time, individuals were regularly lodging complaints about the mistreatment of animals—excessive loads and other abusive behaviors—in public spaces. For example, in October 1882, the mayor of the Tacón district denounced Atiliano Bermúdez, a twenty-two-year-old native of Galicia, for cruelly beating the mule that he saw pulling wagon number 1010, offering a detailed account so that the offender would be penalized.[82]

Another front of the fight against animal cruelty was opposition to bull-fighting and cockfighting, practices that, in the words of Romero Cuyás, constituted "exceedingly immoral" forms of recreation that were without benefit, and upheld "all the roots of evil."[83] Although the symbolic significance of baseball for Cubans at the end of the nineteenth century, as a means of differentiating themselves from Spanish aficionados of bullfighting, has been well documented, the example of contemporaneous animal-protection societies demonstrate that members of both groups—Cubans and Spaniards—actively rejected the bullring.[84] On this issue, animal-protection societies on both sides of the Atlantic, in metropole and colony, coincided, despite the growing popularity of corridas throughout the nineteenth century.[85] The press made visible the activities that took place inside bullrings, inspiring immediate revulsion on the part of those who supported the protection of animals. The press also alerted the public about other gruesome spectacles involving animals: on 25 July 1885 the Diario de la Marina announced that the company of Colonel Pubillones—founder of the famed Cuban circus, el Gran Circo Pubillones—would sponsor a fight between a tiger and a bull as the main event at the Plaza de la Regla. Days later, a battle would ensue between "a magnificent bull, a native citizen of Bolondrón" and the elephant "Romeo," also the property of Pubillones.[86] On some occasions, members of the society acted swiftly in the bullrings, such as on 24 June 1883, when the animal-protection society's president successfully protested against the return of a horse, already injured twice in the belly, to the bullring.[87]

The Cuban animal-protection society received, with elation, the news that in the Spanish metropole, the construction of new bullrings had been prohibited, and such news reinvigorated the struggle for more forceful policies to do away completely with the "national" celebrations of bullfighting. In opposition to that particular brand of nationalism, Cuban supporters of animal protection emphasized the fact that Spain possessed a theater, where its intelligent citizens demonstrated the country's "cult for progress." Although bullfighting represented a deep-rooted tradition with "enthusiastic defenders," the time was ripe to give way to the "progressive project" that opposed it, headed by some branches of the Havana press.[88]

Also on the agenda of the Cuban animal-protection society was the link between animals, diet, health, and hygiene. The society fought, for instance, against the alteration of foodstuffs (such as milk), which it saw as especially harmful to children. Activists criticized the manner in which milk producers infringed on the municipal ordinances: "The majority of the time, the milk was impure . . . and the other times, it was simply watered down, due to the avarice of the owners."[89] Within this context, an unfulfilled "model farm" project sought to achieve an important function, since it planned to sell the milk from cows, asses, or goats directly to stores, where mothers and the infirm could purchase it, "certain of obtaining, with major advantages, so precious a liquid."[90] In the meantime, the members would ensure the purity of the milk, the health of the cows, that they had been well fed, and that they had not recently delivered (hence denying calves their mothers' milk). The society promised, in any case, direct attention to this public-health issue, assuring that a thorough inspection would yield benefits.

In its early years, the Sociedad Cubana Protectora de Animales y Plantas carried out other projects more closely tied to plants. In March 1884, only two years after the society had been founded, an article published in *El Triunfo* refuted its supposed dissolution, countering the claims of some "ill-intentioned individuals." On the contrary, the author announced, the board of directors found itself enmeshed in reforming its ordinances, "in light of the many modifications relating to the protection of animals [that have been] implemented by superior authority."[91] Provisionally, the bylaws of Madrid's Sociedad Madrileña Protectora de los Animales y de las Plantas would govern the Cuban society, while maintaining their stance on not consenting to "the mistreatment of animals, the continued decline of the agricultural industry, or the utter abandonment of the precepts of public hygiene."[92]

In April 1885 authorities approved a series of internal and regulatory reforms. What first stands out is a change in the name of the animal-protection society, now officially called the Sociedad Protectora de Animales y Plantas de la Isla de Cuba (the Society for the Protection of Animals and Plants of the Island of Cuba). Although this may have been a simple formality, it could also indicate a moderation of the initial goals of the society. The new clauses displayed more precision in terms of institutional functioning, and they introduced other important modifications. Article 1 reaffirmed the nature of the society in terms of charity and industry, but it did so more specifically: "to protect man against his ignorance and poor sentiments, to protect useful animals [*animales útiles*], keeping them from all mistreatment and suffering, and to foster the livestock and agricultural industries."[93] The society also redefined and reprioritized many of its goals: to influence primary education through teachers, who would also be encouraged to join the society; to involve the society more in the enforcement of animal-cruelty laws and the collection of fines; and to attract the attention of the government and local authorities "with the purpose of obtaining their effective cooperation in achieving the noble goals that have been proposed."[94] Yet that which the Sociedad Protectora de Animales y Plantas de la Isla de Cuba emphasized most in this new phase was the creation, in February 1886, of an asylum for beggars, named La Misericordia, presumably as a response to the usual critiques of the society's focus on animals instead of on humans. While the existence and goals of this institution go far beyond the objectives of this essay, it is worth noting that, in the end, the society gave rise to a serious conflict that, in the middle of 1890, drove the protection society to its extinction.[95]

Of the most important accomplishments that took place after the reforms of 1885, in relation to the protection of animals, there is little news. One of the reasons for this can be gleaned in the society's minutes from 1885 and 1886: "Multiple efforts [dedicated] for the asylum [La Misericordia] have caused the society to become weak in regards to its original goals."[96] Nevertheless, those same records reveal the dispatch of several commissions to the bullfights, an effort that lasted until the promotion company of the corridas opposed the proposal that agents of the Sociedad Protectora de Animales y Plantas de la Isla de Cuba medically inspect the horses selected to participate in the fight. Another article, which appeared in the September issue of *El Hogar* from 1889, praised the society for compassionately collecting stray dogs, thereby saving people "from the repugnant spectacle of dogs being slaughtered on the main roads by the *salchicha municipal*."[97]

During that time, the Sociedad Protectora de Animales y Plantas de la Isla de Cuba actively involved itself in the struggle against glanders (*muermo*)—an infectious disease that occurs primarily in horses, mules, and donkeys, but is also communicable to humans—which affected many victims in Havana. The society's president, Oscar Conill, and several other members formed part of a special commission established by the government to study the disease. The society also offered to establish an animal hospital where sick or suspicious horses could be deposited, but the proposal failed to receive official backing. One of the most dramatic moments in the struggle against glanders took place on 15 November 1889, with the death of the society's secretary, Pedro Fernández Díaz, a native of Galicia who contracted the disease during his tenure as secretary for the aforementioned government commission set up to fight that very disease.[98]

Moving into the early twentieth century, the U.S. intervention and the establishment of the republic on 20 May 1902 opened a new legislative and associative chapter in relations with animals. Proof of this change is evident even earlier, in the intervening government's order 217, promulgated on 28 May 1900 at the suggestion of the Secretario de Justicia.[99] This order imposed fines and penalties on the following: those who mistreated, cruelly punished, or tormented any animal; those who hurt, mutilated, and thereby killed any animal (whether or not that animal was their "property"); and those who abandoned sick animals or, out of neglect, let them die in public spaces. In addition, the government set up regulations dealing with the transport of animals by train or other means, and established fines for those who "in any way assisted, contributed to, or collaborated in the staging of cockfights or contests between other birds, bullfights, or any other staged fights between animals that, with premeditation, are proposed by the animals' owners or caretakers."[100]

Under more direct influence of the United States, in 1902 the Sociedad Humanitaria, Protectora de los Niños y contra la Crueldad con los Animales (the Cuban Humane Society for the Protection of Children and Against the Cruelty of Animals) was established, following the model of the American Humane Association.[101] The goals of the animal-protection society established by García Villarraza and of the Sociedad Protectora de los Niños de la Isla de Cuba (the Society for the Protection of Children of the Island of Cuba), established in Havana in 1884, thus officially merged by way of an ordinance that was approved on 16 September 1902.[102] During the inauguration of the Sociedad Humanitaria, its long-standing president, Juan Santos Fernández, recalled his old affiliation with the animal-protection society in the 1880s, and he challenged the notion that society need not concern itself with the welfare of animals while there were humans in need of support, especially if one considered the services that animals provided as well as their capacity to suffer and experience pain. He also reaffirmed his opposition to bullfighting and cockfighting. Drawing explicit parallels between the situation of animals and that of enslaved human beings throughout the island of Cuba only a few years earlier, he commented, "This nation that is just born and that enters into a life of equality—[a nation] that only yesterday shook off the enslavement of blacks and achieved liberty for all—must not give in to irrationality and deny the justice and equality bestowed upon all: the doing away with the useless suffering of those beings that support us in our daily labors, and do so until they are slaughtered . . . [is] to preserve the life of man, to make it long-lasting and full of advantages and satisfactions."[103]

Conclusion

In this chapter I have historicized elements of the multifaceted and constantly changing relationship between humans and animals in Cuba throughout the nineteenth century. These factors—changes in technology, animal labor, and attitudes of charity—cannot be excluded from the study of environmental history, since nonhuman animals have played such a significant role in human life as a driving force, means of transport, and source of food and clothing. The slave-run sugar plantations were the scene of (and impetus for) great technological changes in the manufacturing sector and railway expansion, which signified a veritable revolution in terms of transporting merchandise to the ports and, later, raw materials and commodities to the sugar mills. These breakthroughs increased production capacity while decreasing the overall need for slave labor and animal force. Starting in the 1840s, however, the possibility of expanding the scale of production with the help of steam engines directly affected the expansion in the number of oxen yokes used on semimechanized and fully mechanized mills, although in absolute terms the number of oxen was not the decisive factor in the sugar boom. Beginning in the 1870s, new technological advances also began to affect the agricultural sector in terms of reducing dependence on animal draft. However, this was a long process that, while reaching its apex with the agricultural mechanization of the twentieth century, took place over the course of several decades and varied according to a number of local factors.

Nineteenth-century sugar-production manuals reflect the central role that oxen continued to play (just as the ox-driver remained among the highest-paid employees on the plantation) and demonstrate the increasingly common calls to treat the animals more humanely. In this sense, we can establish an undeniable link between these pleas and the emergence, years later, of an animal-protection society within the context of the decline of slavery in Cuba. This chronological coincidence is revealing since it indicates that major social changes, such as the abolition of slavery, tend to induce a reconsideration of mentalities and the ways in which human beings interact with the other constituents of nature.

At present, at least one organization exists in Cuba with goals similar to those described here: the Asociación Cubana para la Protección de Animales y Plantas (The Cuban Association for the Protection of Animals and Plants) was approved and entered in the register of associations of the Ministerio de Justicia de la República de Cuba on 19 February 1987. Its influence on Cuban society, however, appears to be significantly less than that which similar institutions have in other countries, and the scope of its work seems largely to involve protecting domesticated animals and household pets such as dogs and cats. This sensibility coincides with that of public personae who demonstrate a growing preoccupation with recurring scenes of mistreatment and abandonment of animals in the urban sphere.

Although, as a result of changing geopolitical and economic factors, the dawn of the new millennium in Cuba has brought about improved accessibility to petroleum and its byproducts, the recent tendency toward the greater use of animals represents a viable alternative for the country in terms of diversifying its energy sources amid the current environmental crisis and the likely depletion of fossil fuels in the coming decades. Given the numbers cited at the beginning of this essay, it is therefore possible to imagine that Cubans might, in the future, experience the inverse of those changes—namely, the gradual decline of animal draft—that occurred throughout the course of the nineteenth and twentieth centuries. A return to animal labor would, of course, have profound economic, social, and environmental implications, depending on the manner in which Cubans manage the decrease of fossil fuels, and if no other viable alternatives to petroleum are found. Ideas and actions with respect to the protection of animals, or even regarding the defense of their "rights," could play a decisive role in determining the way in which humans will assume such changes.

Notes

An earlier version of this article was published in Spanish as "Facetas de la interacción con los animales en Cuba durante el siglo XIX: Los bueyes en la plantación esclavista y la Sociedad Protectora de Animales y Plantas," *Signos Históricos* 16 (July–December 2006): 80–110. English translation by Alexander Hidalgo and Zeb Tortorici.

1. Arcadio Ríos and Félix Ponce, "Tracción animal, mecanización y agricultura sostenible," in *Transformando el campo cubano: Avances de la agricultura sostenible*, ed. Fernando Funes et al. (Havana: Asociación Cubana de Técnicos Agrícolas y Forestales, 2001), 159–66. See also José Suárez, Arcadio Ríos, and Pedro Soto, "El tractor y la tracción animal," *Revista de Ciencias Técnicas Agropecuarias* 14.2 (2005): 40–43.

2. Humberto Valdés Ríos, "Servicio de transporte público utilizando tracción animal en Cuba," in *La tracción animal en Cuba*, ed. Paul Starkey and Brian Sims, available at the Red Cubana de Tracción Animal website, http://www.recta.org.

3. "Abrecartas, a cargo de Guillermo Cabrera Álvarez," *Granma*, 24 September 1996, 2.

4. Two historic works that touch on the formation of nationalism and Cuban culture in relation to animals are Louis A. Pérez Jr., "Between Baseball and Bullfighting: The Quest for Nationality in Cuba, 1868–1898," *Journal of American History* 81.2 (1994): 493–517; and Pablo Riaño, *Gallos y toros en Cuba* (Havana: Fundación Fernando Ortiz, 2002). For an exploration of some aspects of material culture related to animals, such as wagons and food, see Ismael Sarmiento Ramírez, *Cuba entre la opulencia y la pobreza: Población, economía y cultura material en los primeros 68 años del siglo XIX* (Madrid: Agualarga Editores, 2004). Though studies on cattle breeding are scarce, one could cite, for example, certain passages in Leví Marrero, *Cuba: Economía y sociedad* (Madrid: Editorial Playor, 1974–84). For a panoramic view of animal drafting, see Arcadio Ríos and Jesús Cárdenas, "La tracción animal en Cuba: Una perspectiva histórica," in *La tracción animal en Cuba*, ed. Paul Starkey and Brian Sims (2003), available at the Red Cubana de Tracción Animal website, http://www.recta.org.

5. Edward Wrigley, *Cambio, continuidad y azar: Carácter de la revolución industrial inglesa* (Barcelona: Crítica, 1993); John R. McNeill, *Algo nuevo bajo el sol: Historia medioambiental del mundo en el siglo XX* (Barcelona: Alianza Editorial, 2003), see 36–43. For an insightful work on that Spanish agricultural transition, see Manuel González de Molina and Gloria Guzmán, *Tras los pasos de la insustentabilidad: Agricultura y medio ambiente en perspectiva histórica (siglo XVIII–XX)* (Barcelona: Icaria, 2006).

6. On the role of oxen within the sugar economy, see Manuel Moreno Fraginals, *El ingenio: Complejo económico social cubano del azúcar* (Havana: Editorial de Ciencias Sociales, 1978 [1964]), 1:201–3; and Roland Taylor Ely, *Cuando reinaba su majestad el azúcar* (Havana: Imagen Contemporánea, 2001 [1963]), 618–25. Infor-

mation on the eighteenth century can be found in Mercedes García, *Entre haciendas y plantaciones: Orígenes de la manufactura azucarera en La Habana* (Editorial de Ciencias Sociales: Havana, 2007), 197–98.

7. Gilberto Freyre, *Nordeste: Aspectos da influência da cana sobre a vida e a paisagem do nordeste do Brasil* (São Paulo: Global Editora, 2004), 14.

8. An *arroba* is a weight measurement of about twenty-five pounds; a *caballería de tierra* is a unit of land measure equal to 33.2 acres.

9. José Ricardo O'Farril, "Exposición que hace . . . a la sociedad del método observado en la Isla de Cuba, en el cultivo de la caña dulce y la elaboración de su jugo," *Memorias de la Sociedad Económica de La Habana* 1 (1793): 145.

10. Antonio Morejón and Gato, Archivo Nacional de Cuba (ANC), Fondo Real Consulado y Junta de Fomento (RC y JF), no. 3934, *Memoria sobre la cría de ganados en Cuba*, 1797.

11. Only six plantations achieved the use of water as a source of energy in the Güines zone, near the Mayabeque River, the greatest flowing river in the sugar frontier at the end of the eighteenth century. The use of wind failed due to lack of constancy. See, for example, ANC, RC y JF, Leg. 92, no. 3943, *Experimento de Pedro Diago con un trapiche de viento*, 1800. The first attempt in Cuba to apply steam power to the sugar mills took place in 1797 with a machine purchased in London by the count of Jaruco for the Sybabo plantation.

12. José Francisco de Frías y Jacott (conde de Pozos Dulces), *Memoria sobre la industria pecuaria en la Isla de Cuba* (Havana: Imprenta del Diario de la Marina, 1849), 5.

13. Alejandro Dumont, *Guía de ingenios que trata de la caña de azúcar* (Matanzas: Imprenta del Gobierno a cargo de Campe, 1832), 29.

14. Frías y Jacott, *Memoria sobre la industria pecuaria en las Isla de Cuba*, 10.

15. Antonio Bachiller y Morales, "Memoria sobre el número y valor de los ganados de la Isla, obstáculos que se oponen a su producto y medios de fomentar su consumo y el de las pesquerías," *Memorias de la Sociedad Económica de La Habana* 33 (1846): 347.

16. In comparison, the same region registered a notable increase in the number of horses, from 21,151 in 1846 to 35,426 in 1862, a difference that merits a closer examination in the future.

17. Andrés de Zayas, "Observaciones sobre los ingenios de esta Isla," *Memorias de la Sociedad Económica de La Habana* 12 (1836): 179.

18. Zayas, "Observaciones sobre los ingenios de esta Isla," 180.

19. Zayas, "Observaciones sobre los ingenios de esta Isla," 180.

20. José Montalvo y Castillo, *Tratado general de escuela teórico-práctica para el gobierno de los ingenios de la Isla de Cuba en todos sus ramos* (Matanzas: Imprenta de la Aurora, 1856), 9.

21. Montalvo y Castillo, *Tratado general de escuela teórico-práctica para el gobierno de los ingenios de la Isla de Cuba en todos sus ramos*, 10.

22. Montalvo y Castillo, *Tratado general de escuela teórico-práctica para el gobierno de los ingenios de la Isla de Cuba en todos sus ramos*, 10.

23. Fermín Rosillo y Alquiler, *Noticia de dos ingenios y datos sobre la producción azucarera de la Isla de Cuba para la Exposición de Viena* (Havana: El Iris, 1873).

24. Antonio de Landa, *El administrador de ingenio* (Havana: Imprenta La Fortuna, 1866), 48.

25. Wrigley, *Cambio, continuidad y azar*, 54–55.

26. Landa, *El administrador de ingenio*, 48.

27. Landa, *El administrador de ingenio*, 51.

28. Fraginals, *El ingenio*, 201.

29. Francisco de Paula Serrano, *Agricultura cubana o tratado sobre los ramos principales de su industria rural* (Havana: Oficina del Gobierno y Capitanía General, 1837); José Jacinto de Frías y Jacott, *Ensayo sobre la cría de ganados en Cuba* (Havana: Oficina del Faro Industrial, 1844); Manuel Monteverde, *Estudios prácticos de las condiciones económicas de la industria pecuaria en el distrito de Puerto Príncipe* (Puerto Príncipe [Camagüey]: Imprenta El Fanal, 1856).

30. Ramón de Arozarena and Pedro Bauduy, *Informe presentado a la junta de gobierno del Real Consulado de la siempre fiel Isla de Cuba sobre el estado de la agricultura, elaboración y beneficio de los frutos coloniales en la de Jamaica* (Havana: Imprenta Fraternal de los Díaz de Castro, 1828).

31. The term *bagasse* refers to the dry, fibrous residue remaining after the extraction of juice from crushed sugarcane stalks.

32. Juan Tatjer y Risqué, *Fabricación del azúcar de caña tal como se practica con los aparatos más modernos en la Isla de Cuba* (Havana: Imprenta del Avisador Comercial, 1887), 195–98.

33. *Hornos de quemar bagazo verde. Perfeccionados sin rival. Aplicados a toda clase de calderas de vapor para Ingenios. Con real privilegio para España y sus posesiones de ultramar de Samuel Fiske* (Havana, 1890). Along these lines and taking into consideration that the 1,000 plantations in Cuba used, on average, 40 workers for the extraction of sugarcane remains (some used up to 100), the green bagasse burners could represent a savings of over 20,000 men "who utilizing their strength and labor in other activities could forcefully expand agricultural development among us."

34. *Carros ballena: Privilegio concedido a Felipe Viera Montes de Oca. E. Courtillier y Cía* (Havana: Imprenta del Directorio, 1877).

35. Patria Cook Márquez, "La introducción de los ferrocarriles portátiles en la industria azucarera, 1870–1880," *Santiago* 41 (1981): 137–47.

36. Miguel Rodríguez Ferrer, *Naturaleza y civilización de la grandiosa Isla de Cuba* (Madrid: Imprenta de J. Noguera a cargo de M. Martínez, 1876), 1:801–2.

37. For a general discussion on the formation of civil society during this period, see José Antonio Piqueras, *Sociedad civil y poder en Cuba: Colonia y poscolonia* (Madrid: Siglo XXI, 2005), chaps. 4 and 5. On specific aspects, see Oilda Hevia, *El directorio central de las sociedades negras de Cuba, 1886–1894* (Havana: Editorial de Ciencias Sociales, 1996); Joan Casanovas, *¡O pan, o plomo! Los trabajadores urbanos y el colonialismo español en Cuba, 1850–1898* (Madrid: Siglo XXI, 2000);

Reinaldo Funes Monzote, *El despertar del asociacionismo científico en Cuba* (Madrid: Consejo Superior de Investigaciones Científicas, 2004).

38. See Rebecca J. Scott, *Slave Emancipation in Cuba: The Transition to Free Labor, 1860–1899* (Princeton: Princeton University Press, 1985).

39. Brian Harrison, "Animals and the State in Nineteenth-Century England," *English Historical Review* 88.349 (1873): 786–820.

40. An additional seven cities establish similar associations from 1853 to 1890. Patrick Matagne, "The Politics of Conservation in France in the 19th Century," *Environment and History* 4.3 (1998): 359–66.

41. Roderick F. Nash, *The Rights of Nature: A History of Environmental Ethics* (Madison: University of Wisconsin Press, 1989), 38–53.

42. Sociedad Madrileña Protectora de los Animales y de las Plantas (Unión Internacional de Sociedades Protectoras de Animales), *Exposición de 1881 bajo el patronato de SM la Reina, Jardines del Parterre, Parque de Madrid* (Madrid: Imprenta de G. Juste, 1881).

43. Nash, *The Rights of Nature*; Jesús Mosterín and Jorge Riechmann, *Animales y ciudadanos: Indagación sobre el lugar de los animales en la moral y el derecho de las sociedades industrializadas* (Madrid: Talasa Ediciones, 1995); Jason Hribal, "Pythagoreanism, 1640–1825: The Creation of the Animal Rights Movement," unpublished manuscript.

44. Maurice Agulhon, "Le sang des bêtes: Le problème de la protection des animaux en France au XIXème siècle," in *Histoire vagabonde: Ethnologie et politique dans la France contemporaine* (Paris: Editions Gallimard, 1988), 1:243–82.

45. Jason Hribal, "'Animals Are Part of the Working Class': A Challenge to Labor History," *Labor History* 44.4 (2003): 435–53.

46. Harrison, "Animals and the State in Nineteenth-Century England," 791.

47. Nash, *The Rights of Nature*. See also Arthur F. McEvoy, "Animal Liberation: The Moral Equivalent of Abolitionism?," *Reviews in American History* 18.3 (1990): 425–29.

48. Frías y Jacott, *Memoria sobre la industria pecuaria en la Isla de Cuba*, 13.

49. "Beneficencia para con los animales," *Memorias de la Sociedad Económica de La Habana* 18 (1844): 141–52. Only the author's last name, De Weiss, appears. Among other ideas, he criticized the fact that humans considered animals to be inferior beings, as did "tyrants their subordinates: they believe they exist to fulfill their wishes or to serve their needs, without imagining that reciprocal obligations exist between one and the other."

50. *Aurora del Yumurí* (Matanzas), 11 May 1865, 2.

51. *Aurora del Yumurí* (Matanzas), 11 May 1865, 2.

52. "Maltrato a los animales," *Revista Económica*, no. 17 (27 November 1877): 135.

53. "Maltrato a los animales," *Revista Económica*, no. 17 (27 November 1877): 135.

54. "Maltrato a los animales," *Revista Económica*, no. 17 (27 November 1877): 135.

55. "Importancia de la agricultura," *El Ingenio*, no. 36 (15 December 1878): 575.

56. "Sociedad benéfica," *Revista de Agricultura*, no. 1 (January 1879): 22.

57. "Protección para los animales," *La Nueva Era*, no. 19 (1 August 1882): 300.

58. "Sociedad Protectora de Animales," *Crónica Médico Quirúrgica de la Habana* (July 1882): 349–50.

59. Esteban Borrero Echevarría, "Sociedad Protectora de Animales y Plantas," *La Semana*, no. 12 (19 March 1888): 7–8. Signed in Puentes Grandes, Havana, on 10 January 1883.

60. Esteban Borrero Echevarría, "Sociedad Protectora de Animales y Plantas," *La Semana*, no. 12 (19 March 1888): 8.

61. *Ordenanzas Municipales de Policía Urbana y Rural del Término Municipal de La Habana* (Havana: Fernández y Cía, 1881), 12–13.

62. *Reglamento de la Sociedad Cubana Protectora de Animales y Plantas* (Havana: La Propaganda Literaria, 1882).

63. In the subsequent months, the list expanded to include other colonial administration figures and other powerful actors from the island and the metropolis.

64. José Martí, *Obras Completas* (Havana: Editorial Lex, 1946), 1:1245.

65. Letter from Peter Bergh to Juan García Villarraza, *Boletín de la Sociedad Cubana Protectora de Animales y Plantas* (*Boletín*), 1 February 1883. Bergh's letter was in response to a letter sent by García Villarraza informing him of the foundation of the Sociedad Cubana Protectora de Animales y Plantas.

66. Proposal made by Enrique Lecerff, *Boletín*, 15 February 1883, 36. The issue of *Boletín* that appeared on 1 April 1883 included the list of the publications that the society had received from abroad (56).

67. "Protección en los Estados Unidos," *Boletín*, 15 January 1883, 24.

68. "Sección de noticias generales," *Boletín*, 1 April 1883, 55.

69. Letter from Kuhtmann, Bremen, 5 April 1883, *Boletín*, 1 June 1883, 88.

70. Letter from Kuhtmann, Bremen, 5 April 1883, *Boletín*, 1 June 1883, 88.

71. "Al hombre y al animal," *Boletín*, 1 October 1884, 1.

72. "Nuestros deseos," *Boletín*, 31 January 1883, 26–28.

73. José Romero Cuyás, "¡Moralidad!," *Boletín*, 1 January 1883, 10–11.

74. José Romero Cuyás, "¡Moralidad!," *Boletín*, 1 January 1883, 10–11.

75. Felipe Poey, "¡Indulgencia para los animales!," *Boletín*, 15 January 1883, 21.

76. Juan Cristóbal Gundlach, "Rehabilitación de algunos animales cubanos, persegui-dos y maltratados por preocupación vulgar," *Boletín*, 15 February 1883, 33–34; *Boletín*, 15 March 1883, 47–48; and *Boletín*, 1 April 1883, 53–54.

77. Juan Cristóbal Gundlach, "Rehabilitación de algunos animales cubanos, per-seguidos y maltratados por preocupación vulgar," *Boletín*, 15 March 1883, 48. Due to the damages caused by rats on the plantations, Cristóbal Gundlach suggested, "a solution to increase these creatures [barn owls] rather than destroy them" should be sought.

78. Juan Cristóbal Gundlach, "Rehabilitación de algunos animales cubanos, per-seguidos y maltratados por preocupación vulgar," *Boletín*, 15 February 1883, 33–34; *Boletín*, 15 March 1883, 47–48; and *Boletín*, 1 April 1883, 54. "We have seen some estates take advantage of the maja as mice and rat hunters where maize and other

grain deposits exist; their utility has been so high that those destructive rodents have disappeared."

79. Juan Cristóbal Gundlach, "Rehabilitación de algunos animales cubanos, perseguidos y maltratados por preocupación vulgar," *Boletín*, 15 February 1883, 34.

80. "Sección de noticias locales: Infracciones," *Boletín*, 1 January 1883, 16.

81. "Sección oficial," *Boletín*, 1 February 1883, 28.

82. "Sección de noticias locales: Infracciones," *Boletín*, 1 January 1883, 16.

83. José Romero Cuyás, "¡Moralidad!," *Boletín*, 1 January 1883, 10.

84. Pérez, "Between Baseball and Bullfighting"; and Riaño, *Gallos y toros en Cuba*.

85. Adrian Shubert, *A las cinco de la tarde: Una historia social del toreo* (Madrid: Turner Publicaciones, 2002). Bullfights increased in Spain from about 400 in 1860 to over 700 in 1895. Shubert documents the reaction of the protection societies, in particular, the Plant and Animal Protection Society of Cadiz, founded in 1872 (*A las cinco de la tarde*, 203–5).

86. See *Diario de la Marina*, "Gacetillas" section for those dates. For more on the Pubillones circus, see Francisco José Pantín Fernández, "El Gran Circo Pubillones," *Boletín de las fiestas de Nuestra Señora* (Corao: Asociación Cultural y Recreativa "El Castañéu," 2005).

87. "¡Horror!," *Boletín*, 1 July 1883, 101. It was rumored that horse had lent its services for thirty years.

88. "Las corridas de toros," *Boletín*, 1 April 1883, 49.

89. "Salubridad pública," *Boletín*, 15 January 1883, 18–19.

90. "Salubridad pública," *Boletín*, 15 January 1883, 19.

91. *El Triunfo*, 5 March 1884.

92. *El Triunfo*, 5 March 1884.

93. *Estatutos de la Sociedad Protectora de Animales y Plantas de la Isla de Cuba* (Havana: Imprenta de Soler, Álvarez y Cía, 1885).

94. *Estatutos de la Sociedad Protectora de Animales y Plantas de la Isla de Cuba.*

95. The asylum, in spite of the dissolution of the society, maintained its activities. *Estatutos y reglamento de los servicios interiores del Asilo General para Mendigos La Misericordia* (Havana: Imprenta El Fígaro, 1896).

96. José Romero Cuyás, *Memoria de la Sociedad Protectora de Animales y Plantas de la Isla de Cuba* (Havana: Establecimiento Tipográfico de Álvarez, 1887).

97. *El Hogar*, 8 September 1889, 431. The term *salchicha municipal* as used here appears to refer to the municipal mechanism of collecting stray animals.

98. Book with newspaper clippings, *El muermo en La Habana*, prepared by Juan Santos Fernández, with news dated from 1889 and 1893, Archivo del Museo Nacional de las Ciencias Carlos J. Finlay, La Habana.

99. Leonard Word, *Civil Report of Mayor General Leonardo Wood, Military Governor of Cuba: 1900*, vol. 2. of *Civil Orders and Circulars* (Washington: Government Printing Office, 1900), 406–7.

100. Article 8 authorized agents from legally established animal-protection societies to detain individuals who violated those regulations and to turn them over to au-

thorities to apply the penalties accorded by law. See *Colleción legislativa de la Isla de Cuba: Recopilación de todas las disposiciones publicadas en la "Gazeta de la Habana"* (Havana: Establecimiento Tipográfico, Teniente Rey 23, 1899), 2:369.

101. Dr. Juan Santos Fernández, "Necesidad de leyes protectoras de la infancia en Cuba," in *Memoria oficial de la Cuarta Conferencia Nacional de Beneficencia y Correción de la Isla de Cuba* (Havana: La Moderna Poesía, 1905), 147–52. The full name of its U.S. counterpart was the American Humane Association Societies of the United States Organized for the Prevention of Cruelty to Animals and Children.

102. Cuban Humanitarian Society, *Ordenes generales* (Havana: Imprenta de Cerdeira y Cía, 1913).

103. Juan Santos Fernández, "Sociedad Cubana Protectora de Animales," *Crónica Médico Quirúrgica de La Habana* 28 (1902): 277–84.

8

On Edge

Fur Seals and Hunters along the Patagonian Littoral, 1860–1930

JOHN SOLURI

> Employing the club and rifle, this "civilized" cruelty kills not only
> male seals but also females and pups, the latter being left by the way-
> side because they lack the fine furs that are most valued. . . . Such a
> system of killing, one can easily comprehend, tends to destroy the
> species; those that escape migrate in search of a more secure haven for
> the future.
> —"REPORT . . . ON SEAL HUNTING" (1883)

In 1889 the Chilean president José Manuel Balmaceda signed a decree that
conceded 180,000 hectares of land in Tierra del Fuego to José Nogueira, a
wealthy Portuguese immigrant based in Punta Arenas, the Chilean port on
the Straits of Magellan. A few months later, Mauricio Braun received a
concession to 170,000 hectares of land contiguous to that of Nogueira—
Braun's brother-in-law. These two concessions established the legal foundation for sprawling sheep ranches that would dominate the economy of
southern Patagonia during much of the twentieth century. Nogueira expressed his gratitude to President Balmaceda by presenting him with a
sealskin vest and, for First Lady Emilia Toro, a fur-lined cape made from sea
otter. The garments, crafted in France from marine mammals native to
Patagonia, reportedly pleased the presidential couple.[1] The choice of gifts
was fitting considering that Nogueira had accumulated a small fortune
hunting and trading fur seals along South America's southernmost littoral
zones. Before the first sheep ranchers, missionaries, gold seekers, or tourists

reached Patagonia, both indigenous and foreign visitors forged livelihoods from the riches of the region's coastal ecosystems that were home to a variety of fish, mollusks, crustaceans, waterfowl, and carnivorous marine mammals including South American fur seals (*Arctocephalus australis*). In this chapter I place fur seals and their human hunters at the center of the modern history of southern Patagonia.[2]

The relationship between people and fur seals in southern Patagonia and Tierra del Fuego stretches millennia. Archaeological evidence unearthed in Tierra del Fuego indicates that human groups began regularly hunting seals no later than six thousand years ago, and they continued to do so when seafarers from North Atlantic ports began traversing the region in the sixteenth century.[3] The Spanish began hunting seals in the Plate River region as early as the sixteenth century, but North Atlantic hunters did not begin to operate in earnest along the Patagonia littoral until the late eighteenth century.[4] Between 1780 and 1830, fur seal hunting in the southern hemisphere intensified in conjunction with the rise of trade networks between New England and China.[5] One historian estimated that between 1792 and 1812 as many as 2.5 million sealskins entered Canton, where they were used for producing clothing and felt.[6] Most of these sealskins came from animals that had lived in the southern hemisphere.[7] Fur seal hunting off the coasts of Argentina and Chile declined in the mid-nineteenth century before resuming on a smaller scale in the late nineteenth century and continuing through the first third of the twentieth century.

The long history of fur seal hunting notwithstanding, the historiography on Patagonia tends to focus on human societies and economies based on hunting or herding terrestrial animals, including native guanacos, rheas, and pumas, in addition to introduced domesticates like sheep and horses.[8] Moreover, English-language scholarship on fur seal hunting has focused, with a few notable exceptions, on the seals, hunters, and markets of the northern hemisphere.[9] To be sure, the scale of late nineteenth-century seal hunting that took place on the Pribilof Islands in the Bering Sea dwarfed the contemporaneous activities of hunters in southern oceans. Nevertheless, the fur seal trade in southern Patagonia played a key role in transforming Punta Arenas from a penal colony into a dynamic hub of international trade that was home to a small but powerful group of economic elites (fig. 8.1).[10] The trade also brought the coastal-dwelling indigenous Yámana and Kawéskar into frequent contact with nonindigenous hunting parties. These encounters, characterized by a mix of curiosity, fear, exchange, and violence, would coincide with a decline in Yámana and Kawéskar population levels

8.1 Fur traders aboard a steamship bound for Punta Arenas, Chile. SOURCE: THEODORE OHLSEN, *DÜRCH SUD-AMERIKA* (HAMBURG: LEIPZIG: L. BOCK AND SOHN, 1894). COURTESY OF THE LIBRARY OF THE ARNOLD ARBORETUM, HARVARD UNIVERSITY, CAMBRIDGE, MASSACHUSETTS.

over the course of the nineteenth century, from a few thousand to a few hundred.

Writing a history of fur seals and hunters presents unique methodological challenges. Historical sources on fur seal hunting in the southern hemisphere tend to be fragmentary and scattered. While there are thousands of whaling ship logs archived in New England repositories, my research to date has turned up only a small number of logs for vessels whose stated objective was fur seal hunting. Unlike domesticated animals such as cattle, horses, or dogs, marine mammals are seldom the subject of literature, legal disputes, or probate records. Of course, this is true of most noncaptive animal populations, but marine mammals are particularly hard to "track" because they spend most of their time in underwater environments. Moreover, whales, sea lions, and seals tend to have very large ranges that rarely correspond to the regional or national boundaries that frame most historical research. Centering fur seals thus presents a daunting set of challenges for landlubber historians whose units of analysis and lived experiences tend to be literally grounded.[11]

Methodological challenges may only be the tip of the iceberg. Dorothee Brantz has recently argued that the problem with writing animals into history may be due less to a dearth of sources than to a "question of agency."[12] If agency implies an ability to transform social relations to some degree, Brantz reasons, incorporating animals into historical narratives requires a "radical rethinking of the project of history" that rejects the "epistemological separation of nature and culture."[13] Brantz largely echoes Erica Fudge's call to challenge humanist traditions that situate people "in splendid isolation" from animals and other nonhuman organisms.[14] However, when she calls for embedding people in a *natural order*, Fudge perhaps inadvertently reinforces notions of a stable Nature, even as she works to destabilize its components. I seek to avoid this potential pitfall by focusing on the *relationships* between fur seals and hunters in specific times and places, rather than concentrating on the singular discursive subject "animals."[15] I am hesitant to reject liberal notions of "human rights" or to affirm "animal rights" where animals are implicitly understood to be individuals. The historical and ecological meanings of animals are always contextual—conditioned by when and where they live.

My approach is best described as ecological history whose objective is to highlight the historicity of the relationships between groups of people and groups of animals. After providing a brief overview of South American fur seals' ecology and lifecycles informed by contemporary marine biology, I trace fur seal–human relationships in southern Patagonia over the course of the nineteenth and early twentieth centuries. In so doing, I pay close attention to hunting processes because I want to consider not only how the practices of indigenous and North Atlantic hunters diverged in critical ways, but also the extent to which documents (e.g., ships' logs) produced by hunters can be read "against the current" to reveal clues about the relationship's ecological dynamism. I also trace the creation of conservationist policies in twentieth-century Argentina and Chile before offering some reflections on integrating seals into historical narratives.

The "Double Lives" of Fur Seals

South American fur seals belong to a group of nineteen species of marine mammals that are endemic to the coastal waters or river systems of South America. In the southernmost portion of the continent, a current flowing from Antarctica divides into the Humboldt Current in the Pacific, reaching nearly to the equator, and the Malvinas Current in the Atlantic. These currents help to define the world of South American fur seals, whose range extends along the long coastlines of Argentina and Chile.[16] However, the idea of marine mammal populations existing in defined water masses (i.e., currents) is too simplistic. A handful of seal species that breed on sub-Antarctic islands or on Antarctic ice occasionally move to the coastal waters of South America and have even established colonies in Patagonia, blurring the boundaries of ecosystems and species. Finally, species themselves are largely the products of human scientists and subject to change on account of shifting practices among both scientific communities and the organisms that they study.[17]

Taxonomists classify South American fur seals and sea lions as otariids or "eared seals."[18] The popular Spanish names *lobo de dos pelos* (fur seals) and *lobo de un pelo* (sea lion) reflect the presence or absence of the characteristic that commercial hunters most valued: fur seals possess a dense, soft undercoat in addition to a coarse outer coat. In contrast, sea lions possess a single, coarse layer of fur.[19] All otariids lead "double lives": they spend extended periods of time foraging in open waters, punctuated by stints on land to give birth, mate, socialize, shed fur, and rest.[20] From an evolutionary perspective, otariid anatomy reflects their amphibious character: their ears, eyes, and limbs all function in both marine and terrestrial environments. For example, otariids can walk and run on all four limbs and scale steep, slippery rocks. They are also agile swimmers capable of leaping out of the water to breathe when swimming fast.

South American fur seals give birth and mate on land between October and February. During the birthing and mating season, male fur seals establish territories and seek to mate with females—who tend to greatly outnumber males—in that territory. Female fur seals choose their birthing site and subsequently mate with a nearby male. Unlike most land-dwelling placental mammals, marine mammals produce a single offspring; twins are extremely rare.[21] After giving birth, female otariids continue to feed while they are lactating, moving back and forth between sea and rookery for about four months, at which point most young are weaned. Female fur seals live twenty to thirty years and begin reproducing around age three or four. Male seals have shorter life spans than females and are reproductively active for as little as two years, due to social dynamics among males.

Laboratory tests indicate that otariids can learn complicated tasks and possess excellent memories, cognitive abilities that presumably are helpful for living in marine environments where sources of food are unevenly distributed. Otariids vocalize both on land and underwater, sounds that appear to serve primarily as a form of intragroup communication.[22] Although these findings, like all research, are provisional, they compel historians and other humanists to confront the possibility that at least some animals have "experiences" and may be able to communicate them via a vocalized language (leaving aside other sensory exchanges).

This biological understanding of fur seals reveals four important characteristics that conditioned the modern history of fur seal hunting in Patagonia: the annual formation of large, land-based colonies for birthing and mating enabled mass hunting during the late spring and early summer; very high ratios of females to males (10:1) on rookeries may have diminished hunters' concerns about killing the former; and fur seals' tendency to produce a single offspring per year limited the rate at which their populations could reproduce themselves, particularly when subject to mass hunting pressures. Finally, fur seal biogeography also played a crucial role: the rocky outcroppings at the very edge of land and sea on which fur seals established temporary colonies posed enormous obstacles for human hunters and may possibly explain why fur seals have persisted in far larger numbers than many terrestrial mammals. These characteristics greatly influenced, but did not wholly determine, fur seals' changing relationships with humans. Indeed, it is the historicity of social and ecological phenomena with which I grapple here.

Seals and Indigenous Hunters

Nineteenth-century travel accounts of Tierra del Fuego often refer to contact with "canoe people"—indigenous Kawéskar and Yámana who navigated the region's many channels in canoes made from rough-hewn wood planks lashed together with a variety of materials, including whale whiskers, strips of sealskin, and plant fibers. The amount of time that the Yámana and Kawéskar spent in their canoes reflected the importance of marine ecosystems to their daily lives. They fished and foraged along the coastlines, where patches of giant kelp (*Macrocystis pyrifera*) weakened ocean tides and provided habitats for fish, crustaceans, and other sources of food. The fauna found in kelp beds, or *cachiyuyos*, attracted predators, including seals and otters. Yámana hunters' flat-bottomed canoes reportedly glided with ease over the dense entanglements of kelp, guided by women who did all of the rowing and much of the fishing (which at times involved diving into the cold seawater to haul out their catch).[23] In addition to fish and marine mammals, Yámana consumed large amounts of shellfish along with smaller quantities of otters, birds, eggs, mushrooms, and guanacos. Although there is no evidence that they hunted whales, Yámana organized impromptu celebrations that could last for several days on discovery of a beached whale.

The food resources available to the Yámana were thus not as impoverished as many nineteenth-century travel accounts—including that of Charles Darwin—might lead one to believe. In fact, some scholars argue that marine resources enabled the population density of Yámana settlements along the Beagle Channel to exceed that of the terrestrial forager-hunter groups in the arid steppe regions of Patagonia.[24] Paradoxically, twentieth-century studies of Tierra del Fuego's inner canals and channels indicate they are not highly productive aquatic environments. What, then, enabled the Yámana population to become unusually dense for a foraging society? Adrian Schiavini has argued that the key to explaining the paradox lies in the behavioral ecology of otariids whose movements during the winter months effectively linked up the interior waters of Tierra del Fuego to the much more biologically productive ecosystems that surrounded the archipelago. In other words, the limited resources generally found in the Beagle Channel were offset by the "import" of energy in the form of seals and sea lions whose seasonal movements connected the Beagle Channel—the center of Yámana settlement—with more productive ocean waters.[25] Schiavini's argument is indirectly supported by archaeological studies that indicate that fur seals and sea lions constituted the "most important nutritional resource" for Yámana settlements prior to the nineteenth century.[26]

Historical sources are far less conclusive than is the archaeological record with regard to the importance of seal flesh in Yámana foodways. Some late nineteenth-century observers reported seeing "Fuegians" hunting seals.[27] For example, in 1870 one missionary stated, "At present, the only way to obtain food is to spend [their] days rowing their canoes in search of fish, seals, and dolphins."[28] As this description suggests, the Yámana engaged in pelagic hunting, pursuing individual seals in the water by using long spears outfitted with a single large bone barb.[29] There are no descriptions of Yámana hunting seals or sea lions on land during breeding season. However, many travel accounts published between 1820 and 1890 describe the Yámana eating mussels and other kinds of shellfish. Rich in proteins, shellfish contain very little fat and few carbohydrates, making it virtually impossible to base a diet primarily on their consumption. In contrast, fur seals provided far more calories from fats than would have been obtainable from other sources including terrestrial mammals such as guanacos. Luís Orquera and Ernesto Piana contend that readily accessible supplies of shellfish complemented other foods and provided a safety valve during times of scarcity, but they acknowledge that some key historical sources do not entirely support their argument.[30]

The relationship between nineteenth-century Yámana and the marine animals that they reportedly called *töpöra* therefore remains rather fuzzy.[31] What is not disputed is that Yámana did not prey on fur seal breeding colonies. Their pelagic hunting practices combined with low levels of consumption and accumulation greatly reduced the likelihood that Yámana seal hunters produced long-lasting, large-scale ecological effects on populations of South American fur seals. However, this does not imply that the relationship between Yámana hunters and seals was one of static equilibrium; the archaeological evidence does not capture annual fluctuations in seal availability. The arrival of North Atlantic sealers added a critical new dynamic to the relationship between human hunters and fur seals in the region by introducing new methods and motives for hunting marine mammals that would quickly undermine the capacity of fur seal populations to reproduce themselves and draw the Yámana into new worlds of biological, cultural, and economic exchange.

Seals and Sea Hunters

If their fatty tissues were primarily what made seals desirable to Yámana hunters, their dense, soft fur was responsible for turning seals into valuable commodities in parts of Asia and the North Atlantic. The first wave of commercial sealing in the southern hemisphere began in the late eighteenth century when North Atlantic "sea hunters" began venturing into southern waters in pursuit of marine mammals including whales, sea elephants (*Mirounga leonine*), and sea lions (*Otaria flavescens*) in addition to fur seals.[32] One of the first English-language writers to call attention to the marine mammals found along South America's littoral zones was Alexander Dalrymple who reported in the late eighteenth century that fur seals existed in "such numbers that they killed eight or nine hundred in a day with bludgeons on one small Islot" near the Falklands. George Foster, a naturalist who accompanied James Cook on his voyage toward the South Pole, also made reference to hunting marine mammals along the Patagonian littoral in 1775.[33]

Subsequently, investors in England and the United States organized whaling and sealing expeditions to the Falkland/Malvinas Islands and Isla de los Estados (an island separated from Tierra del Fuego by the Straits of Lemaire). Fearful of incursions into its nominal imperial territory, the Spanish Crown created the Real Compañía Marítima de Pesca. The enterprise, based in Puerto Deseado, hunted tens of thousands of sea lions along the Patagonia littoral during the 1790s.[34] During this first wave of seal hunting in the southern hemisphere, the coastline of southern Patagonia— with the important exception of Isla de los Estados—was not a major hunting ground. However, many sealing expeditions passed through the region on their way to hunting territories in the eastern Pacific, raising the likelihood that some vessels paused to hunt marine mammals and trade with indigenous people (fig. 8.2).[35]

By 1830, few fur seals remained on the Juan Fernández, Malvinas, or Isla de los Estados islands. A new generation of commercial fur seal hunters,

8.2 Indigenous hunters seek to trade furs with passengers on approaching steamship. SOURCE: THEODORE OHLSEN.
DÜRCH SUD-AMERIKA (HAMBURG: LEIPZIG: L. BOCK AND SOHN, 1894). COURTESY OF THE LIBRARY OF THE ARNOLD ARBORETUM,
HARVARD UNIVERSITY, CAMBRIDGE, MASSACHUSETTS.

including the Patagonian native Luís Piedra Buena and José Nogueira, emerged in the 1860s and concentrated its hunting activities in the waters of southern Patagonia. Government officials registered the exportation of more than 25,000 sealskins between 1877 and 1880 from Punta Arenas, a total that did not include those taken by foreign sealers.[36] At its peak, the Punta Arenas fur seal fleet consisted of 28 boats and 600 workers.[37] A London-based merchant testified that the number of sealskins harvested in the general vicinity of Cape Horn between 1876 and 1892 surpassed 113,000, a volume consistent with the Chilean historian Mateo Martinic's finding that Punta Arenas-based crews killed an average of 6,000 seals per year during the 1880s.[38] Although historical sources on the number of fur seals hunted are not entirely consistent, nor do they tend to be precise about the location of hunting territories, one can safely assume that, on average, late nineteenth-century sealers killed several thousand fur seals per year in southern Patagonia.[39]

Sealing remained a lucrative livelihood in the late nineteenth century, at least for the owners and captains of the vessels. In 1879 Carlos Wood, governor of Punta Arenas, reported that the activities of fewer than 200 sealers had generated more than 300,000 pesos of revenue in five or six months.[40] Three years later, two New England-based vessels, the *Thomas*

Hunt and the *Express*, returned from the "coast of Patagonia" with a total of 4,300 sealskins. The *Whalemen's Shipping List and Merchant's Transcript* reported, "The owners will receive a handsome dividend and probably the vessels will be refitted and sent out again. Stonington has been fortunate in the seal fishery for several years."[41] The money generated by sealing expeditions was divided very unequally. An account of an Antarctic fur-seal voyage indicates that the vessel's owners retained approximately two-thirds of the $30,000 netted by the sale of 1,400 fur seals in London. The captain earned $2,000 (6.7 percent of the net proceeds); the first mate earned $1,500 (5 percent); and six other individuals earned between $218 and $874. The twenty-one-member crew each earned $165.40, from which the owners deducted advances for food, clothing, insurance, and other expenses.[42]

The contracts that José Nogueira offered to potential crewmembers in Punta Arenas between the years 1873 and 1883 suggest that prevailing wages declined during that period. For example, in 1873 Nogueira enlisted a ten-person crew that divided two-thirds of the net proceeds "without distinction." However, a contract from the following year stipulated that the captain, first mate, and second mate would receive a significantly higher percentage of the net proceeds than the crew. An 1883 contract provided for two-thirds of net earnings to pass into the hands of the owners and captain of the vessel; the remaining third was to be divided equally among the crew.[43] Moreover, Nogueira routinely provided advances to cover expenses incurred during the two or more months spent sealing, during which time he profited from the sale of provisions to his crews. Ship owners and sea captains thus enjoyed a disproportionate share of the capital accumulated from sealing.

In addition to shedding light on social relations among people, the distribution of profits generated by sealing expeditions may help us to understand the persistence of hunting operations in the face of declining seal populations in the second half of the nineteenth century. If late nineteenth-century investors were able to lower their labor costs, then they might have been able to hunt fur seals under conditions of relative scarcity, delaying the onset of "economic extinction," while placing additional pressures on already dwindling seal populations.[44] Furthermore, since the earnings of everyone on the vessel, from the captain to the greenest crewmember, derived from a percentage of the net proceeds, there was a strong incentive to kill as many of the most valuable fur seals as possible. More research is needed to be able to determine the strength of the historical relationship between changing labor costs, the profitability of sealing voyages, and rates of fur seal hunting, but on a conceptual level, historians looking to center animals must keep in mind that relationships among people can exert important if indirect force on animals.[45]

Labor history intersects with the history of fur seals on a far more intimate level: in order to kill fur seals, hunters first had to find them and negotiate the edge environments in which seals congregated on a seasonal basis. Rocky shorelines, high winds, and powerful waves made seal hunting a dangerous occupation. Sealers tested their patience and risked their lives to reach seal colonies. Both published sources and logbooks indicate that hunting gangs often waited days or weeks before being able to land safely. The journal of William Henry Appleman, a North American sealer who worked in Tierra del Fuego in 1875, offers an example of the trials that faced sealers when they tried to approach their prey. In one instance, Appleman spent two weeks trying to land on a rocky outcropping where a thousand seals lay just out of reach. On another occasion he witnessed the death of several companions when their launches smashed into a rocky shoreline.[46] Trade journals in New England also reported on the death of fur seal hunters. For example, the *Whalemen's Shipping List and Merchant's Transcript* reported in 1876 that four seamen were "washed overboard from the rocks and lost while engaged in sealing" in Tierra del Fuego.[47]

Crews that successfully landed on hunting grounds faced the prospect of spending weeks or even months separated from their crewmates.[48] Hunting gangs established their base camps at a considerable distance from rookeries in order to avoid being detected by their prey, whose sense of smell was believed to be acute. When approaching a fur seal colony, hunters tried to position themselves between the animals and the sea in order to prevent the seals from rushing into the water. Then, as sea captain Charles Scammon described, the "knock down" began: "In former times, when fur-seals abounded, they were captured in large numbers by the ordinary seal-club in the hands of the sealer who would slay the animals right and left by one or two blows upon the head. A large party would cautiously land leeward of the rookery if possible; then, when in readiness, at a given signal, all hands would approach them shouting and using their clubs to the best advantage in the conflict. Many hundreds were frequently taken in one of these 'knock downs' as they were called."[49] This description of sealing is generally consistent with those found in early nineteenth-century sources. In some cases, sealers would deliver a stunning blow and then plunge a knife into their prey so that it would bleed to death. When the killing was over, workers immediately began to beam the skins and preserve them with coarse salt.

By the 1870s, the method of sealing had changed considerably according

8.3 Hunters approach fur seals. Note the depiction of firearms. SOURCE: THEODORE OHLSEN, *DÜRCH SUD-AMERIKA* (HAMBURG: LEIPZIG: L. BOCK AND SOHN, 1894). COURTESY OF THE LIBRARY OF THE ARNOLD ARBORETUM HARVARD UNIVERSITY, CAMBRIDGE, MASSACHUSETTS.

to Scammon: "At the present day the animals have become so scarce and shy at the once favorite resorts that the hunter often has to watch and wait for them singly, and it is frequently difficult to approach near enough to dispatch them with the club, so that a rifle must be used. Where new rookeries are found, the seals are quite tame and are easily approached and clubbed. The rifle is never used unless absolutely necessary, for it makes holes in the skin that greatly reduce their value."[50] A source from 1909 described a similar transformation in seal hunting: "Formerly the seals were captured by getting between them and the shore and knocking them down with clubs. Now, however, the constant warfare waged against them has made them wild and they are shot at long range, it being impossible to approach near enough to insure their capture in any other way."[51] When the German artist Theodor Ohlsen visited Punta Arenas on his voyage to Valparaíso in the late nineteenth century, he drew a picture depicting *loberos* (sealers) crouched upon a rocky ledge, with rifles in hand, preparing to attack a small group of hauled-out seals (fig. 8.3).[52]

These historical narratives are noteworthy for at least two reasons: they lend a certain agency to fur seals insofar as the animals are described as reacting to intense hunting by becoming "shy" or "wild," a perceived change that compelled hunters to resort to firearms that reduced the value of skin. Such narratives can be read as self-serving or perhaps self-deluding insofar as the hunters' seem to deny their role in depleting fur seal populations. However, if the frequently invoked image of seals seeking refuge from hunters cannot be taken at face value, it suggests that hunters' intimate contact with furs seals engendered respect if not empathy for their preys' capacity to adapt to a changing environment. Secondly, hunters' acknowledgment that shooting fur seals reduced their market value alludes to the "social lives" of the commodity that resulted from the hunt: North Atlantic people generally did not prize seals for their flesh, but rather for their fur which was transformed into garments. North Atlantic markets paid different prices depending on the size, age, color, origin, and condition of the fur. Historicizing the fashionableness of seal fur would shed light not only on the economic forces driving the trade but also would help to reveal the symbolic meanings of animals in the rapidly transforming agro-industrial societies of the North Atlantic.

A primer on seal hunting written by George F. Athearn, a late nineteenth-century sea captain based in New England, further illustrates the extent to which hunters viewed their prey as social beings. Remarking on "the habits" of Cape Horn fur seals, Athearn wrote:

> About the 25th of November the young clapmatches [female seals] of four or five years, that are to have their first pups, come on shore . . . but the main herd of the old clapmatches do not begin to haul in any great number before the 5th of December and from that to the 25th of December they come in fast. I don't think it is a good plan to commence killing until they get well into pupping. Don't kill any of the old wigs [male seals] until you have worked off most of the clapmatches, or near the end of the pupping season, as they hold the other seals and will not let them go off the rocks if they can prevent it. If you should want all the seals that are on the rocks to make up your cargo, you can all through the pupping season be working off the young wigs, which are always hauled in small rookeries near the pupping seal, driven there by the old wigs.[53]

In this account Athearn recommends exploiting the fur seals' social dynamics during the birthing and mating season in order to maximize the number of skins to be obtained. Old male seals possessed little economic value per se, but Athearn considered them important because they "held" the seals on the rookeries. His recommendations, which included killing pups and female seals as well as young males, reveal no interest in conserving fur seal populations.

The logbooks that I have reviewed seldom provide enough details to judge how closely Athearn's recommendations corresponded to actual practice. Nevertheless, they convey a sense of the challenges of hunting fur seals and the hunters' undeniable persistence. For example, the schooner *Thomas Hunt* departed Stonington, Connecticut, on 23 July 1874, bound for

Cape Horn. The vessel reached Tierra del Fuego in mid-October. The crew found what they were looking for on some small islets known as the Evouts: "When a short distance from the rock saw about 150 seal. . . . Sent [crew-member] on board for more guns and ammunition. Captain returned in the boat and together we pulled and on getting in close found the rock covered with seal from the base to the top and the outer rocks covered. . . . The ground swell was so heavy that we could not land at this gulch or at hardly any point on the rock. . . . Fearing that if we attempted to land at another point than at the gulch we should loose [sic] the most of them, we returned on board."[54]

Ten days passed before the crew of the *Thomas Hunt*, now accompanied by the crew of the *Charles Shearer*, was able to land and begin shooting fur seals. By 26 October, the two crews had beamed and salted 461 fur sealskins. Three days later, the *Thomas Hunt* set a course for the Diego Ramírez Islands. The log entry for 31 October states that a "number of wigs were on the rocks where we landed but too near the water to club. Climbed up and took seven seal on the top of the rock passed over and saw plenty of seal but mostly on the lower rock on which was a heavy sea."[55] The next day, the crew found "plenty of seal on weather side of the island but mostly on lower rocks. Knocked down and shot 124." For the remainder of November, the sealers, struggling in "rough conditions," secured only a small number of skins.

In December the *Thomas Hunt* headed for the South Shetland Islands, but the crew found far more ice and snow than seals in the sub-Antarctic region. They returned to the Diego Ramírez Islands after burying a crew-member who had fractured his skull when he fell while repairing a topsail. For the remainder of December and January, the sealers visited dozens of islands, rocks, and canals searching for seals. During this period, the crew enlisted the assistance of an "Indian pilot" and traded tobacco for sealskins with local inhabitants.[56] The *Thomas Hunt* and the *Charles Shearer* began the return voyage to New England in early March with a combined cargo of 3,149 skins—the output of nearly five months of active hunting.

The profits made from the voyage apparently were sufficient to prompt the owners of the *Thomas Hunt* to finance another voyage in 1878. The log entries provide few details, but repeated references to both "beaming skins" and "boiling blubber" indicate that the crew hunted fur seals and sea lions.[57] From September to November, the crew encountered only modest num-bers of seals. Then, on 19 December, at a place identified as "North Rock," the sealers obtained 5,000 skins. The brief log entry for that day does not remark on this extraordinary quantity, but on the following two days, "all hands" participated in dressing the skins for storage. Hunting resumed on 23 December in a different location but yielded only two skins. Returning to North Rock on 29 December, the sealing crew killed more than 1,000 seals. In early February the *Thomas Hunt* began its return voyage to Stonington laden with approximately 7,900 sealskins.[58]

The experiences of the *Thomas Hunt* in the 1870s point to the importance of timing: seal hunters knew from experience when and where fur seals generally hauled out for breeding, but variations in the precise timing and location of fur seal colonies in any given year created considerable uncertainty. Fur seal hunters at times tried to overcome this problem by depositing crewmembers on islands where they would remain for weeks or even months in the hope of hunting arriving fur seals. For example, on 3 November 1875 the schooner *Golden West* landed a party of eight men on the Diego Ramírez after taking a mere 23 seals. The vessel did not return for more than four months, when it removed the landing gang and 451 skins.[59]

This sample of fur sealing vessels' logs, combined with other historical sources, provide some clues about the relationship between fur seals and North Atlantic hunters. Although images of men clubbing seal pups would become a late twentieth-century symbol of animal cruelty, fur seal hunting in southern Patagonia was marked by violence and death for seals and men alike. Crewmembers of sealing vessels experienced hunger, scurvy, and exposure to severe weather. They endured prolonged periods of social isolation and extremely hazardous working conditions. The distribution of profits generated from the exchange of fur sealskins appears to have become increasingly skewed over the course of the nineteenth century in favor of those who ventured capital but not their bodily labor to hunt seals. The work was highly seasonal and the results uncertain; seal hunters and their dependents probably lived on credit and faced extended periods of time when they were forced to find other forms of livelihood. Moreover, the association of clubs with "primitive" societies belies the historical reality that modern markets and elite fashion aesthetics compelled hunters to make clubs the weapon of choice even when it put the hunters themselves at greater risk of bodily harm.

Assuming that the above description of the organization of labor and work processes associated with fur seal hunting is reasonably accurate, there can be little surprise that seal hunters' descriptions of Patagonian animals—be they fur seals, penguins, or guanacos—are unsentimental. On the rare occasion when a sealer expressed a sentiment toward an animal, the object of affection was invariably a mascot. For example, describing a period of solitude aboard a vessel anchored in the Falkland Islands in the 1790s, Eben Townsend wrote, "I was then captain, mate, and all hands. As I had enough to do I was not so lonesome as you may imagine. I was left with a dog, a cat, and five kittens, but the dog killed the cat, and the kittens being but a day old, died also. I never felt the loss of a cat so much."[60] Moreover, in the only examples I have come across in which authors—retrospectively—convey a sentiment about fur seals, they do so by drawing analogies to kittens or canine puppies to express affection for, or remorse about, encountering a young seal "pup" whose mother had been killed.[61]

This fragmentary evidence hints at a wild / domesticated binary that may have been as important in the past as a human-animal dyad. Also, the fact that English-speaking hunters referred to juvenile seals as "pups" strongly suggests that scholars should approach hunters not as "inhumane" or "cold-blooded" killers, but rather as people inhabiting (and sometimes dying) in the contradictions of a historical moment in which notions of animal welfare emerged side-by-side with expanding global trade networks and empires that transformed ecosystems on a vast scale and in so doing threatened a wide range of animal species.

Fur Seals and Conservationists

The contradictory notions toward animals in general and fur seals in particular are also apparent in late nineteenth-century efforts to conserve fur seal populations. In 1880 the government of Argentina gave official warning that any vessel engaged in hunting whales, seals, sea lions, or waterfowl along the coast of Patagonia or adjacent islands would be seized. One year later, the Falkland Islands government issued an ordinance prohibiting fur seal hunting during the breeding season.[62] In 1882 a visitor to Tierra del Fuego stated that fur seals would "soon disappear" because the hunters "kill all that they can get their hands on paying no attention to the age of the seals.... With proper regulation sealing could be a source of national wealth but at present it is truly destructive."[63]

The following year, the Chilean minister of foreign relations named a three-person commission to draft a report on regulating the sealing industry. The commission warned that if the government continued to allow hunters to operate without regulations, the results would be a loss of revenue for the state and the disappearance of a livelihood for the "self-sacrificing" settlers in those "distant lands."[64] The report also called attention to the illegal activities of foreign seal hunters, citing the example of the U.S. schooner

Florence, which reportedly took 13,000 sealskins from Chilean waters. The Chilean committee's recommendations included prohibiting the use of fire-arms, banning hunting during mating season, and limiting the total annual kill to 20,000 seals. The committee also suggested that the best way to enforce these regulations would be by granting concessions to hunting grounds on the assumption that the concessionaires would have a self-interest in preventing unauthorized sealing. The set of recommendations was not adopted; three years later, Oscar Viel, one of the report's authors and a former governor of Magallanes, again brought the threat of declining seal populations to the attention of the minister of colonization.[65]

In 1891, following a year of rising tensions between President Balmaceda and the Chilean Congress, a civil war erupted in Chile. When the fighting terminated, Balmaceda was dead and a new government formed in which the executive branch ceded considerable power to the congress.[66] Signifi-cantly, concerns about fur seal hunting persisted throughout this period of political upheaval. Balmaceda's successor, President Jorge Montt, decreed a one-year ban on fur seal hunting in Chilean territory in 1892. The following year, the Chilean Congress extended the ban for four years. The brief debate held in the Cámara de Diputados focused primarily on the eco-nomic effects of the ban and the challenges of ensuring that foreign sealers complied with it. The terms of the debate were almost entirely economic or "utilitarian"; no one expressed concern about the possibility of biological extinction. In fact, one vocal opponent of the measure opined that since fur seals had little potential for domestication, they did not merit effort to conserve them.[67]

Once in place, the ban came under criticism for "favoring foreigners who, due to the lack of vigilance along our[Chile's] coasts, have been able to dedicate themselves to the industry without any risk of any kind."[68] The government responded by awarding Chilean nationals exclusive rights to hunt fur seals in designated regions. Between 1905 and 1908, hunters took some 50,000 fur seals from the "Cape Horn" region, most of which were shipped from Punta Arenas to London.[69] Around this time, some sealers turned to pelagic hunting. For example, between 1902 and 1911, at least twenty-one Canadian schooners voyaged to the southern hemisphere, where they hunted seals in open water and shipped the skins via Port Stanley (Falk-lands) or Punta Arenas.[70] Sporadic sealing voyages continued as late as 1915, but the number of seals killed was small.[71]

Between 1921 and 1929, the governments of the Falklands, Argentina, and Chile legislated new protections for fur seals including the establishment of national reserves and complete bans on hunting and commerce.[72] The efficacy of these measures is not clear. In 1937 the Argentine government, acknowledging that earlier efforts to establish reserves had failed to result in a significant increase in the population of fur seals, joined its neighbor Chile in banning the hunting, commerce, and shipping of fur seals in its national territory. Chilean government statistics indicate that only two fur sealskins were exported from Punta Arenas in 1939.[73]

The decline in the Patagonian fur seal trade did not mean the end of hunting marine mammals in the region. Both nutrias (*Myocastor coypus*) and sea lions (*Otaria flavescens*) were hunted intensively during the mid-twentieth century. Between 1920 and 1954, traders exported approximately 325,000 sea lion pelts from Chile.[74] In Argentine Tierra del Fuego, hunters killed more than 128,000 sea lions between 1937 and 1947.[75] The incomplete records of Chilean naval authorities stationed on Navarino Island indicate that local people—including individuals identified as Yámana—continued to hunt nutrias and "lobos" (probably sea lions) through the 1940s.[76]

Edgewise: Fur Seals, Hunters, and Historians

Today, populations of South American fur seals can be found along South American coastlines from Uruguay to Peru. Scientists estimate the total population to be between 200,000 and 250,000 and increasing. In fact, the International Union for the Conservation of Nature (IUCN) considers South American fur seals to be a species of "least concern" for extinction.[77] In Patagonia, fur seals, along with sea lions, penguins, guanacos, and other native fauna have become signs of wild nature used to lure tourists to the region. How and why fur seal populations have rebounded, and whether they will continue to increase, are questions that remain unanswered by this study, whose exploratory nature has served primarily to sharpen questions central to the relationship between fur seals and hunters in the nineteenth and twentieth centuries.

One set of questions revolves around the geographies of fur seal hunting. The fur seals' breeding grounds on the literal edges of land and sea appear to have enabled both mass hunting activities and the survival of seal populations in the region. This paradox results from fur seals' practice of giving birth on land. Historically, North Atlantic-based commercial hunters exploited the fur seals' seasonal "hauling out" to kill them en masse. At the same time, the rocky, wave-swept places where seals established their temporary colonies created formidable barriers to hunters and were inhospitable to long-term human settlement. Human activities therefore have not destroyed the critical habitats of fur seals to nearly the extent that they have reduced habitat for large terrestrial carnivores. The extent to which conservationist legislation enacted by national or regional governments has protected these habitats remains to be researched, but one suspects that the locations of fur seal breeding grounds have played a significant role.

The importance of geographical contexts in human-animal relationships is further revealed by contrasting the historical meanings of "sea wolves" (*lobos marinos*) with those associated with terrestrial wolves, or in the case of southern Patagonia, pumas (*leones*). In many pastoral societies, wolves and pumas conjure images of cunning and dangerous predators.[78] In contrast, carnivorous marine mammals have seldom been perceived as a serious threat to human livelihoods. However, this may be changing in light of the rapid expansion of fish farming in southern Chile. Sea pens filled with salmon or other kinds of fish sometimes lure hungry sea lions, prompting aquaculture firms to erect underwater barriers. Reports have also circulated about aquaculture employees killing sea lions.[79] If aquaculture continues to expand in southern Patagonia, there is a good possibility that the meanings of "sea wolves" will become more akin to those of terrestrial wolves in ranching regions.

The edge environment is also an important consideration for thinking about the historical meanings of killing fur seals. I suspect that romantic notions of "the hunt" can influence in unexpected ways how historians view fur seal hunters: fur seals do not possess reputations for ferocity; in contrast to a lion or a bear, fur seals appear rather defenseless. However, the places that they inhabited created treacherous and bleak conditions that appear to have claimed the lives of a fair number of hunters. Moreover, the methods and scale of killing reflected less the bloodthirsty desires of the hunters than a changing global economy. Fur seal hunting coincided with the rise of industrial, mass-consumer societies in the North Atlantic, whose economies of scale and aesthetic sensibilities influenced the practices of fur seal hunters, including choice of weapons. When viewed as part of a long-distance commodity flow, both fur seals and their human hunters are seen as acting under constraints, suggesting that it might make more sense to talk about relative power relations than about agency.

Far-flung places of consumption are important to study not only for the ways that they influenced hunting techniques; recovering the ways in which people used and displayed sealskins can reveal clues about the symbolic meanings of animals as well.[80] How and why sealskins became objects of desire in Chinese and North Atlantic societies remains to be explained, as does the twentieth-century shift that led many consumers to reject garments fashioned from wild animals as inhumane. In order to answer these questions, we need to understand more about how different people classified fur seals over time. The historian Harriet Ritvo has documented the confusion that seals, or "talking fish," presented to English naturalists in the eighteenth and nineteenth centuries.[81] How did this confusion get resolved? How did hunters and traders in Punta Arenas, Santiago, London, or Canton fit fur seals into their mental frameworks? Was there a connection between the designation of baby seals as "pups," movements to prevent cruelty to domesticated animals, and a turning away from sealskins as fashion? Asking these kinds of questions about specific animals may help scholars to move beyond a human/animal binary.

Placing fur seals at the center of southern Patagonia's history also brings into view indigenous hunters, including the Yámana women and men who used canoes and spears to hunt seals in open water. Archaeologists, whose excavations in Yámana settlement sites indicate a sharp decrease in seal consumption beginning in the nineteenth century and a concomitant increase in shellfish consumption, have argued that the arrival of North Atlantic hunters disrupted seasonal patterns of seal movements and in the process created a fat deficiency in Yámana diets. This hypothesis appears to rest on an unstated assumption that Yámana lacked a collective capacity to adapt to socioenvironmental changes. However, historical sources suggest that some Yámana responded to the arrival of North Atlantic interlopers by engaging them in trade or serving as their pilots, raising the possibility that at least some Yámana improvised alternative ways to get access to food.[82] More research is needed to examine to what degree, if any, fur sealing created opportunities for a "middle ground" similar to the dynamics of the North American fur trade described by the historian Richard White.[83]

The shifting relations among fur seals, Yámana, and colonizers is important because even as some national legislators in Chile and Argentina expressed concerns about dwindling fur seal populations, state incentives to colonize southern Patagonia contributed to the destruction of indigenous societies via violence, resettlement, and disease. In fact, the removal of indigenous hunting societies such as the Yámana was often justified in the name of protecting cattle and sheep—the domesticated animals that would form the basis of settler society in the region. Here I suggest that there was a discursive linking of native people and wild animals that stood in opposition to the settlers and domesticated animals that state agents and missionaries understood to be the basis of civilization.

Finally, we remain with the vexing question of whether fur seals played a role in this history beyond that of passive victim. Can historical sources be read "against the current" to reveal the agency of seals? For example, the last line of this chapter's epigraph describes seals "emigrating in search of a safer haven for the future." Other late nineteenth-century sources, when justifying the use of firearms, often refer to "wild" or "shy" seals.[84] W. Nigel Bonner argued that this was little more than denial: "Clearly what had happened in these cases was that the seals, remaining faithful to their now-insecure breeding grounds, returned to be greeted with a sealer's club and death."[85] However, the small number of ships' logs that I reviewed indicates that, by the 1870s, there was little certainty from one season to the next about where large seal colonies would be found, which potentially undermines Bonner's claim that seals remained "faithful" to their breeding grounds. This limited, indirect evidence, combined with the ability of captive seals to learn reactions to human commands, leaves open the possibility that some fur seals responded to intensive hunting by seeking out new breeding grounds.[86]

That said, the dynamism of marine mammals should not be conflated with agency or power. Moreover, the traditional toolkit of historians may not be capable of identifying the historicity of fur seals: did late nineteenth-century seals respond differently to hunters than late eighteenth-century ones? This is an intriguing question whose answer at least for the moment remains beyond reach. Centering fur seals in the history of Patagonia exposes the edges of historical knowledge.

Notes

The epigraph is from Alfredo de Rodt, H. A. Howland, and Oscar Viel, "Informe pasado al Sr. Ministro de Relaciones Exteriores sobre la caza de lobos marinos," 17 May 1883, Chile, Ministerio de Relaciones Exteriores, Archivo General Histórico, Fondo: Histórico, vol. 107.

1. Mateo Martinic, *Nogueira el pionero* (Punta Arenas: Ediciones de la Universidad de Magallanes, 1986), 128–29.
2. For my purposes, "southern Patagonia" includes Tierra del Fuego, Cape Horn, and a number of other islets lying in the national territories of Argentina or Chile.
3. Luis Abel Orquera and Ernesto Luis Piana, *La vida material y social de los Yámana* (Buenos Aires: Editorial de la Universidad de Buenos Aires, 1999); and Adrian Schiavini, "Los lobos marinos como recurso para cazadores-recolectores: El caso de Tierra del Fuego," *Latin American Antiquity* 4 (1993): 346–66, 358.
4. W. Nigel Bonner, *Seals and Man: A Study of Interactions* (Seattle: University of Washington Press, 1982), 59.
5. For estimates of eighteenth-century seal kills, see W. Nigel Bonner and Richard

M. Laws, "Seals and Sealing," in *Antarctic Research*, ed. Sir Raymond Priestly, Raymond J. Adie, and G. de Q. Robin (London: Butterworths, 1964), 163–90; and Briton Cooper Busch, *The War against the Seals* (Kingston: McGill-Queen's University Press, 1985). Also see Richard M. Jones, "Sealing and Stonington: A Short-Lived Bonanza," *Log of Mystic Seaport* 28 (January 1977): 119–26; Edouard Stackpole, *The Sea-Hunters: The New England Whalemen during Two Centuries, 1635–1835* (Philadelphia: J. B. Lippincott, 1953); A. G. E. Jones, "The British Southern Whale and Seal Fisheries," *Great Circle* 3 (April 1981): 20–29; and Charles H. Townsend, "Fur Seals and the Seal Fisheries," *Bulletin of the Bureau of Fisheries* 28 (February 1910): 315–22.

6. James Kirker, *Adventurers to China: Americans in the Southern Oceans* (New York: Oxford University Press, 1970).

7. New England–based seal hunters killed several hundred thousand fur seals (*Arctocephalus philippii*) on the Juan Fernández Islands, situated in the Humboldt Current, 670 kilometers off the coast of central Chile. In one documented example, the crew of the Salem, Massachusetts–based *Concord* stowed more than 20,000 fur sealskins between May and July 1800. Log of *Concord* (Salem, Massachusetts), Phillips Library, Salem, Massachusetts.

8. There are important exceptions; in addition to his biography of Jose Nogueira, see Mateo Martinic, "Actividad lobera y ballenera en litorales y aguas de Magallanes y Antartica, 1868–1916," *Revista de estudios del Pacífico* (1973): 7–26; and Orquera and Piana, *La vida material y social de los Yámana*.

9. Most English-language accounts of sealing along the Patagonian littoral appear as individual chapters in reports focused on seals and seal hunting in the northern hemisphere or the Antarctic region. Several British and North American scholars discuss sealing voyages to the Falkland Islands and the South Shetland Islands in the context of the "discovery" of Antarctica. During the Cold War, the British Antarctic Survey funded studies of seals (primarily *Arctocephalus gazella*) in the South Georgia and South Shetlands Islands. Compare W. Nigel Bonner, *The Fur Seal of South Georgia* (London: British Antarctic Survey, 1968).

10. The Chilean government established Punta Arenas as a penal colony in 1843. In the 1860s the town's status changed to that of a minor port (*puerto menor*).

11. Animals are the subject of a growing number of works by historians and other social scientists, but to date most of this scholarship has focused on Europe, the United States, or the British empire (Africa and South Asia). Excepting salmon in the Pacific Northwest of the United States and whales at an international scale, fish and marine mammals have received little attention. For a brief overview of trends in animal studies, see Harriet Ritvo, "Animal Planet," *Environmental History* 9.2 (April 2004): 204–20. Also see Adrian Franklin, *Animals and Modern Cultures: A Sociology of Human-Animal Relations in Modernity* (London: Sage, 1999). On animals and empire, see Virginia DeJohn Anderson, *Creatures of Empire: How Domestic Animals Transformed Early America* (Oxford: Oxford University Press, 2004); and John M. MacKenzie, *The Empire of Nature* (Manchester:

Manchester University Press, 1988). Some notable works that begin to center animals in Latin American contexts include Robert W. Wilcox, "Zebu's Elbows: Cattle Breeding and the Environment in Central Brazil, 1890–1960," in *Territories, Commodities, and Knowledges: Latin American Environmental Histories in the Nineteenth and Twentieth Centuries*, ed. Christian Brannstrom (London: Institute for Latin American Studies, 2003), 218–46; Gregory Cushman, "'The Most Valuable Birds in the World': International Conservation Science and the Revival of Peru's Guano Industry, 1909–1965," *Environmental History* 10 (July 2005): 477–509; and Regina Horta Duarte, "Pássaros e cientistas no Brasil: Em busca de proteção, 1894–1938," *Latin American Research Review* 41.1 (spring 2006): 3–26.

12. Dorothee Brantz, *Beastly Natures: Animals, Humans, and the Study of History* (Charlottesville: University of Virginia Press, 2010), 2.

13. Brantz, *Beastly Natures*, 3.

14. Erica Fudge, "A Left-Handed Blow: Writing the History of Animals," in *Representing Animals*, ed. Nigel Rothfels (Bloomington: University of Indiana Press, 2002), 3–18.

15. Fudge, "A Left-Handed Blow," 15; Philippe Descola and Gísli Pálsson, "Introduction," in *Nature and Society: Anthropological Perspectives*, ed. Philippe Descola and Gísli Pálsson (London: Routledge, 1996), 1–21.

16. Biological evidence suggests that seals inhabiting the Atlantic coast of South America seldom mate with fur seals on the Pacific side. See J. I. Túnez, Daniela Centrón, H. L. Cappozzo, and M. H. Cassini, "Geographic Distribution and Diversity of Mitochondrial DNA Haplotypes in South American Sea Lions (*Otaria flavescens*) and Fur Seals (*Arctocephalus australis*)," *Mammalian Biology* 72 (2007): 193–203.

17. Enrique A. Crespo, "South American Aquatic Mammals," *Encyclopedia of Marine Mammals*, ed. William F. Perrin, Bernd Würsig, and J. G. M. Thewissen (San Diego: Academic Press, 2002), 1138–43.

18. Otariids possess external ear flaps, hence the term "eared" seals. Along with "earless" seals (phocids) and walruses (odobenids), otariids constitute a group of marine mammals known as pinnipeds. Although pinnipeds are among the more visible sea creatures, scientific understanding of their behaviors and ecological roles is surprisingly limited due to the challenges presented by the large spatial and temporal scales at which pinnipeds interact. For useful introductions to seals, see National Audubon Society, *Guide to Marine Mammals of the World* (New York: Alfred A. Knopf, 2002); William F. Perrin, Bernd Würsig, and J. G. M. Thewissen, eds., *Encyclopedia of Marine Mammals* (New York: Academic Press, 2002); Ronald W. Nowak, *Walker's Marine Mammals of the World* (Baltimore: Johns Hopkins University Press, 2003); and Walter Sielfeld, *Mamíferos marinos de Chile* (Santiago: Ediciones de la Universidad de Chile, 1983).

19. Fur seals and sea otters rely on their fur to trap a layer of air that serves as their primary insulation. The density of hair follicles on fur seals is approximately fifty times greater than that of terrestrial mammals. Sea lions possess a single layer of

coarse hair that international markets did not value; they were hunted primarily for their blubber, which was reduced to oil. Pamela K. Yochem and Brent S. Stewart, "Hair and Fur," in *Encyclopedia of Marine Mammals*, ed. William F. Perrin, Bernd Würsig, and J. G. M. Thewissen (New York: Academic Press, 2002), 548–49.

20. Fur seals eat a wide range of fish, cephalopods (octopus, squid, cuttlefish), and crustaceans (krill, shrimp, rock lobster). Studies of northern hemisphere fur seals suggest that their diets can vary based on age, sex, and the changing availability of prey. Roger L. Gentry, "Eared Seals," in *Encyclopedia of Marine Mammals*, ed. William F. Perrin, Bernd Würsig, and J. G. M. Thewissen (New York: Academic Press, 2002), 348–51.

21. Fur seals and sea lions tend to mate during the four- to eleven-day perinatal nursing period. Physiologically, this is possible due to female otariids' Y-shaped uterus, one part of which holds the developed fetus, while the other part is able to host a new embryo soon after birth. The embryo enters a state of suspended animation for a few months before implanting. From that point, fetal growth takes place for seven to nine months.

22. Gentry, "Eared Seals," 348–51.

23. Orquera and Piana, *La vida material y social de los Yámana*, 133.

24. Schiavini argues that Yámana population densities were thirty to forty times higher than those of other indigenous groups in Patagonia ("Los lobos marinos como recursos para cazadores-recolectores marinos," 358).

25. Schiavini, "Los lobos marinos como recuros para cazadores-recolectores marinos," 358.

26. Luis Abel Orquera, "The Late-Nineteenth-Century Crisis in the Survival of the Magellan-Fuegian Littoral Natives," in *Archaeological and Anthropological Perspectives on the Native Peoples of Pampa, Patagonia, and Tierra del Fuego to the Nineteenth Century*, ed. Claudia Briones and José Luís Lanata (Westport, Conn.: Greenwood, 2002), 146.

27. See J. A. Allen, "Fur-Seal Hunting in the Southern Hemisphere," in *The Fur Seals and Fur-Seal Islands of the North Pacific Ocean*, ed. David Starr Jordan (Washington: Government Printing Office, 1900), 3:307–19; A. Howard Clark, "The Antarctic Fur-Seal and Sea-Elephant Industry," in *The Fur Seals and Fur-Seal Islands of the North Pacific Ocean*, ed. George Brown Goode (Washington: Government Printing Office, 1884–87), 2:400–67, sec. 5; Bartolomé Bossi, *Exploración de la Tierra del Fuego* (Montevideo: La España, 1882); and R. W. Coppinger, *Cruise of the "Alert": Four Years in Patagonian, Polynesian, and Mascarene Waters* (London: W. Swan Sonnenschein, 1883).

28. Orquera and Piana, *La vida material y social de los Yámana*, 235.

29. E. Lucas Bridges, *Uttermost Part of the Earth* (New York: E. P. Dutton, 1950), 97.

30. Orquera and Piana, *La vida material y social de los Yámana*, 162–65.

31. Bridges, *Uttermost Part of the Earth.*

32. Stackpole, *The Sea-Hunters*; Allen, "Fur-Seal Hunting in the Southern Hemisphere"; and Busch, *The War against the Seals*, 9.

33. Robert Cushman Murphy, "The Status of Sealing in the Subantarctic Atlantic," *Scientific Monthly* 7 (1918): 112–19.

34. Martinic, *Nogueira el pionero*, 24.

35. Kirker, *Adventurers to China*, 12.

36. Alfredo de Rodt, H. A. Howland, and Oscar Viel, "Informe pasado al Sr. Ministro de Relaciones Exteriores sobre la caza de lobos marinos," 17 May 1883, Chile, Ministerio de Relaciones Exteriores, Archivo General Histórico, Fondo: Histórico, vol. 107.

37. Mariano Guerrero Bascuñan, *Memoria que el delegado del Supremo Gobierno en el territorio de Magallanes, don Mariano Guerrero Bascuñan, presenta al señor Ministro de Colonización* (Santiago de Chile: Imprenta y Librería Ercilla, 1897), 407.

38. J. A. Allen, "Mammalia of Southern Patagonia," in *Reports of the Princeton University Expeditions to Patagonia, 1896–1899*, ed. William B. Scout (Princeton: Princeton University Press, 1899), 3:313; and Martinic, "Actividad lobera y ballenera en litorales y aguas de Magallanes y Antartica, 1868–1916," 9.

39. This estimate is conservative. The number of seals killed was likely higher because seals that fell into the sea when shot sank and were not recovered. Furthermore, the secrecy with which sealers conducted their business, combined with the limited regulatory structures in place, created conditions under which official export figures were unlikely to be complete. For additional sources on the scale of hunting activities in the region, see Clark, "The Antarctic Fur-Seal and Sea-Elephant Industry," 402; Federico Albert, "Los Lobos Marinos de Chile," *Revista Chilena de Historia Natural* 5 (1901): 33–41; Bossi, *Exploración de la Tierra del Fuego*, 53; Barnard L. Colby, *For Oil and Buggy Whips* (Mystic, Conn.: Mystic Seaport Museum, 1990), 166; and Allen, "Fur-Seal Hunting in the Southern Hemisphere," 312.

40. Martinic, *Nogueira el pionero*, 28.

41. *Whalemen's Shipping List and Merchant's Transcript*, 9 May 1882.

42. The source did not indicate the length of the voyage. Clark, "The Antarctic Fur-Seal and Sea-Elephant Industry," 428.

43. Martinic, *Nogueira el pionero*, 30–33. Also see Coppinger, *Cruise of the "Alert,"* 114.

44. The tonnage of New England sealing vessels declined over the course of the nineteenth century in conjunction with smaller per-voyage cargos of sealskins. This presumably lowered costs. An important missing piece of information is a time series of fur prices in the London markets.

45. If one factors in social reproduction, the equation becomes more complex. Martinic refers to sealers' spouses receiving advances to buy food and other items to sustain families left behind in Punta Arenas. Income-generating activities initiated by spouses or other family members may have further "subsidized" the male-dominated sealing industry.

46. Busch, *The War against the Seals*, 209.

47. "Letters," *Whalemen's Shipping List and Merchant's Transcript*, 9 May 1876.

48. Crews often found themselves stranded and resorted to eating seal flesh and bird eggs to survive while awaiting rescue from a passing vessel. For examples, see

Clark, "The Antarctic Fur-Seal and Sea-Elephant Industry," 429; Kirker, *Adventurers to China*, 44–45; and Busch, *The War against the Seals*, 207.

49. Quoted in Clark, "The Antarctic Fur-Seal and Sea-Elephant Industry," 426.
50. Quoted in Clark, "The Antarctic Fur-Seal and Sea-Elephant Industry," 426.
51. Edwin Swift Balch, "Stonington Antarctic Explorers," *Bulletin of the American Geographical Society* 41 (1909): 473–92.
52. Mateo Martinic Beros, *Recorriendo Magallanes antiguo con Theodor Ohlsen*, 2d edn. (Punta Arenas: La Prensa Austral, 2005), 81.
53. Quoted in Clark, "The Antarctic Fur-Seal and Sea-Elephant Industry," 431.
54. Log of schooner *Thomas Hunt* (Stonington), entry dated 13 October 1874, Log 841, New Bedford Whaling Museum, New Bedford, Rhode Island.
55. Log of schooner *Thomas Hunt* (Stonington), entry dated 31 October 1874, Log 841, New Bedford Whaling Museum, New Bedford, Rhode Island.
56. This is the only reference to indigenous guides that I have found. See Rubén Stehberg, *Arqueología histórica Antártica: Aborígenes sudamericanos en los mares subantárticos en el siglo XIX* (Santiago, Chile: Dirección de Bibliotecas, Archivos y Museos, 2003).
57. Log of schooner *Thomas Hunt* (Stonington), entries dated 6 September 1878 and 16–18 September 1878, Log 841, New Bedford Whaling Museum, New Bedford, Rhode Island.
58. Due to ambiguities in the log entries, I was unable to distinguish fur seals from sea lions.
59. Log of *Golden West*, Log 32, Center for Archival Records Research, Mystic Seaport, Mystic, Connecticut.
60. Quoted in Clark, "The Antarctic Fur-Seal and Sea-Elephant Industry," 462.
61. Colby, *For Oil and Buggy Whips*, 86; and Coppinger, *Cruise of the "Alert,"* 116.
62. On Argentina, see *Whalemen's Shipping List and Merchant's Transcript*, 28 December 1880; on the Falklands, see A. B. Dickinson, "Southern Hemisphere Fur Sealing from Atlantic Canada," *American Neptune* 49 (fall 1989): 281.
63. Bossi, *Exploración de la Tierra del Fuego*, 53.
64. Rodt, Howland, and Viel, "Informe pasado al Sr. Ministro de Relaciones Exteriores sobre la caza de lobos marinos."
65. Oscar Viel to Minister of Colonization, 16 November 1886, Chile, Ministerio de Relaciones Exteriores, Archivo General Histórico, Fondo: Histórico, vol. 107.
66. Patrick Barr-Melej, *Reforming Chile: Cultural Politics, Nationalism, and the Rise of the Middle Class* (Chapel Hill: University of North Carolina Press, 2001).
67. Sesión Ordinaria, 10 August 1893, Cámara de Diputados, in *Boletín de las Sesiones Ordinarias en 1893* (Santiago, Chile: Imprenta Nacional, 1893), 372.
68. Guerrero Bascuñan, *Memoria*, 408; and Albert, "Los Lobos Marinos de Chile," 41.
69. Townsend, "Fur Seals and the Seal Fisheries," 319; and Balch, "Stonington Antarctic Explorers."
70. At least 60,000 skins were shipped from Port Stanley during this time period. Dickinson, "Southern Hemisphere Fur Sealing from Atlantic Canada," 284.

71. Martinic, "Actividad lobera y ballenera en litorales y agues de Magallanes y Antartica, 1868–1916," 10–13.
72. Dickinson, "Southern Hemisphere Fur Sealing from Atlantic Canada," 289; J. Agustín Iriarte and Fabian M. Jaksic, "The Fur Trade in Chile: An Overview of Seventy-Five Years of Export Data (1910–1984)," *Biological Conservation* 38 (1986): 243–53, 244–45; and "Disposciones en vigor que reglamentan la pesca general," 12 May 1926, Chile, Ministerio de Relaciones Exteriores, Archivo General Histórico, Fondo: Argentina, vol. 370.
73. "Producción del territorio de Magallanes durante el año 1939," Chile, Ministerio de Relaciones Exteriores, Archivo General Histórico, Fondo: Argentina, vol. 674.
74. Iriarte and Jaksic, "The Fur Trade in Chile."
75. Pedro H. Bruno Videla, *Recursos naturales de la Tierra del Fuego: Lobos Marinos* (Ushuaia, Argentina: Gobernación del Territorio Nacional de la Tierra del Fuego, Antártida e Islas del Atlántico Sur, 1980), 2–7; and José Viera-Gallo B., "La pesca en la República Argentina," enclosure in Germán Vergara to Ministro de Relaciones Exteriores, 22 July 1950, Chile, Ministerio de Relaciones Exteriores, Archivo General Histórico, Fondo: Argentina, vol. 1019.
76. Stehberg, *Aqueología histórica antartica*, 185–97.
77. "Arctocephalus austris," IUCN Red List of Threatened Species, http://www.iucn redlist.org/apps/redlist/details/2055/0, last updated 4 September 2012.
78. On wolves in two European contexts, see Garry Marvin, "Wolves in Sheep's (and Others') Clothing," in *Beastly Natures: Animals, Humans, and the Study of History*, ed. Dorothee Brantz (Charlottesville: University of Virginia Press, 2010), 59–78.
79. See Maritza Sepúlveda and Doris Oliva, "Interactions between South American Sea Lions *Otaria flavescens* (Shaw) and Salmon Farms in Southern Chile," *Aquaculture Research* 36.11 (July 2005): 1062–68.
80. On the many uses of animal skins and furs in the early twentieth century, see Charles H. Stevenson, *Utilization of the Skins of Aquatic Animals*, extracted from U.S. Fish Commission Report 1902.
81. Harriet Ritvo, *The Platypus and the Mermaid, and Other Figments of the Classifying Imagination* (Cambridge: Harvard University Press, 1998), 46–47.
82. Bridges, *Uttermost Part of the Earth*, 115.
83. Richard White, *The Middle Ground: Indians, Empires, and Republics in the Great Lakes Region 1650–1815* (Cambridge: Cambridge University Press, 1991).
84. Early nineteenth-century whalers used similar language to describe their "wary" prey. See Lance E. Davis, Robert E. Gallman, and Karin Gleiter, *In Pursuit of Leviathan: Technology, Institutions, Productivity, and Profits in American Whaling, 1816–1906* (Chicago: University of Chicago Press, 1997), 148–49.
85. Bonner, *Seals and Man*, 60.
86. For a similar argument about whales, see Davis, Gallman, and Gleiter, *In Pursuit of Leviathan*, 148–49.

9

Birds and Scientists in Brazil

In Search of Protection, 1894–1938

REGINA HORTA DUARTE
Translated by Zeb Tortorici
and Roger Arthur Cough

During the first decades of the twentieth century, the processes of constructing national identities in various Latin American countries were decisively linked to the sciences of the natural world. Brazil's renewed contact with international markets through the export of coffee, bananas, cocoa, sugar, and tobacco meant that natural history and biology stood out as the sources of rational solutions to problems such as soil exhaustion, changes in rainfall patterns due to deforestation, and, above all, the pests that were attacking the plantations—all consequences of the abrupt changes in environmental conditions caused by the expansion of territorial occupation. Various institutional models and scientific practices were borrowed, but the United States was, without doubt, the most marked influence on the types of biological thought that merged with nationalistic processes. In that country, biology also constituted a strong component of nationality, and the study of nature and hygiene was highlighted as an indispensable component in the formation of full citizens.[1]

In Brazil the experience of the republic after 1889 was concerned largely with the search for institutions which would serve the country's entry into the modern world. In this context, scientists and intellectuals constantly appealed to the state for more direct intervention in the law and for the exploitation of nature to be contained. Among preservationist discourse, the denunciation of the extermination of birds was a recurring theme that was certainly in accord with international movements then on the rise: after all, the biologists active in Brazil included in their curricula vitae participation in renowned worldwide scientific associations and events, visits to institutions abroad, and publications and editing of specialized periodicals, in addition to maintaining active correspondence with members of the international scientific community.

In 1894 the publication of Emílio Goeldi's *As Aves do Brasil* (The Birds of Brazil) revolutionized ornithological knowledge in Brazil, relating it to the demands of constituting nationality. In 1938 the self-taught Eurico Santos published books on the exuberant Brazilian avifauna and the disastrous consequences of its probable extinction due to predatory hunting for trade in feathers or food for impoverished populations. Between these dates, and prior to them, the Brazilian political process suffered radical changes. Since its independence from Portugal in 1822, Brazil had been instituted as a constitutional monarchy. In 1889 the republic was proclaimed by a group of army personnel, and with its coming, Dom Pedro II, emperor since 1840, was exiled with his whole family. The first constitution of the republic dates from 1891 and marks the foundation of a constitutional federalist state with a decentralized administration and a high degree of legislative and executive autonomy for the various states of the federation. In addition to the dominant military presence, the group linked to the export of commodities (especially coffee) exercised strong political power in a republic constituted along oligarchic lines. The great majority of the population was excluded from the effective exercise of political citizenship. In 1930 dissident elites and members of the emerging middle class organized the so-called 1930 Revolution and proposed the construction of a centralized nation. Above all, they criticized the power of the local oligarchies in distant backlands of Brazil, which acted on the margins of legality and exercised immense private power, much to the detriment of the public institutions. The dissident project championed the administrative, political, and cultural integration of the most distant regions of the country's enormous territory, and sought the construction of a modern state. A principal leader of the revolution, Getúlio Vargas, became the president of Brazil soon after the movement's victory. Over the years, the authoritarian character of this political project increased, and in 1937 Vargas installed a dictatorship supported by the army. The "New State" (Estado Novo, 1930–45) was a strongly centralist political régime, with a populist, anticommunist, and nationalist appeal.

Brazilian society's relationship to birds assumes interesting aspects throughout these turbulent decades of Brazilian history. Decisive contacts with North American culture become especially evident at the dawn of the twentieth century. The curious cultural and historical construction of the prevalent images of nature in Brazil are revealed on analysis, reinforcing the conception that the way in which humans describe and comprehend the natural world is inseparable from the values and conflicts that they have lived through.[2] Historical inquiry makes explicit the impasses suffered by the scientific community in Brazil at the hands of the instituted powers, their attempts at intervention, and appeals to authorities largely responsible for delineating a common identification between scientists and birds. Pointing to the utility and the role of birds in constituting a prosperous nation, scientists also clamored for their place in the spotlight, in the name of Brazilian progress. In focusing on such debates, I aim in this chapter to confront a forgotten tradition, that is, the adoption by scientists of important practices in defense of nature in the first decades of the Brazilian Republic.

Luxury and Vanity

In 1908 a high-class shop in the city of Rio de Janeiro, the Maison Blanche, advertised in the magazine *Fon-Fon!* that they offered the most beautiful hats, under which "all physiognomies become Romanesque, mystical, and of a voluptuous, feminine haughtiness." The advertisement was accompanied by an illustration of an elegantly dressed woman wearing a large hat decorated with numerous plumes and feathers. In the pages of the same magazine, over the two first decades of the twentieth century, many other shops offered scarves, furs, fans made with feathers, assorted plumes and feathers, ornaments for the hair made from stuffed birds, in addition to hats brought "by the last steamers from Europe" (fig. 9.1).[3]

The magazine *Fon-Fon!*, founded in Rio de Janeiro in 1907, printed social columns as well as sections on literature and political, cultural, and sporting news. Its title evoked ideals of modernity, consumption, and desire for sophistication of the urban Brazilian elite of that time, visible also in the plethora of objects, medicines, cosmetics, watches, clothes, accessories, cars, and weapons announced in the copiously illustrated pages of the magazine. In those years, the city of Rio de Janeiro underwent staggering transformations: the city's center—the primary target of hygiene experts—was remodeled; the thoroughfare Avenida Beira-Mar opened; botanical gardens were planted; the port was outfitted; and electric trams began to circulate on many main roads. An elegant, "French" space was mapped out with promenades on chic streets and car rides through the new postcard

9.1 Fashion, women, and birds in a Brazilian shop's advertisement.
SOURCE: *FON-FON!* 13.30 (1919): 9.

city. Consumerism dictated the acts of an elite obsessed with the latest fashion and the *dernier bateau* articles.[4]

The social columns cultivated extravagant habits: in the hot streets of Rio de Janeiro, ladies paraded with enormous hats and furs. Famous actresses had themselves photographed with plumes and feather fans. Particular types of hats were associated with different traits: ornaments made of wings denoted sentimental, dreamy women; plumes were used by fanciful women; and so on. Much attention was given in the news to the largest hat in the world, one-and-a-half meters in diameter, the feathers of which fell to the waist of the woman, a well-known European actress, who donned the hat. Magazine articles about the São Paulo elite emphasized women in their grandiose plumed headpieces.[5] Such adherence to European customs and styles occurred simultaneously with the extensive penetration of foreign capital in Brazil and the increase in internal wealth generated by the export of coffee, cocoa, and rubber. The country entered into a market internationalized by European and North American commercial expansion, in which the hegemony of French fashion imposed ostentatious standards, the tone of which was dictated by the promise of belonging to high society and to the chic world.[6]

Among the symbolic habits of acquiring status, hunting for sport gained prominence, and among the countless new consumer goods available, weapons constituted a major attraction for the male public. Advertisements —such as those of the representatives of the Winchester Repeating Arms Company, Standard and Hunt—ran photos and prints of hunts carried out by appropriately dressed men, accompanied by beautifully groomed dogs, in bucolic scenarios of skies lined with birds ready to be shot. Detailing the caliber and properties of each product, representatives of the foreign firearms factories boasted that European hunters had a predilection for their weapons in particular (fig. 9.2). These images emphasize a type of masculine sociability that was practiced in the context of the hunts. Frequented exclusively by males, the hunts provided an opportunity for friendly competition among social equals. Hunting clubs organized game hunts on vast estates and ranches (*fazendas*). One report showed the large landed estate (*latifundio*) where the president of the republic at the time, Wenceslau Braz, was in the habit of practicing this sport. Also in the news were the results of competitions, such as hunts of white-tailed doves (*juriti*), that were carried out during picnics on the outskirts of Rio de Janeiro, in which the most refined families participated.[7]

Hunting manuals of the time described memorable adventures through the forests of Brazil and included instructions on shooting game as well as descriptions of the weapons and ammunition used in hunting. One book published in 1924 criticized the restrictions that São Paulo's municipal authorities placed on the hunting of partridges during certain times of the year, which caused "a certain antipathy," because "in fact it is no small thing to deny us our favorite pleasure [*nosso prazer predileto*] for seven long months a year."[8] However, "to indulge the vice [of the hunt], every now and again we can become the furtive hunter, which we vouch is a delight [*o que afirmamos ser delicioso*]."[9] A delight also, that text guaranteed, was the flesh of the rufous-bellied thrush (*sabiá*), whose sweet song so many poets had praised without ever knowing its taste, as meat capable of "perfectly satisfying even the most exigent palate."[10] In detailed tables, the reader was told

2ᴷ 650 gr.

Espingarda de 3 canos (2 para chumbo, 1 para bala) marca HUNT
dos afamados fabricantes :
Manufactura allemã de Armas, H. Burgo Müller & Söhne.

A **HUNT** não deixa o caçador sem caça. Na Europa é a arma predilecta hoje.

Vendas a prestações e a dinheiro

Com os amigos representantes para o Brazil: QUACKEBEKE & ROCHA

ℭASA GARANTIA Rua do Theatro 3, R
Caixa 337 – Telephone 3

9.2 "HUNT never leaves the hunter without the hunt. In Europe, it is the preferred gun of today." Advertisement for the weapons manufacturer HUNT. SOURCE: *FON-FON!* 3.33 (1909): 6.

that for hunting this delicate bird the best weapon was a thirty-six caliber, with one gram of gunpowder and twelve grams of number nine shot lead.[11]

In fact, plumed hats and sportive hunting were merely the tip of the iceberg of a genuine massacre of Brazilian avifauna that took place between the end of the nineteenth century and the first decades of the twentieth century in the Amazon, along the coasts, in the vast arid backlands of Brazil (the *sertão*), and in diverse rural areas spread throughout the country. The trade in feathers flourished throughout these decades. Taking into consideration the data of legal exports, between 1901 and 1905 Brazil sold around six thousand kilos of feathers to Germany, England, France, and the United States. Between 1910 and 1914, legal exports totaled some twenty thousand kilos. The most prized specimens were feathers from flightless rheas (*emas*), herons, scarlet ibises (*guarás*), parrots, parakeets, macaws, violaceous and chestnut-bellied euphonias (*gaturamos*), toucans, hummingbirds, and various types of wood-rail (*saracuras*). Two factors should be taken into consideration in the analysis of such data. First, each individual animal contributed only a few meager grams of feathers. In the second place, contraband in the most remote corners of the country—where the bird hunters acted due to the inefficacy of government control—makes these figures a very poor estimation of the number of birds that were actually killed during those years. Furthermore, the increase in quantity throughout those years cannot be said to indicate greater economic gain, as exports reached a value of 849,192,000 réis between 1901 and 1905, but in an evident devaluation, scarcely 543,274,000 réis between 1910 and 1914.[12] There was, moreover, trade in leather made from the skin of birds. In Rio de Janeiro, just one fazenda was responsible for annually exporting 20,000 hummingbird skins to France. In countless Brazilian coastal cities and towns, hundreds of bird skins were sold at absurdly inexpensive prices, later to be auctioned in London.[13]

In spite of the popularity of the use of feathers in Brazil until the mid-1920s, the acceptance of these products declined in some countries due to the rise of movements against the extermination of birds. One of the greatest importers of feathers, the United States of America—where there was the same craze for hats and decorative objects made with birds or parts of these animals—was precisely the country that led the way in contesting this trend. As part of Woodrow Wilson's environmentally active presidency, in 1913 the Federal Tariff Act was passed, prohibiting the importation of the feathers of many types of wild birds, as well as their heads, wings, tails, and skins or parts of the skins (manufactured or not), unless they were intended for proven scientific or educational use. Such prohibition was hailed as one of the many successes of campaigns by conservationist groups like the National Association of Audubon Societies. Founded in 1896 in honor of the famous ornithologist, naturalist, and artist John James Audubon (1785–1851), the National Audubon Society enjoyed the participation of doctors, attorneys, naturalists, sportsmen, and women who were appalled by fashion trends they considered cruel and unacceptable. The group's activists raised awareness in schools, organized meetings, and pressured Congress to prohibit the sale of hats and other objects which utilized feathers, thus achieving many successes and conquering public opinion.[14] Such actions certainly found a resonance in the events organized around Bird Day, which was commemorated for the first time in 1894 through the initiative of Charles

Almanzo Babcock, superintendent of schools in Oil City, Pennsylvania, who sought to raise awareness of bird protection and integrate bird conservation education into school curriculum. In the following years, the commemoration of Bird Day became popular in various American cities. Supported by the U.S. Department of Agriculture, in 1901 Babcock published a book, *Bird Day: How to Prepare for It*, with curriculum suggestions and instructions for schools on how to prepare for the event, all with the aim of engendering a love of birds among young Americans.[15] In 1911 the American Game Protection Association also started to defend hunting practices that were compatible with conservation.

There were protests by conservationists and the consequent mobilization of public opinion in other countries also. Throughout the whole of the nineteenth century, a new sensibility around the relationships between humans and animals, one which condemned cruelty to animals, was created and put into practice on a global scale. Many European countries promulgated laws on the matter. It is curious, however, that this new sensibility did not impede the rise of trends in fashion for decorations made from bird feathers and skins, nor did it deter the expansion of the breeding of songbirds in captivity. The Society for the Protection of Birds—founded in London in 1891 and incorporated by Royal Charter in 1904—worked against the use of plumes throughout the entire British empire. Acts such as Queen Alexandra's declaration in 1906 that she would never again use feathered hats or Queen Mary's decision in 1911 to discard her adornments made from birds before a visit to India had a profound influence. In 1884 the first meeting of the International Ornithological Congress was held in Vienna. The Ligue pour la Protection des Oiseaux started its activities in 1912, in France. In Latin America the Sociedad Cubana Protectora de Animales y Plantas was founded in 1882, and the Asociación Ornitológica del Plata was formed in 1916 in Argentina. Notwithstanding this, such movements succeeded in regulating the trade in feathers only at a much later period. In Canada the importation of some products was banned in 1915, whereas in England such prohibition was decreed only in 1921. In Argentina legal measures would become effective only after 1936.[16]

An article that appeared in 1902 in the Australian magazine *Emu*, published by the Royal Australasian Ornithologists Union (founded in 1901), drew attention to the international dimension of the extermination of birds, as migratory species visited various countries, crossed oceans, ignored frontiers, and interconnected nations. In Latin America herons flew over the Amazon rainforest and entered territories belonging to Colombia, Venezuela, and Brazil. Between 1890 and 1913, around ten thousand herons were shot down in Venezuela. The poor mestizo populations of Honduras hunted partridges, pigeons, peacocks, and pheasants in transit. At the same time, the passenger pigeon (*Ectopistes migratorius*), which inhabited parts of the United States, Mexico, Colombia, and Venezuela, became extinct. In this matter, too, the United States led the way with the Migratory Bird Treaty Act of 1918, which implemented the convention signed in 1916 with England and Canada for the protection of migratory birds. A 1936 statute implemented the convention between the United States and Mexico for the protection of migratory birds and game mammals.[17]

In Brazil, a great supplier of feathers and plumes, legislation was established much later, in 1934. However, many dissonant voices fashioned conservationist practices that, while aware of the specific nature of the problems faced in Brazil, had as constant models the events, struggles, and successes of various movements against the extinction of birds in other countries. Many intellectuals criticized the destruction of nature at the same time that they questioned the liberal model predominant in Brazilian politics during the first decades of the republic. They did not share in the euphoric climate of "progress" about which the government boasted (pointing to the rise in trade, the entrance of foreign capital, the growth of cities, and the sophistication of urban life). One of the most important representatives of this critical tide was Alberto Torres (1865–1917). An essayist, jurist, and politician who served as a reference for innumerable intellectuals and scientists of the 1920s and 1930s, Torres affirmed that "the mansions of our avenues, the carriages, the automobiles, the jewels, easy and bright letters, the clothes, [and] the fashions" obfuscated the real Brazil, that of "virgin forests and mines," a potential that had been imprudently "mined, scraped, pulverized, [and] dried out," leaving behind only "the shimmering spangles of our cities," defined as the "center of agitation and luxury."[18]

With a desire to integrate the vast backlands of the country's interior, various scientists accompanied the commission of Cândido Rondon in order to install telegraph lines in the sertão. Situated among the impressive stories of these expeditions was the visit of Theodore Roosevelt, between 1913 and 1914, guided by Rondon through the continental immensity of Brazil. As president, Roosevelt took decisive action for the conservation of nature, just as he had acted for the valorization of love for one's home, for one's mother, for country, and for God. He joined his great enthusiasm for the West and for the history of the pioneers with a defense of simple, agrarian society in which men overcame hardship through their capacity and their efforts, and in the process renewed their virtues.[19] Roosevelt's presence left an indelible mark on the Brazilian participants of the expeditions and established an important contact with North American cultural values at the beginning of the twentieth century, thus becoming a reference point for those who defended the preservation of nature in Brazil.

In an effort to incorporate isolated lands and populations, the state installed telegraph lines, which constituted one strand in the network of strategies designed to make the government more present and visible in the northwest of the country. Rondon, consistent with his positivist conceptions, fought for the protection of the indigenous and in defense of their land against the exploitative interests of local rubber barons and proprietors. With the official establishment of the Serviço de Proteção ao Índio (Indian Protection Service) in 1910, Rondon and his allies experienced tense situations in which the inefficacy of the federal government and an evident lack of authority in facing up to special interests became evident. The difficulties of actually carrying out republican public practices in the vast Brazilian territory became apparent.[20]

Among the participants, Edgar Roquette-Pinto (1884–1954), an anthropologist and future director of the National Museum, stressed the violence and lawlessness in the regions traversed. In Corumbá, Mato Grosso, he'd heard it said that there the law in force for the resolution of conflicts was "article forty-four, paragraph twenty-two"—a reference to the caliber of the Winchester rifle and the barrel of an automatic pistol. In the same region, he'd witnessed the illegal sale of kilos of feathers. And it reminded him, with sorrow, of the women who would use them, and he asked himself how many would feel remorse "if they saw those white flakes flying through the blue sky" and how many would prize their adornment even more "simply because it cost the happy life of the herons."[21] At the heart of this confrontation, centered around the harsh reality of the sertão—that remote, arid backcountry in Brazil's interior, where birds and Indians were being decimated—and criticism of the luxury and vanity characteristic of the urban Brazilian belle époque, a dual configuration of new rationalities and new sensibilities came to be articulated.

Sense and Sensibilities

The first protests against the extermination of birds in Brazil came from certain eminent foreign scientists who were initially contracted by the government of the Empire of Brazil, the political entity founded in 1822 and replaced by a republic in 1889. Hermann von Ihering (1850–1930), a German who soon became a Brazilian citizen, and Emil August Göldi (1859–1917), a Swiss scientist, were contracted as zoologists of the Imperial Museum and arrived in Brazil in 1880 and 1885, respectively. Shortly after the founding of the Republic of Brazil in 1889, Ihering was appointed director of the Museu Paulista in São Paulo, a position he held from 1894 to 1915, and Göldi headed the Museu Paraense, a research institution and museum in the northern state of Pará, between 1891 and 1907.[22]

The *Revista do Museu Paulista* and the *Boletim do Museu Paraense*, publications founded respectively by Ihering and Göldi, both printed articles in defense of birds in Brazil. In addition to using their scientific authority, Ihering and Göldi systematized political arguments—reaffirming their love for Brazil, despite being foreigners—that were openly directed at governmental authorities, who they urged to take steps they considered pressing. They questioned the model of liberalism adopted at that time and, to the detriment of a liberty which they considered excessive and damaging to the morals and economy of Brazil, stood up for greater emphasis on the common good. Over the first quarter of the twentieth century, such ideas would encounter ramifications in a wider context of new political positions that became current in Brazilian society. In the face of a radically federalist régime, the disorganization of the country and the incapacity of instituted power to attend to national interests came under fire.

In two appeals to the governor of Pará in 1895, Emil August Göldi (or, as he began to sign his name, Emilio Augusto Goeldi), using his position as a dedicated naturalist who had studied Brazilian wildlife for twelve years and as director of the Museu Paraense de História Natural e Etnografia, protested against the "barbarous destruction of herons and Scarlet Ibises." These birds, which were among "the most fascinating natural adornments on the majestic Amazon River," were undergoing a genuine massacre. In order "to pluck out the few feathers of which thousands are necessary to yield one kilogram of the heinous merchandise," the hunters carried out killings in the laying areas—such as the island of Marajó at the mouth of the Amazon River—which led to the abandonment of eggs and chicks, a fatal reduction in the numbers of these birds, and the risk of their extinction within a few years.[23]

The same tone of complicity on subjects relative to Brazil—which Ihering certainly assumed to prevent his denunciation being ignored because it was coming from a foreigner—was the keynote of Ihering's protests. If Goeldi insisted on "Brazilianizing" his name, Ihering, in addition to becoming a naturalized Brazilian citizen, always spoke of Brazilian society using the inclusive pronoun *we*. Pointing to the persecution suffered by birds—the animals most affected by unrestrained hunting, the expansion of railways, and the aggressive use of the soil—Ihering stressed the emergence of an economic ornithology, "a subject that, until now, has been almost completely ignored among us," and the urgency of following the examples given "by the great nations of the old and new world."[24]

The two scientists, in addition to protesting, drew up proposals for the instituted powers. Goeldi requested that the governor of Pará, who had invited him to be director of the Museu Paraense, put pressure on Congress to approve new hunting laws. In particular, he advocated for the absolute prohibition of hunting herons and Scarlet Ibises between June and the end of January, the protection of nesting groups on private and unoccupied government land, and the stipulation of prohibitive taxes on the trade in feathers. Furthermore, he argued, resolute propaganda against the fashion of feathers should be instituted. Goeldi declared himself willing to help and, providing the utmost publicity to the question in various European magazines, sought out "well-intentioned ladies on both sides of the ocean" who would understand the gravity of the situation.[25] Such measures, he felt, should cover the entire territory of Brazil.

Furthermore, Ihering advocated the following: a federal law against predatory hunting, with a precise definition of which animals could be hunted and in exactly which periods of the year; the explicit prohibition of hunting numerous bird species regardless of the time of year; the end of hunting for trade; the establishment of specific fines proportional to the gravity of the infraction; and the prohibition of children from hunts, given that "the boys kill the little birds simply for fun [*os meninos matam os passarinhos simplesmente por divertimento*]."[26] In São Paulo the wealth generated by coffee and growing industrialization did not prevent the spread of poverty among various segments of the population, contributing still further to the devastation of nature, according to the renowned zoologist. Echoing Goeldi's accusations—for whom the primary agents of the extermination in Pará were "individuals disinclined to take up any serious occupation," those whose profits were spent on *aguardente* (distilled spirits) in the nearest tavern—Ihering pointed out, in a classic classist attitude, how many wretched men, "idle people of the lower classes and scoundrels," used to sell birds in the streets and markets, tied up with string in groups of three to five.

On the coast of São Paulo, Artur Neiva (1880–1943) was one witness to the extermination of birds. A student of tropical diseases and sanitation, he traveled in 1912 in the service of the Instituto Oswaldo Cruz to campaign for sanitation and education in hygiene. He belonged to a generation of intellectuals who privileged discussions about the poor populations of Brazil's interior—largely characterized as sick, malnourished, abandoned by the public authorities, and morally reprehensible—and who offered themselves as the only agents capable of redeeming the poor by their civilizing mission. In this view, poverty, ignorance, violence, immorality, sickness, and the destruction of nature merged. In the modern nation yet to be constructed, there would be no place for the debased trade in birds that had been plucked, salted, and sold by the dozens, for insignificant prices, in poor fishing villages whose inhabitants rarely had the opportunity to eat meat and were desperate for any "sum that mitigated their immense poverty."[27] Ihering worked for several years as a traveling naturalist for the Museu Nacional. In addition to publishing a study on the nests and eggs of birds in Brazil, he published works on birds in São Paulo, Rio Grande do Sul, and Rio de Janeiro. In 1894 he pointed out the clear reduction in the avifauna in the city of São Paulo, as well as a significant increase in insect pests (given that birds were powerful and untiring insectivores) in the whole of that region which had been greatly affected by the advance of agriculture and the railways. The question, however, went beyond the limits of that state, and Ihering defended measures throughout all of the Brazilian territory. The insistence on the economic utility of certain birds made it acceptable to support the pursuit of others, such as the sparrows which were known for destroying crops. Even in this case, a nationalistic perspective prevailed, stressing the fact that they were nonnative birds.[28]

The argumentations of Goeldi and Ihering assumed an important political dimension, as they pointed out the consequences of these practices for Brazil and criticized the then dominant liberal models. They acted with extreme care, certainly fearing the accusation of being monarchists: having initially been contracted by Dom Pedro II, Emperor of Brazil between 1840 and 1889, they suffered pressures soon after the proclamation of the republic by Brazilian military personnel in 1899, and Goeldi himself was dismissed from the National Museum.

To Goeldi, the extermination of herons and scarlet ibises was one of the most "scandalous crimes practiced against nature in this beautiful country." These birds maintained a hydro-hygienic balance, benefiting the whole Amazon Basin. But, for Goeldi, this was not the most serious issue at hand. The trade in feathers, a "sordid business" (*torpe negócio*) moved merely by human greed, morally degraded the Brazilian people, sinking them in barbarism and "base murder" (*vil assassinato*), distancing them from civilization. The ostensibly easy profit of illegal practices devalued the honest and productive earnings of activities that generated taxes, "the result of sweat and honest work," such as tilling the soil and industry. Debasing the "sacred and intangible patrimony" of the nation, the trade in feathers sank its populations into moral degradation and annihilated the assets of future generations.[29] According to Ihering, the attempts at local regulation, when they existed, proved to be ineffective. The absence of inspection and of punishments, which really acted to deter the abuses, revealed the true dimension of the problem. Those acts were crimes against the nation, "against future wealth and against the country."[30]

Anticipating the possible criticisms that restrictive federal laws would exceed the functions of the liberal and federalist republic, Ihering affirmed that the needs of a modern state were incompatible with orthodox liberalism. After all, did not public hygiene, undeniably necessary, require certain regulations, under pain of the population succumbing to epidemics? Did not the country still need to demand military service of its citizens, just as it did their political and juridical functions? There too should be restrictions on the few who destroyed for personal gain what belonged to all Brazilian citizens of the present and the future.

To base his example more soundly, he cited the great pioneer in federal laws for the protection of nature, the United States of America, where the protection of birds was seen as a subject of national interest. In that country the ministry of agriculture worked together with eminent ornithologists, including T. S. Palmer, the author of various books about the economic importance of birds to agriculture. An examination of the stomachs of birds, sacrificed for that purpose, proved the utility of the vast majority of North American birds to the plantations. American protection laws were not limited to defining months in which to interrupt hunting, but, in the name of national interest, established permanent prohibitions against hunting the great majority of bird species. In foreign protection laws could be found, according to Ihering, the great superiority of this type of legislation—which was also established in other "civilized" countries such as England, Germany, Canada, Australia, all of them already fully convinced that "to watch over the conservation of the ornithological riches of their countries" was not only a right, but a "duty of governments"—making such legislation an example to be followed. Once foreign protection laws were examined and compared to see "which legislation would be the most recommended for application among us," the convenience of "giving preference to American laws" became evident in the search for references in the constitution of Brazilian laws.[31]

Goeldi also cited the example of North American federal laws. The United States, "where democratic institutions conserve their purest and most genuine character," was set up as a model, as it knew how to "trace the limits of individual liberty" and "cut out certain excrescences with which this liberty wanted to try to invade the terrain of common interest and the well-being of the social collectivity."[32]

In spite of being foreigners, Goeldi and Ihering mobilized nationalist arguments to defend measures against the "absolute carnage" suffered by the birds. They certainly achieved a resonance with the practices of other intellectuals of the epoch, as in the case of Alberto Torres, defender of agriculture as the true moralizing path for the Brazilian economy and critic of European fashions and the excessive decentralization of the Brazilian state. Simultaneously with the objections of Ihering, Torres denounced the "prostitution of our soil" as well as the accumulation of wealth by rapid and easy means, facilitated by the extensive railway network linking several Brazilian states. Following the example of nations such as the United States and Canada, where "the governments are starting to care for their riches," Torres called for the "public authorities to try to halt the devastation of the natural vegetation."[33] Such protection, he argued, should be combined with the genuine agricultural vocation, the evident destiny of Brazil, the opposition of which "would be a crime against its nature and against human interests."[34] In the proposals of Torres, directed to his contemporaries, agriculture and conservation should work together in a regulated exploitation of the soil, guided by study and compliance with natural laws. In a nation of solid wealth, there would no longer be a place for the "multitude of vagabonds" left to their own devices and mired in vice, violence, and barbarity; all workers would be upright and sustained by the sweat of their brows. Furthermore, with the use of native labor, there would be no need for mass waves of immigrants, who, it was assumed, spread dissolute customs and eventually joined marginal ranks and "parasitic professions."

On this last point, Ihering surprisingly would have agreed. In pointing to Italy as a lamentable exception in Europe concerning laws for the protection of birds, he always stressed how the Italian colonists in Brazil gave way to predatory behavior in relation to birds. The ornithologist Olivério Pinto, of the Museu Paulista, also commented on "the traditional voracity of the

Italians in relation to all flying birds, without distinction as to type or size."[35] Artur Neiva, passing through regions in the south of Brazil, had noted that there was not one wild bird in existence, as "the Italian colony established there had eaten up even the last little bird."[36] Alípio de Miranda Ribeiro, one of the members of the Rondon Commission and a zoologist of the National Museum, requested state action against the extermination of the birds by the Italian colonists.[37]

In conjunction with the denunciation of predatory practices on the part of the poor populations and immigrants, the criticisms of decentralization, the idealization of a strong state, and the defense of the "real interests of the Nation" came together in systems of political programs in which the oligarchic republic was criticized as an inappropriate vehicle for creating the basic conditions of life and work for the development of the new Brazilian man, virtuous and great, just like the land itself. From the mythical perspective of an agrarian nation, the preservation of nature, which up to then had been wasted or poorly utilized, should merge with its rational use. The reigning cosmopolitanism and the importation of European models were denied, contradictorily, as many of the references of intellectuals like Goeldi, Ihering, Torres, and Pinto were in fact foreign, such as the international preservationist movements.[38]

In his writings, Alberto Torres outlined the liberal national agenda. As a liberal, he maintained his defense of the preservation of civil rights, on the basis of international laws and of representative political mechanisms. But the nation reigned supreme above all else, requiring the concentration of power and the increase of the state's functions.[39] It is important to note how these were the same presuppositions of Ihering's arguments for the defense of the establishment of federal hunting laws in Brazil.

In these historical conditions, the signs of the formation of new sensibilities, in which birds figure prominently, in relation to the natural environment are endowed with a special sense. In 1912 the Sociedade Brasileira Protetora dos Animais (Brazilian Society for the Protection of Animals) published a booklet on the negative moral effect of the hunts, detailing the "suffering and massacres dictated by fashion." Governed by cruelty and mercantilism, the hunts were transforming victims into merchandise and executioners into merchants. In a sequence of photos, the *via crucis* of the herons was presented visually: the female in the nest, the same bird in search of food for the chicks, its dead body with the feathers plucked, the chicks waiting for the mother, the despair of hunger, the wasting away in silence, and, in the final image, the chicks dying. Appealing to women in their maternal roles, such texts implored women and fashionable ladies to cease being accomplices of such tragedy.[40]

The repudiation of cruelty to birds became quite frequent, always associated with the moral foundations of society. Some books on hunting assumed a tone of denunciation, as in a 1925 work in which the author affirmed that hunts no longer existed: what had previously been practiced was "in cold blood, the most revolting destruction," and as a consequence, the birds had disappeared. Delicate creatures such as thrushes (*sabiás*), orioles (*melros*), plumbeous seedeaters (*patativas*), and the rufous hornero (*joões de barro*) all succumbed, "torn apart by innumerable hunters, in general, immigrants who arm themselves with a twelve caliber gun, [which is] practically a cannon."[41] Ihering protested against the young men who amused themselves by setting traps and going on hunting trips with shotguns. Hermann Luderwaldt, zoologist of the Museu Paulista, denounced the brutalizing effect on the morals of young people who destroyed nests, set traps, or even blinded the birds in order to "tame them." Olivério Pinto, for his part, lamented the "war of extermination" conducted by boys armed with slingshots—easily made because of the availability of rubber from automobile inner tubes—which resulted in bird watching having to be carried out in areas increasingly distant from the populated zones. The praise of intellectuals, such as that by the critic José Veríssimo (1857–1916), for the initiatives of the American government, which had prohibited the importation of animal parts, also had a profound effect. On returning from Paris, elegant American ladies "saw their beautiful aigrettes on their expensive hats pulled out by the hands of the customs officials."[42]

In 1911, public-education authorities in São Paulo sought to establish their own Bird Day, along the lines of what schools in the United States had done, with a view to developing the "delicate and noble feeling of protection for birds" in the hearts of the children. It was imperative to stop the "unpatriotic and perverse" acts, the pitiless extermination with "shotgun, trap, and sticky resin," through "the education of the sentiments."[43] As proposed, Bird Day was to be commemorated in the schools and was to include the participation of the family, with an emphasis on the utility of insectivorous birds for agriculture—"these active and happy little workmen which slave away from morn till night, without demanding pay increases, killing and destroying the myriads of insects"—and on their aesthetic importance, in addition to the moral example which "bird life" provided "for the development of some of the best attributes and impulses of our soul," including its acts of "generosity and altruism, devotion to offspring, as well as other estimable qualities."[44] A book with poems and excerpts from the most diverse authors on birds—such as Victor Hugo, Michelet, Olavo Bilac, Gonçalves Dias, and Rodolpho Ihering (Hermann von Ihering's son and a zoologist of the Museu Paulista)—was also published.[45]

Bearing the title "Protegei os passarinhos" (Protect the Birds) and reproduced from an agricultural magazine, Rodolpho Ihering's article proved the utility of birds.[46] In São Paulo the plague of the leaf-cutting ants, a great torment to agriculture, had as one of its principal causes the disappearance of the birds. During the flight of millions of the winged female ants, the starters of new colonies, there were no longer birds to interfere with the ants' uncontrolled proliferation. Opening up the stomachs of various Brazilian species proved that the vast majority of birds ate hundreds of such insects a day. Scientific data therefore proved, without a doubt, the need to fight for the protection of birds. Nevertheless, the author doubted that this data would be sufficient to change attitudes in a way that would go beyond a simple utilitarian defense of the birds, and that as a result would favorably affect only some species. In the end, would there not exist, among men, a sentiment of piety and a repugnance for the evil deed of snatching "from its nest the little bird which devoted itself, with a mother's love, to the raising of its chicks," leaving "the featherless chicks with their beaks wide open, awaiting the morsels that the mother would bring back"?[47] Such arguments indicate how attention to birds, by some sectors of Brazilian society, became associated with the delineation of new rationalities, new sensibilities, and new values.

Work and Virtue

The insistence of scientists and public authorities on legal measures against the depredation of nature should be considered in the context of the constitution of the identity of the researchers themselves: in defending birds, they felt themselves to be useful to society, yet, at the same time, as disregarded by society as the birds were. Hermann von Ihering lamented the lack of interest in the natural sciences and the state government's abandonment of the Museu Paulista. Although the museum's library was good and zoological exhibitions were open to the general public, they were rarely visited, except by agriculture students and doctors wanting information on parasites and their vectors. In a 1938 work on the topic of biology in Brazil, Cândido de Mello Leitão, zoologist of the Museu Nacional, described his feelings on thumbing through the reports of Agassiz, a visitor to Brazil in the middle of the nineteenth century. Among pages criticizing the precarious maintenance of the collections in what was then the Imperial Museum, he found the following comment in pencil, dated 1885 and signed by Dr. Goeldi: "The same thing, even today" (*Ainda hoje a mesma coisa*).[48] Mello Leitão lamented that any visitor to the exhibitions could still repeat those same words. He urged the constitution of a society in which the objects of the natural sciences as well as their students received value, encouragement, and protection.

In the case of birds, scientists, in addition to addressing governmental authorities, sought a strategic approach toward the agricultural sectors. But it was the example of North American ornithologists that formed the basis of scientific propaganda around the insectivorous nature of birds within a rational agricultural system that was attuned to the preservation of nature. Once again, the speeches of the scientists harmonized with other Brazilian thinkers in attributing an agricultural vocation to Brazil.

This argument made explicit an important modification in ornithological knowledge about Brazil. Unlike in the past, when foreign travelers described and classified various specimens, and sent crates replete with collections to the museums of Europe, scientists of the Museu Nacional, the Museu Paulista, the Museu Paranaense (in Paraná) and the Museu Paraense (in Pará) produced knowledge that went far beyond the bounds of classification and morphology. More than serving as inanimate pieces in museums, the birds became objects of detailed investigations: their lives, feeding habits, songs, and reproductive faculties were studied, as was their integration with nature and their relationships to flora, fauna, and members of the same species.

Goeldi surrounded himself with more than a hundred live birds in his study, for everyday observation. He was also interested in the "animal world in a state of freedom." He explored "mountain ranges and meadows, the wide open spaces and the forests, the heat and the cold" with the aim of "listening to the pulsations of the world of organisms" and, by means of narrative, of "stimulating a taste for animal nature" in each of his readers. The idea grew that Brazilians, possessors of one of the richest avifaunas on the planet, simply did not know anything about this aspect of the greatness of their land. Olivério Pinto, for one, stressed the importance of studying the avifauna and its applicability in the defense of Brazil's economic interests, increasing even more the value of the biological investigations."[49]

Bird watching and the knowledge of birds that it produced accordingly became subjects of nationwide interest and, following Roquette-Pinto's premise that the urgency of studying Brazil was "in order to love it consciously, study the land, the animals, [and] the people," a means of cultivating love for the country. On the importance of biology, Roquette-Pinto further affirmed the excellence of a scientific education in the formation of future Brazilian citizens, stressing that thinkers such as Alberto Torres owed a great deal of their intellectual brilliance to the constant study of biology: "It would not have been possible to write what he had written on social questions without a solid biological basis."[50]

Bird watching became popular in various countries. In the United States fervor for bird watching and ornithology began near the end of the nineteenth century. In England bird watching went beyond the exclusively scientific environment in the 1930s and became important in reinforcing national identity and in the cultural construction of the national space. Internationally, "expressions" and national symbols were set up around birds.[51] In Brazil birds not only served as national symbols but were also considered a guarantee for the success of the economy.

The basis of this argument was founded in scientifically conducted observation, the results of which were widely divulged. Hermann von Ihering published warnings about the sad situation of the "fatherland of the clouds of locusts" in agricultural magazines and in widely circulated newspapers. Rodolpho von Ihering defended commonly pursued birds, such as the rufous-collared sparrow (*tico-tico*) (fig. 9.3). After an examination of the stomachs of these birds, researchers calculated that in a thirty-square-kilometer area of land inhabited by some forty tico-ticos, "these workmen" could destroy 33,600 insects per week. Alípio de Miranda Ribeiro, a great collaborator on the magazine *O Campo* (The Field), repeatedly published articles on the actions of the birds in efforts to defend against insect pests and plagues.[52]

But in addition to publishing these illustrious scientists, such magazines ran articles about the increased support for the defense of birds among people tied to agriculture. Sections on the usefulness of swallows, thrushes, and the Great Kiskadee (*bem-te-vis*) taught readers how to identify the birds by distinguishing their song and habits. Booklets, described as works of agricultural defense in their prefaces, reproduced photos of the useful birds that were to be protected. In the *Jornal do Comercio* an anonymous agriculturalist requested the protection of smooth-billed anis (*anus*), "spontaneous and alert defenders of cattle and crops," and "permanent and effective auxiliaries on which tilling the soil can depend to protect it from the ticks

Em defeza do "tico-tico"

Não ha entre nós quem não conheça o trepego passarinho que o povo denominou "Tico-tico" e que por parte dos naturalistas recebeu primeiro o nome scientifico *Zonotrichia pileata* ou *matutina*, nome que depois teve de ser substituido por *Brachyspiza capensis*.

Excusado será pois descrever a sua pluma-choroso do passaro, como o ouvimos no correr do dia. Cantando assim, elle vae saltitando do chão ao galho mais proximo e dahi ao muro, de onde em curto vôo vae pousar além, sobre o grammado — não para se distrahir ou brincar, mas sempre «cuidando da vida».

Entretanto o nosso passarinho, ainda como musico, completa o seu repertorio com varias outras melodias. Assim de manhã cedinho o canto é um tanto mais alegre e á noite lá para as 8, 9 ou mesmo 10 horas, ás vezes, elle ainda faz ouvir, uma só vez, uma ligeira estrophe, de effeito extremamente poetico.

Tudo isto, porém, os observadores da biologia das nossas aves já registraram em varios trabalhos ornithologicos, sendo interessante o facto de que, segundo a proveniencia, o canto varia muito entre os individuos todos da mesma especie. Com pequenas variantes o mesmo «Tico-tico» occorre desde o Mexico até a Patagonia.

Nada porém nos dizem os livros com relação a um dos pontos mais interessantes da biologia deste passarinho. Por mais que procurasse-mos em toda a literatura ornithologica, allás a mais bem cuidada de todas as secções da nossa fauna, nada encontramos de mais positivo sobre o que constitue o principal alimento dessa avesinha, a mais vulgar de todas. Como adiante mostraremos, não é isto uma pergunta ociosa. Pelo contrario, abstracção feita do interesse esthetico que temos pelas aves, admirando a

O TICO-TICO reconhecido como passarinho util, porque em bôa parte a sua alimentação consiste em insectos.

(Cliché *Chacaras e Quintaes*)

gem, pouco vistosa, é verdade, mas interessante, assim como o seu canto — que o caipira traduz como: «Todo dia assim... assim.... assim!» Essa explicação etymolo-onomatopaica, como diriamos em falta de melhor vocabulo, repete bem as notas ou syllabas do canto um pouco

9.3 "In Defense of the 'Tico-tico.'" SOURCE: RODOLPHO VON IHERING. *CHÁCARAS E QUINTAIS* 4 (1913): 47–49.

and the all-devouring locusts." The constant citation of Hermann von Ihering in scientific and nonscientific print media was evidence of the degree of resonance that was achieved beyond the strictly scientific milieu.[53]

Birds were presented as tireless collaborators in the greatness of the nation, "little workmen" that in seeking their sustenance simultaneously served their country. The recurring image of the bird as a useful worker becomes still more provocative when related to the emergence, in those very same years, of the idea that work was an indispensable virtue of the "new Brazilian man" (*novo homen brasileiro*), in the formation of a citizen-worker who, in struggling for his own individual wealth, would never, even for a moment, cease to fight for the wealth of the nation as well.[54]

The argument for the utility of birds was, accordingly, placed at the border of a defense of moral character. Observation of birds and detailed descriptions of their behavior were conducted within a logic that was absolutely anthropomorphic.[55] Certain values and qualities, through which some sectors of society aimed to oppose the urbanized civilization of the Brazilian belle époque which sheltered luxury feather hats, were attributed to birds. For the defenders of Bird Day, for example, the birds were generous, altruistic, loving, and fully devoted to their offspring and to conjugal life. Their physical beauty educated the aesthetic sense, and their songs inspired love for music and poetry. For many women, female birds came to symbolize conjugal fidelity, motherly love, and unflagging dedication to the home. For the children, these small avian beings demonstrated the value of kindness, love for one's country, honest work, and cooperation and harmony between the members of a collective community. Given that the "nursery of evil is in children . . . and in women, with their bad taste for decorations of plumes," the fatal consequences of the extermination of birds were evaluated alongside the importance of education in combating it.[56] Combating what was considered to be a type of ancestral barbarity in women and children—as well as in hunters, who were, much to the detriment of the nation's interests, solely interested in their own profits—proponents of birds' interests sought to make visible an image of an authentic, national, harmonious, and organic civilization. Even in 1930, for example, the Cruzada de Amor aos Pássaros (Crusade of Love for Birds), on the island of Paquetá, organized the release of some two thousand birds in the commemoration of 13 May, the date that slavery was abolished in Brazil.[57]

This tendency strengthened, as can be seen in the pages of the magazine *Fon-Fon!* in 1930 and in the various articles published by the Sociedade Rural Brasileira. In one such article, the author called on mothers to free their children from the specter of cruelty to animals and to raise good and compassionate moral beings who would become the men of tomorrow. Formed as much by modest individuals as by those of renowned talents, societies for the protection of animals were defended "because the codes of kindness and protection of the weak sum up the principles of a true democracy."[58] Another essay translated into Portuguese an article from a New York newspaper about birds and the importance given to them by scientists and members of the U.S. government, which stressed not only the utility of birds but also the ethical examples provided by the habits of birds.[59]

In 1938 Eurico Santos (1883–1968)—founder of four agricultural magazines and an activist of institutions like the Clube Zoológico, the Comitê Internacional Para a Proteção das Aves (International Committee for the

Protection of Birds), and the Sociedade Nacional de Agricultura—published two books that divulged scientific knowledge of the birds of Brazil, which, for many years, served as important works of reference for those interested in the subject. According to the author, it was essential that there be an intermediary to facilitate dialogue between scientists and the public. In addition to emphasizing the insectivorous action of birds, Santos presented important information on the principal families and species of Brazilian avifauna. His emphasis on the various moral qualities of birds tells us, without a doubt, much more about the values of the society in which the author lived than it did about the lives of the birds which were his focus. Along these lines, he described the rufous hornero (*joão de barro*) as an industrious, pure, and peaceful worker, "as honest as a saint." Swallows (*andorinhas*) were, in his estimation, "living examples of tameness, fraternal love, fidelity and merit," for among them the bonds of conjugal love were so strong that only death could undo them. Some avian colonies, described as "true cooperative[s] of mutual succor," lived in great solidarity. Regarding numerous types of birds, the author affirmed the perfect love and harmony that characterized the couples: when a "barbarous hunter shot down one of the consorts," he could be sure of getting the other, which always stayed faithfully nearby.[60]

Dozens of examples of motherly love, conjugal love, and social cooperation fill the pages of Santos's *Pássaros do Brasil*, showing the birds in their meekness, but also in their tenacity and capacity for survival throughout time. The persecution of birds reminded Santos of the lives of saints, "who went around the world preaching kindness and concord, but who died at the hands of evil, brutish people," hence the necessity and the clamor of sensible people for the protection of these animals.[61] In using scientific and experimental data to talk about the birds, Santos introduced a strategy by which to persuade his readers of the "natural" status of certain virtues. Curiously, his work dates from the early years of the New State (Estado Novo), a great champion of the image of a "new man" centered within the family, "the political cell of society," which was to be supported and protected by the state through social security and housing measures. Preoccupation with the family became a "central question in the protection of the 'Brazilian man' [*homen brasileiro*] and in the actual material and moral progress of the country."[62]

Federal protection of birds became regulated in 1934, in a moment of optimism for the scientific community and for the many associations linked to the defense of natural resources, many of them strongly nationalistic in content. Under the auspices of the government, the Primeira Conferência Brasileira de Proteção à Natureza (the First Brazilian Conference for the Protection of Nature) was held. The decree of 1934 established protection for animals, under the tutelage of the state. It prohibited maltreatment of animals (defined in thirty-one specific situations) and established fines and penalties for such acts. In relation to birds, the decree mandated adequate cages, hygiene, and fresh water and food for birds that were for sale. It prohibited the fighting and illegal transport of birds, and the negotiation and hunting of insectivorous birds, songbirds, hummingbirds, and certain birds of small size during any period of the year, except for those utilized, with due authorization, for the purposes of scientific research. In 1941 this decree would be complemented by the Lei das Contravenções Penais (Law of Penal Contraventions), whose article 64 prohibited cruelty to animals. The Código de Caça (Hunting Code) of 1943 established protection for animals that were useful to agriculture, as well as for small birds and fowl (except for those that were harmful to agriculture), rare species, and homing pigeons.[63]

Of significant interest is a hunting manual, published in 1934, that gave the history of this practice in Brazil, discussed its sociological aspects, warned against the depredation of nature and certain species, commemorated the new law, and urged the lovers of the noble sport to comply with it. Considering the centuries of "unending and useless holocausts" and the self-mutilation of Brazil, it pointed out the dangers of such ecological unbalance and celebrated the new possibilities established in the present.[64]

The ecological vision of nature as an immense, balanced, and stable organism—certainly a tendency in the biological understandings of the first half of the twentieth century—went along with the perspectives of humans who, left to their own devices, acted in ravaging and egoistic ways. Paradoxically, if nature was intrinsically harmonious and cooperative, man in his natural state was potentially destructive, which necessitated containment by outside forces. Hermann von Ihering described plants and animals as a totality, "a complicated mechanism from which parts cannot be removed, just as one cannot try to modify it" without causing serious consequences. For the scientist, the case of Brazil was not a problem of its inhabitants or its culture, especially given that "in all parts [of the world], man is the same, and individual greed aims for immediate profit."[65]

Such a Hobbesian perspective, shared by innumerable sectors of Brazilian society in those decades, pointed to a veritable state of war by human beings among themselves and with nature, a condition that demanded the emergence of a strong state. Furthermore, this project found expression in the constitution of 1937, which elevated Brazil's natural assets to the category of public patrimony. Through the authoritarian intervention of Getúlio Vargas, the dissolution of conflicts and unjust competition was deemed possible by strengthening the apparatus of the nation, in which a "new man," under the purview of the state, would become attuned with nature and with other human beings through cooperation and harmonious coexistence. The organicism of nature, the nation, and individuals thus became intertwined.[66]

Throughout the following years, however, given paltry budgets for forestry authorities, unreliable supervision over an immense territory, and the absence of effective participation on the part of civil society, the limits of governmental action and support for preservation became apparent. Efforts to establish national parks, at the time, did not privilege ecosystems of great biodiversity, but rather favored accessible areas near urban centers—such as Itatiaia National Park, situated between the states of Rio de Janeiro and Minas Gerais, or Serra dos Órgãos, a national park about a one-hour drive from the city of Rio de Janeiro—or strategic centers such as Iguaçu National Park, located at the meeting point of Brazil, Argentina, and Paraguay. The preservation of natural patrimony was undoubtedly important to the projects of the Vargas government. But, beyond its cultural and political symbolism, nature outside the national parks was principally considered to be a source of exploitable wealth for economic development, and industrializing projects emerged as the essential commitment of the New State.[67]

In this new nation to be established, human relationships with birds were invested with innumerable meanings, constituted in new practices and diverse symbolisms, which provided evidence of the sociohistorical character of such visions and concepts of nature. The sudden rise in the use of pesticides obscured the insectivorous activities of birds. In the 1940s agricultural magazines ran innumerable advertisements for insecticides, promising to resolve an already classic dilemma wherein "either Brazil will finish off the leaf-cutting ants, or the leaf-cutting ants will finish off Brazil."[68] The tico-ticos now shared the skies with crop-dusting airplanes, proof of the "incredible technical audacity of the man-bird," in a fever for development and technology that would grip Brazil in the years to come.

The study of the relationship between birds and scientists in the early decades of the twentieth century permits us to reiterate a few significant conclusions. Scientists and intellectuals ardently defended the need to alter practices that exploited nature, envisioning the construction of a solid economy for the nation. They pointed out that contemporary modes of consumption were irresponsible and destructive, and they directed their efforts at lawmakers, calling for legislative changes that would control economic interests and reconcile issues regarding the preservation of nature. These scientists and intellectuals participated significantly in the rise of a new social and historical sensibility, with the ethical condemnation of cruelty against animals. They also represented attacks against Brazil's fauna as crimes against the nation, and, specifically in the defense of birds—migratory creatures—they pointed to the preservation of nature as being in the interest of humanity, beyond the bounds of national borders. In this search for protection, however, obstacles abounded, and the scientists felt repeatedly devalued by the priorities of governmental action. They also recognized the unenthusiastic efforts of the government regarding the regulation and inspection of practices exploitative of nature, and suffered bitter defeats vis-à-vis the euphoria of progress and development that gained momentum in the country from 1940 onward.

At present, scientists and activists are racing against time to save the 117 threatened species of birds in Brazil, which, according to a recent BirdLife International report, is one of the countries worldwide with the highest rate of bird extinctions.[69] This panorama is differentiated from that at the start of the twentieth century in that there is now legislation for the protection of the Brazilian fauna, a systematic policy for the creation of parks, and an active civil society. However, there are still grave problems, including the vibrant illegal traffic of live wild birds, the persistence of widespread hunting of birds and fowl by impoverished rural populations, the destruction of forests, which has been aggravated by the actions of illegal timber and palm dealers, and the serious pollution of Brazil's rivers.[70]

With the creation of new protected areas, protests against the expulsion of the poor inhabitants amass. On the other hand, others point out the impossibility of preserving the already fragile ecosystems existing within populated reservations, indicating the continued presence of romanticized ideas of "traditional populations."[71] The debate on the role of the state in Brazil has become reinvigorated. In a present historical moment that is replete with old and new challenges, the protection of birds remains a decisive theme of Brazilian society.

Notes

An earlier version of this article was published in Portuguese as "Pássaros e cientistas: Em busca de proteção 1894–1938," *Latin American Research Review* 41.1 (2006): 3–26.

The English translation is by Zeb Tortorici and Roger Arthur Cough. I dedicate this work to Marina, who loves the birds.

1. See Stuart McCook, *States of Nature: Science, Agriculture, and Environment in the Spanish Caribbean, 1760–1940* (Austin: University of Texas Press, 2002); Stuart McCook, "Las epidemias liberales," in *Estudios sobre historia y ambiente en América*, ed. Bernardo García (Mexico City: Colegio de México, 2002), 2:223–47; and Philip Pauly, *Biologists and the Promise of American Life* (Princeton: Princeton University Press, 2002).

2. William Cronon, ed., *Uncommon Ground: Toward Reinventing Nature* (New York: W. W. Norton, 1995), 19–56.

3. "Chronica da moda," *Fon-Fon!* 1.1 (1907): 32–35; "Os chapéus da Maison Blanche," *Fon-Fon!* 2.14 (1908): 3; and "A Brasileira," *Fon-Fon!* 13.30 (1919): 9. See also Helmut Schindler, "Plumas como enfeite da moda," *História, Ciências, Saúde—Manguinhos* 8 (2001): 1089–108.

4. José Murilo de Carvalho, *Os bestializados* (São Paulo: Companhia das Letras, 1989), 39–41; Jeffrey Needell, "Rio de Janeiro at the Turn of the Century," *Journal of Interamerican Studies and World Affairs* 25 (1983): 83–103; Nicolau Sevcenko, *Literatura como missão* (São Paulo: Brasiliense, 1983), 25–77; Maria Zanon, "Os galicismos na *Fon-Fon!* Influências lexicais francesas no português do Brasil" (PhD diss., Universidade de São Paulo, 2003).

5. "O chapéu e o caracter feminino," *Fon-Fon!* 7.27 (1913): 59–61; "Estação theatral," *Fon-Fon!* 4.37 (1910): 27; "A elite paulista," *Fon-Fon!* 7.34 (1913): 45; "O maior chapéu do mundo," *Fon-Fon!* 5.26 (1911): 3; "Senhoras e senhoritas," *Fon-Fon!* 4.38 (1910): 7.

6. James Laver, *A roupa e a moda* (São Paulo: Companhia das Letras, 1993), 216; François Boucher, *A History of Costume in the West* (New York: Thames and Hudson, 1987), 388–401; Gilda Mello Souza, *O espírito das roupas* (São Paulo: Companhia das Letras, 1987).

7. "A Hunt não deixa caçador sem caça," *Fon-Fon!* 3.33 (1909): 6; "Standard," *Fon-Fon!* 4.28 (1910); "Winchester," *Fon-Fon!* 13.35 (1919): 9; "Club dos caçadores," *Fon-Fon!* 3.35 (1909): 30; "Em Minas, uma caçada nas margens do Sapucahy," *Fon-Fon!* 8.38 (1914): 34; "Notas sociais," *Fon-Fon!* 13.30 (1919): 26.

8. Alberto de Carvalho, *Manual do caçador* (São Paulo: Edição do autor, 1924), 26.

9. Carvalho, *Manual do caçador*, 26.

10. Carvalho, *Manual do caçador*, 27.

11. Carvalho, *Manual do caçador*, 28. See also Bento Arruda, *Por campos e mattas* (São Paulo: Editora Monteiro Lobato, 1925), 20–22; Bernardo de Castro, *Tiro ao vôo* (Rio de Janeiro: Edição do autor, 1925); Henrique Silva, *Caças e caçadas no Brasil* (Rio de Janeiro: Garnier, n.d.); Pereira da Cunha, *Viagens e caçadas em Mato Grosso* (Rio de Janeiro: Francisco Alves, 1922).

12. A. Redondo, "O comércio das penas no Brasil," *Brotéria: Vulgarização científica* 15 (1917): 33–35; Teresa Urban, *Saudade do Matão* (Curitiba: UFPR/Boticário/Fundação MacArthur, 1998), 55–58; João Menegheti, "Curiosidades sobre a caça à

perdiz," *Natureza em Revista* 9 (1982): 28–30; Warren Dean, *A ferro e fogo* (São Paulo: Companhia das Letras, 2000), 264–65; Nancy Leys Stepan, *Picturing Tropical Nature* (New York: Cornell University Press, 2001), 51. In 1905, 840,192 thousand réis were worth $271,741.44. In 1914, 543,274 thousand réis were worth $157,549.46 (Thomas Holloway, *Imigrantes para o café* [Rio de Janeiro: Paz e Terra, 1984], 268).

13. Emilio Augusto Goeldi, *As aves do Brasil* (Rio de Janeiro: Livraria Clássica de Alves e Cia, 1894), 214–43; Hermann von Ihering, "Necessidade de uma lei federal de caça e proteção das aves," *Revista do Museu Paulista* 3 (1902): 228–60.

14. See the Smithsonian's National Museum of American History website, http://www.americanhistory.si.edu/feather/fthcex.htm; Robin W. Doughty, *Feather Fashions and Bird Preservation: A Study in Nature Protection* (Los Angeles: University of California Press, 1975).

15. Charles Babcock, *Bird Day: How to Prepare for It* (New York: Silver, Burdett, 1901); F. G. Blair, *Illinois Arbor and Bird Days* (Springfield, Ill.: Superintendence of Public Instruction / Schnepp Barnes State Printers, 1915).

16. Keith Thomas, *O homem e o mundo natural* (São Paulo: Companhia das Letras, 1996), 217, 289. The author highlights Italy and Spain as exceptions in that they did not have movements against animal cruelty in the nineteenth century. On the Royal Society for the Protection of Birds, see the society's website, http://www.rspb.org.uk; and "Hats Off to Birds," at the Nova Scotia Museum website, http://museum.gov.ns.ca/mnh/nature/nsbirds/feat05.htm. Reinaldo Funes Monzote, "Las orígenes del asociacionismo ambientalista en Cuba: La Sociedad Protectora de Animales y Plantas (1882–1890)," in *Naturaleza en declive: Miradas a la historia ambiental de América Latina y el Caribe*, ed. Reinaldo Funes Mozonte (Valencia: Centro Francisco Tomás y Valiente UNED Alzira-Valencia / Fundación Instituto de Historia Social, 2008), 267–309; http://avesargentinas.org.ar.

17. W. W. Allen, "Notes on the Albatross," *Emu* 2 (1902): 100; Santiago Olivier, *Ecologia y subdesarrollo em América Latina* (Mexico City: Siglo XXI, 1983), 203–23; Pedro Cunnil, "Movimientos pioneros y deterioro ambiental y paisajístico en en siglo XIX venezolano," in *Estudios sobre historia y ambiente en América II: Norteamérica, Sudamérica, y el Pacífico*, ed. Bernardo García Martínez and María del Rosario Prieto (Mexico City: El Colegio de México, 2002), 141–59; "Migratory Bird Program," U.S. Fish and Wildlife Service website, http://www.fws.gov/migratorybirds.

18. Alberto Torres, *O problema nacional brasileiro*, 3d edn. (São Paulo: Companhia Editora Nacional, 1978), 94–95, 104; Adalberto Marson, *A ideologia nacionalista em Alberto Torres* (São Paulo: Duas Cidades, 1979).

19. Lúcia Oliveira, *Americanos* (Belo Horizonte: Universidade Federal de Minas Gerais, 2000), 74–75, 97, 131–43; Theodore Roosevelt, *Através do sertão do Brasil* (São Paulo: Companhia Editora Nacional, 1944).

20. Todd Diacon, *Stringing Together a Nation: Candido Mariano da Silva Condon and the Construction of a Modern Brazil, 1906–1930* (Durham: Duke University Press,

2004), 10, 110. See also Antônio Lima, *Um grande cerco de paz* (São Paulo: Vozes, 1995).

21. Edgar Roquette-Pinto, *Rondônia*, 3d edn. (São Paulo: Cia Editora Nacional, 1935), 88–89.

22. "Hermann von Ihering," *Natureza em Revista* 2 (1971): 6–10; Maria Lopes, *O Brasil descobre a pesquisa científica* (São Paulo: Hucitec, 1997), 158–212, 248–91.

23. Emilio Goeldi, "Destruição das garças e guarás," *Boletim do Museu Paraense* 2 (1898): 27–42; Goeldi, *As aves do Brasil*, 242–43.

24. Ihering, "Necessidade de uma lei federal de caça e proteção das aves," *Revista do Museu Paulista* 5 (1902): 228–60.

25. Goeldi, "Destruição das garças e guarás," 31–39.

26. Ihering, "Necessidade de uma lei," 255.

27. Artur Neiva, "Prefácio," in Eurico Santos, *Pássaros do Brasil*, 2d edn. (Rio de Janeiro: F. Briguiet e Cia, 1948), 8, 9; Nísia Trindade Lima, *Um sertão chamado Brasil* (São Paulo: Revan, 1999), 93–121. On the relationship between nature and poor inhabitants of Amazônia, see Candace Slater, *Entangled Edens: Visions of the Amazon* (Berkeley: University of California Press, 2003), 183–204.

28. The following are a few of Hermann von Ihering's works: "As aves do Estado de São Paulo," *Revista do Museu Paulista* 3 (1898): 113–476; "As aves do Estado do Rio Grande do Sul," *Annuario do Estado do RGS para o anno 1900* (Porto Alegre: Graciano Azambuja, 1899); "Aves observadas em Cantagalo e Nova Friburgo," *Revista do Museu Paulista* 4 (1900): 149–64; "Catálogo crítico-comparativo dos ninhos e ovos das aves do Brasil," *Revista do Museu Paulista* 4 (1900): 191–300; "Contribuições para o conhecimento da ornithologia de São Paulo," *Revista do Museu Paulista* 4 (1902): 261–329; "As aves do Paraguay em comparação com as de São Paulo," *Revista do Museu Paulista* 6 (1904): 310–84; *As aves do Brasil* (São Paulo: Museu Paulista, 1907); "Novas contribuições para a ornithologia do Brasil," *Revista do Museu Paulista* 9 (1914): 411–48.

29. Goeldi, "Destruição das garças e guarás," 36.

30. Hermann von Ihering, "Devastação e conservação das matas," *Revista do Museu Paulista* 8 (1911): 485–500, 485.

31. Ihering, "Necessidade de uma lei," 244–49; Hermann von Ihering, "Proteção às aves," *Revista do Museu Paulista* 9 (1914): 316–32; Ihering, "Devastação e conservação das matas."

32. Goeldi, "Destruição das garças e guarás," 30.

33. Torres, *O problema nacional brasileira*, 92–93, 100–101.

34. Marson, *A ideologia nacionalista em Alberto Torres*, 136–37, 149, 162.

35. Mário Olivério Pinto, "Resultados ornithológicos de uma excursão pelo oeste de São Paulo e sul de Mato Grosso," *Revista do Museu Paulista* 17 (1932): 689–826.

36. Quoted in Santos, *Pássaros do Brasil*, 169

37. Alípio de Miranda Ribeiro, cited in Santos, *Pássaros do Brasil*, 91; Ihering, *Proteção às aves*, 316; Ihering, "Necessidade de uma lei," 249.

38. Lúcia Oliveira, Mônica Velloso, and Ângela Gomes, *Estado Novo* (Rio de Janeiro:

Zahar, 1982), 109–50; Lúcia Oliveira, *Elite intelectual e debate político nos anos 30* (Rio de Janeiro: Fundação Getulio Vargas, 1980); Simon Schwartzman, Helena Bomeny, and Vanda Costa, *Tempos de Capanema* (Rio de Janeiro: Paz e Terra, 2000).

39. Marson, *A ideologia nacionalista em Alberto Torres*, 60; José Luiz Franco, "Proteção à natureza e identidade nacional: 1930–1940" (PhD diss., Universidade de Brasília, 2002).

40. Eugênio George, *As caçadas, o que elas exprimem moralmente* (Rio de Janeiro: Cattaneo and Borseti, 1912), 1–11; "As Caçadas ou os dramas dos ninhos," *Chácaras e quintais* 6 (1912): 13–15.

41. Bento Arruda, *Caça, caçadas e caçadores* (São Paulo: s/e, 1925), 20–22.

42. Ihering, *Proteção às aves*, 322; Hermann Luederwaldt, "Algumas considerações sobre a natureza do Brasil," *Revista do Museu Paulista* 16 (1929): 317–27; Pinto, "Resultados ornitológicos," 693; José Veríssimo, cited in Edgar Schneider, "Aspectos sociológicos da caça," in C. F. Buys, *Armas e munições de caça* (Porto Alegre: Livraria Globo, 1934), 225.

43. Arnaldo Barreto, Ramon Roca, and Theodoro de Morais, eds., *Festa das aves: Prova e verso, collectanea organizada* (São Paulo: Diário Official / Directoria Geral da Instrução Pública, 1911), 3–4

44. Barreto, Roca, and Morais, *Festa das aves*, 5.

45. Barreto, Roca, and Morais, *Festa das aves*, 14–16.

46. Rodolpho Ihering, "Protegei os passarinhos," in *Festa das aves: Prova e verso, collectanea organizada*, ed. Arnaldo Barreto, Ramon Roca, and Theodoro de Morais (São Paulo: Diário Official / Directoria Geral da Instrução Pública, 1911), 141–46.

47. Rodolpho Ihering, "Protegei os passarinhos," 146.

48. Cândido de Mello Leitão, *A biologia no Brasil* (São Paulo: Companhia Editora Nacional, 1938), 164–65; Ihering, *Proteção às aves*, 325.

49. Pinto, "Resultados ornitológicos," 691; José Alves, *A ornitologia no Brasil* (Rio de Janeiro: EdVerj, 2000), 327–42; Goeldi, *As aves do Brasil*, 3; Fernando Straube and Alberto Urben-Filho, "Tadeusz Chrostowski (1878–1923): Biografia e perfil do patrono da ornitologia paranaense," *Boletim do Instituto Histórico e Geográfico do Paraná* 52 (2002): 35–52.

50. Edgar Roquette-Pinto, "Alberto Torres," *Revista Nacional de Educação* 2.18–19 (1934): 1–8.

51. Helen Macdonald, "What Makes You a Scientist Is the Way You Look at Things: Ornithology and the Observer, 1930–1955," *Studies in History and Philosophy of Biological and Biomedical Sciences* 33 (2002): 53–77.

52. Hermann von Ihering, "A pátria das nuvens de gafanhotos," *Chácaras e Quintais* 5 (1911): 21–23; Rodolpho von Ihering, "Em defesa do tico-tico," *Chácaras e Quintais* 4 (1913): 47–49; Hitoshi Nomura, "A colaboração de Miranda Ribeiro para o conhecimento da zoologia brasileira na época da Comissão Rondon," *Revista de Ornitologia Paranaense* 4 (2000): 26–29; Santos, *Pássaros do Brasil*, 104.

53. A Pessoa, "Ainda sobre nossos pássaros cantores," *Chácaras e Quintais* 2 (1929): 173–74; Wilson Costa, *Os pequenos amigos da agricultura* (São Paulo: A Capital, 1914); "Carta Interessante," in *Festa das aves: Prova e verso, collectanea organizada*, ed. Arnaldo Barreto, Ramon Roca, and Theodoro de Morais (São Paulo: Diário Official / Directoria Geral da Instrução Pública, 1911), 162–63; Ernesto Niemeyer, "As andorinhas e os mosquitos," *Chácaras e Quintais* 53.6 (1936): 698–702; "Uma ave útil ao criador," *Sítios e Fazendas* 2.4 (1937): 32.

54. Oliveira, *Estado Novo*, 151–66.

55. On anthropomorphism, biological knowledge, and social practices, see Marshall Sahlins, *The Use and Abuse of Biology* (Ann Arbor: University of Michigan Press, 2003), 93–108.

56. Barreto, Roca, and Morais, *Festa das aves*, 4–6, 109–10, 118.

57. Schneider, "Aspectos sociológicos da caça," 214.

58. "Proteção aos irracionais," *Fon-Fon!* 24.9 (1930): 65–66.

59. "Conselhos," *Fon-Fon!* 24.7 (1930): 67; "A vida dos pássaros," *Fon-Fon!* 24.42 (1930): 42.

60. Santos, *Pássaros do Brasil*, 13–18, 25, 55, 127.

61. Eurico Santos, *Da ema ao beija flor*, 2d edn. (Rio de Janeiro: Briguiet, 1952).

62. Oliveira, *Estado Novo*, 156.

63. Decreto 24.645 de 10 Julho 1934, *Coleção das leis da República dos Estados Unidos do Brasil* (Rio de Janeiro: Imprensa Nacional, 1936), 720–23; Edna Dias, *Tutela jurídica animal* (Belo Horizonte: Mandamentos, 2001); Eurico Santos, *Caças e caçadas* (Rio de Janeiro: Briguiet, 1950).

64. Schneider, "Aspectos sociológicos da caça," 267.

65. Ihering, "Devastação e conservação das matas," 485–86, 493. On ecology and "organicism," see Michael Barbour, "Ecological Fragmentation in the Fifties," in *Uncommon Ground: Toward Reinventing Nature*, ed. William Cronon (New York: W. W. Norton, 1995), 233–55.

66. Oliveira, *Estado Novo*, 114; Schwartzman, Bomeny, and Costa, *Tempos de Capanema*, 183; Alcir Lenharo, *A sacralização da política* (Campinas: Papirus, 1986), 18; Franco, "Proteção à natureza e identidade nacional," 6–13.

67. Seth Garfield, "A Nationalist Environment: Indians, Nature, and the Construction of the Xingu National Park in Brazil," *Luso-Brazilian Review* 41.1 (2004): 139–67, see 142–49; José Drummond, *Devastação e preservação ambiental no Rio de Janeiro* (Rio de Janeiro: Eduff, 1997), 141–208; "Primeira Conferência Brasileira de Proteção à Natureza," *Boletim do Museu Nacional* 11.1 (1935): 54–61; Franco, "Proteção à natureza e identidade nacional," 14–34; Schwartzman, Bomeny, and Costa, *Tempos de Capanema*, 17.

68. "Combate às pragas," *Sítios e fazendas* 7.12 (1942): 5–7.

69. "State of the World's Birds 2004: Indicators for Our Changing World," Case Studies, BirdLife International, http://www.birdlife.org/sowb.

70. Fábio Olmos, "Correção política e biodiversidade," in *Ornitologia e Conservação*, ed. Jorge Albuquerque (Tubarão: Unisul, 2001), 291–99.

71. Candace Slater, *Entangled Edens: Visions of the Amazon* (Berkeley: University of California Press, 2002); Paul Little, *Amazonia: Territorial Struggles on Perennial Frontiers* (Baltimore: Johns Hopkins University Press, 2001); Antônio Diegues, *O mito moderno da natureza intocada* (São Paulo: Hucitec, 2000); Olmos, "Correção política e biodiversidade," 279–311; Lúcia Ferreira, "Conflitos sociais em áreas protegidas no Brasil," *Idéias* 8 (2001): 115–49.

10

Trujillo, the Goat

Of Beasts, Men, and Politics in the Dominican Republic

LAUREN DERBY

> They killed the goat
> On the highway,
> Let me see him,
> Let me see him!
>
> They killed the goat,
> But they didn't let me see him.
> —ANTONIO MOREL, "LA MUERTE DEL CHIVO"

On the first anniversary of the death of Rafael Trujillo, the dictator who ruled the Dominican Republic with an iron fist for three decades (1930–61), celebratory *antitrujillistas* formed a new popular fête in his honor. Called *la fiesta del chivo*, the feast of the goat, these rites invoked the custom of *rezos* or prayers on the anniversary of a loved-one's death, while inverting them into a subversive ritual of exorcism of the Trujillato. In Hostos Park in the capital city, a large image of a goat with Trujillo's face sporting a goatee was erected, then burned and shredded in an orgy of jubilation; afterward there was dancing in the streets to *merengue típico*. Dominicans devoured loads of goat meat, and awarded some fifty goats along with liters of rum to the hundreds of people who turned out for the event, with prizes for best poetry on the meaning of Trujillo's death, as well as best dance performance to the now-famous rural merengue "Mataron al Chivo." A parade commenced at Las Damas street in the colonial zone, marching through the downtown area, ending at the site of Trujillo's assassination. The events were accompanied by several bands, fireworks, plenty of alcohol, and dancing in the streets. The celebrants even raised a goat into a chair from one of Trujillo's notorious torture centers, La Cuarenta, and carried it to the obelisk, a monument constructed in honor of the city's name change to Ciudad Trujillo in 1937, where the animal was then shot. Other festivities riffed on the goat theme as well: the air force adorned its jubilee with goat heads from the beasts used to prepare the quintessentially creole soup *sancocho*, and the armed forces' band played "La Muerte del Chivo" over and over.[1] The central parade and festivities were organized by the antitrujillista Eduardo Jiménez Martínez in the city center and at the site of Trujillo's assassination, but the idea spread like wildfire and was enthusiastically picked up by the marines and armed forces, as well as by private clubs and families in the major cities, which hosted their own festivities in the capital and in major towns such as Santiago, San Pedro, San Francisco, and Salcedo. Even today many antitrujillistas celebrate 30 de Mayo with a goat fête.[2]

In contrast to the ferocious spirit of revenge that prevailed on Trujillo's death, when Dominicans in New York had rioted at the consulate, tearing down Trujillo's ubiquitous uniformed portrait from the wall and otherwise wreaking havoc, the events on the first anniversary of his death were raucous yet gleeful, as in the spirit of those who mischievously invaded the private beaches and farms of the Trujillo family as if to claim them as part of the public realm. After thirty years of life in a police state, Dominicans experienced exhilaration as well as intense anxiety at the death of Trujillo, like the experience of the death of God, an unspeakable joy mixed with dread.[3] But why was the goat chosen to symbolize Trujillo?

In this chapter I seek to explain why Rafael Trujillo, the ruthless dictator of the Dominican Republic, was transformed into a goat by those who despised him. Trujillo's sobriquet first emerged among the underground opposition to Trujillo, who called him "the goat" while they were conspiring to assassinate him to camouflage the plot from authorities, a moniker which became public after his death when crowds dressed effigies of Trujillo as a devil and burned them as they would Judas figures during carnival in acts called "la fiesta del chivo."[4] I locate his *apodo* (nickname) within several contexts, including peasant lore about goat behavior, as well as everyday male verbal arts, since animal monikers such as tiger and lion are popular labels of Dominican masculinity connoting cunning and sexual predation; the goat also draws on popular shape-shifter narratives as well as animal sacrifice in Afro-Catholic popular religiosity, or *vodú*.

Importantly, rendering Trujillo a goat was a gesture of popular humor, in the spirit of *choteo*, a form of verbal punning and oblique scoffing at figures of power characteristic of a Hispanic Caribbean creole sensibility, one which has a proclivity to convert rage into laughter.[5] Within the domain of popular religiosity, goats are marked as beasts of sacrifice, as preeminent gifts to the *misterios* (deities); they are animals which reside between nature and culture as proto-pets kept in the patio and even in the house, notwithstanding their status as farm animals, and as such they are "both 'man' and animal, friendly and hostile . . . threshold creatures . . . alarmingly imbricated with the forms of life which [betoken] civility."[6] Like pigs, they are ubiquitous as the banks of the rural poor, held in reserve for hard times, and free access to feral goats has been an important feature of rural life since the colonial period.[7] As a symbol, the goat also conveyed Trujillo's political location, since the goatee became an emblem of liberal politicians in the 1920s.[8] This chapter moves from the barnyard to the street to excavate the meanings conveyed by making Trujillo into a goat and then eating him.

Beasts of Memory

Yet why would an animal have been chosen to symbolize Trujillo by those who loathed him?[9] The fiesta del chivo was first and foremost a speech act, one which invoked the goat as a potent sign of creole identity and resistance to the Trujillo dictatorship. Yet, as an emblem, its "symbolic weight" arose from its role as a food product, which as Claude Lévi-Strauss revealed, constitutes a core element of culture as the place where the natural world is made social.[10] Yet the goat in this case is also a mnemonic which recalls the "embodied history" of a particular mode of production, one which formed the basis of a "masterless" peasantry in the mountains for centuries in the Dominican Republic.[11] Unlike in Cuba and Puerto Rico, sugar came late to the Dominican Republic, and for centuries rural denizens relied on feral boar and goat as part of an itinerant swidden economy of which agriculture was only a minor component.[12] The goat is a beast invested with dense sociocultural meaning and value due to its centrality in the Dominican rural economy and everyday life, both as a source of sustenance for a protopeasantry of hunters and as the privileged food of sacrifice for the gods, and thus was a key component of rural lived experience.[13] The patio goat could be used as a lien for acquiring loans where credit was scarce, and its offspring provided access to cash when needed; it thus linked the domestic economy with the market. A constant presence around the family, it also provided meat for special occasions.[14] Since the goat and the pig were ubiquitous in

10.1 Creole pig, Bánica, Elias Piña, Dominican Republic. PHOTO BY LAUREN DERBY.

the Dominican *campo* and thus infrequently sold, they were marked as residing primarily outside the domain of market exchange, unlike cattle, which formed the basis of the eighteenth-century contraband economy to Haiti.[15]

But most important, wild goat and pig serve as embodied memories of life before state subjection, which conjure up a *cimarrón* (runaway slave) history of resistance to the state in the form of open-range *montería*, or free access to feral pork, cattle, and goat in the *monte* (bush) within an economy of hunting and gathering which had been the norm before the Trujillo regime from the seventeenth century until the early part of the twentieth century (see fig. 10.1). Liberal state builders excoriated this *crianza libre* (open range) economy, eventually passing fencing laws in 1908 which began to curtail that which had been termed as "the greatest obstacle and worst enemy" for the nation's prosperity and development, largely because it sustained a nonmarket peasantry, the nexus of a vibrant economy outside of state control and one which failed to contribute to state formation and national progress.[16] Implicitly, the goat and the pig—the most beloved creole meats in *platos típicos*—are contrasted with those products associated with the market economy and processes of commodification of which sugar is the central signifier.[17] The highly prized goat and pig—said to be the tastiest of meats—thus resemble the beloved Andean guinea pig which connotes the nourishment and warmth of hearth and family, and thus contrasts with rice, the symbol of "racial superiority and commodity fetishism," proletarianized labor and empty calories, a food which is "glamorous but insubstantial."[18]

In a process of what Anton Blok would describe as "metonymical totemism," in which the species' characteristics epitomize certain features of the person, Trujillo was nicknamed the goat after his death by those who loathed him.[19] Partly this was a means by which to camouflage critical talk about Trujillo in the tense, uncertain, and violent period after his assassination, when his eldest son, Ramfis, took political control along with the notorious hit man Johnny Abbes, director of the intelligence agency which had controlled the apparatus of everyday terror under the regime.[20] During this period, the press was a mouthpiece for the regime, falsely declaring, for example, that people deplored Trujillo's assassination and stating incontrovertibly that "the Era of Trujillo will not disappear as long as the Dominican Republic exists because it is an integral part of it."[21] By the autumn, however, things had changed, and resistance to these deplorable conditions could be voiced. The newspaper *Unión Cívica* placed popular repression front and center, showcasing large photos of dead bodies in the streets, with headlines such as "The Government's Terror Unleashed Lets Loose Machine Gunfire on the People."[22] Meanwhile, in provincial towns such as La Vega, thugs armed with sticks attacked protesting crowds.[23]

The first anniversary of Trujillo's death in May fell within a short window between the rule of Ramfis Trujillo and the savage repression of the military secret service (SIM), and that of the Consejo del Estado, when the democratic movement Unión Civica came out of hiding and the fiesta del chivo could take place safely in broad daylight, organized in part by the Fidel Castro-inspired Movimiento 14 de Junio, named for one of the invasions intended to unseat Trujillo from power during his reign.[24] Trujillo had been known by his enemies by a series of apodos, or nicknames, such as *el jefe* and *chapita* (bottle cap), which alluded to Trujillo's penchant for plastering his chest with medals and military honors.

But Trujillo the goat was a post-Trujillato invention with several layers of meaning. The billy goat was intended to convey Trujillo's volatile temperament and erotic appetite, as well as his inability to be controlled by others, since goats like cats are notorious for their difficulty to herd. According to one livestock trader specializing in goats, Trujillo was "bronco, vivo; brincaba dondequiera" (lively, astute, untamed).[25] Just as goats dominate territory and cannot be reined in, they can also signify the unruly transmission of information: *boca de chivo* means gossip, and Trujillo's spies were called his *chivato*. Trujillo survived many attempts on his life, and was thus an *hombre chivo*, eternally suspicious. Even after his death, Trujillo, it seems, lurked in the bushes like an ornery goat waiting to turn back into a man and take the reins of power again at the right moment.[26] Just as Trujillo was known for his meticulous attention to appearance, in the world of farm animals goats are uniquely fastidious and are thus allowed to wander freely into the home because they are considered clean. Like rabbits, they do not defecate indiscriminately, and like people, they refuse to eat previously nibbled food.[27]

Like a goat, Trujillo was also renowned for his fierceness and virility. Possessing large testicles, the billy goat is notoriously promiscuous due to the fact that it resists pairing, representing "unrestrained nature"; indeed, *chivería* means "flirtation" in Dominican Spanish, and *chivatica* means "strong body odor."[28] The goat, of course, is also horned, and *cornudo*, or being horned, refers to "cuckolds," since "like deceived husbands billy goats tolerate the sexual access of other males to females in their domain."[29] Since Trujillo was famous for his sexual avarice, and for his penchant for deflowering damsels, this apodo was an inverted comment on his reputation for sexual predation and for thus rendering men around him cuckolds.[30] Perhaps due to the uniquely inexpressive quality of the goat visage, the goat also connotes dissimulation, as seen in several popular Dominican sayings such as "hazte el chivo loco"—pretend you don't know, that is, feign innocence.[31] As a figure of resistance, Trujillo as goat was thus a "public secret"— "that which is generally known but cannot be articulated"—which revealed his uncouth savagery in the eyes of Dominicans who had suffered so much under his rule.[32] Trujillo as goat was popularized in a merengue written to celebrate Trujillo's demise, which disguised popular glee of his assassination in a celebratory lyric about a goat slaughter.

As a ribald ritual, the feast of the goat drew on a popular carnivalesque rite which marked the end of Lent and the exultation of Jesus' resurrection throughout the Americas. Rendering an effigy of Judas and then jubilantly destroying it while officially sanctioned by the church was a long-standing Spanish and Latin American custom, one in which, as Bill Beezley has said, celebrants reveled in the "riotous opportunity to turn the world upside down and set aside customary rules of social hierarchy and convention."[33] As such the ritual drew on the repertoire of popular satanic theatrical traditions in Carnival and Easter as African-descended peoples appropriated and turned on its head an abiding colonial demonic view of African-

10.2 Devil masquerade on Good Friday, Bánica, Elías Piña, Dominican Republic. PHOTO BY LAUREN DERBY.

derived paganism.[34] For example, the Dominican carnival character the *diablo cojuelo* represents the moors who were defeated by St. James the patron saint of Spain in an eleventh-century battle and who eventually became demons that chased infidels back to church in seventeenth-century Spain. In the Dominican frontier town of Bánica, on Good Friday youths impersonate the devil with grotesque masks that combine bits of garbage such as packing materials, old animal masks, and wigs (often several); some wear long gowns of white, black, and red, and ferociously whip each other; others blacken their faces to resemble monstrous slaves (see fig. 10.2). Good Friday is also the day when Gagá secret societies (Rara in Haiti) come out of hiding to engage in mock battles over territory in the sugar zones. As Elizabeth McAlister notes, the Rara conduct the spiritual work that is necessary when Jesus, along with the angels, saints, Judas, and the recently dead, disappears into the underworld on Good Friday.[35]

While drawing on a Catholic script, these rites are thoroughly grassroots performance events, as can be seen in the mixture of European imagery of the devil and popular cultural references such as Batman, who as a gothic superhuman fits comfortably within the rubric of demonic representation, yet updates it. Popular representations of the devil draw on a range of satanic images, from the figure of the Jew, which in Navarrete carnival has a bull face with human features, to the *lechón* (lit., suckling pig), which dresses like Satan but has a porcine face, wears brightly colored and fringed attire, and carries a bladder on a string with which he whips anyone in his way.[36] The vilification of African-derived religious forms also helps explain Carnival's trope of the "black rogue," who at times appears as a pretty black boy.[37] However, the devil's image can only be approximated since he is by definition changeable, and in Mexico can even appear as a dapper handsome *güero* (white guy) walking into a bar.[38] The common thread in these diabolical representations is the unification of opposites: an animal face with human features, the inside being outside, or a man dressed as a woman, the latter figure of which is said to have come from Haiti.[39] Yet the oldest Dominican rendition of Mephistopheles has a goat's beard; in Mexico the tell-tale sign of the devil is hoofed feet.[40]

The lechón is a reminder that the devil was often imaged in the colonial period as a hybrid shape-shifting beast which appeared in various guises— as a dog, pig, goat, or bird, or as a monstrous half-man, half-goat. Indeed, today Satan is often euphemized as a "cabrito sin cuernos" (a hornless goat), so as not to conjure him.[41] Tales of shape-shifting and monstrous hybrids were canonized in the early conquest narratives of marvels, and as James Arnold reminds us, monsters have been frequently adopted by Latin Americans, "occupying a necessary, liminal position at the edge of any culture's conceptual field where alien others must be dealt with."[42] The northern European werewolf, for one, made its way to the New World. The sixteenth-century Italian navigator Antonio Pigafetta described a pitiful creature in the New World with the ears and head of a mule, the body of a camel, a deer's legs, and the whinny of a horse.[43] But Africa is also replete with shape-shifters such as the man-leopards of Nigeria, the Tabwa lion terrorists of Central Congo, and the Banyang were-animals of Cameroon; so Caribbean lycanthropy is overdetermined via several sources, most likely including Amerindians, since the Otomi and Nahua of Mexico also shared shape-shifting beliefs.[44] Indeed, the Haitian countryside is rife with such nocturnal monsters, from the vampiric *lougarou* and the *galipote* to the *bizongo*, which appears as a dog in packs seeking victims at night, and the *Lansetcòd*, which turns people into cattle with a crack of his whip.[45] But over time shape-shifting has creolized, becoming less evil phenomenon than endearing trickster. Today a common genre of rural haunting narratives concerns *bacás*, small demonic hybrid creatures—goats that meow, bulls with gold teeth, pigs with cat's tails, cows that cry like babies—that are identified through their anomalous animal behavior, such as sheep that do not leave a road when a car approaches. The hidden protagonist of devil pact narratives, bacás are said to bring wealth to their owners by stealing others' livestock and crops. These changeable creatures thus evoke the dread of the goblin-foxes of Japan, the sinister Bedouin hyena with its uncanny human yelp, or the rabbit of southern African-American lore, which is a heinous witch in disguise.[46]

Yet if the fiesta del chivo was a Dominican *dechoukaj* (the term used for the mixture of rejoicing and rioting that exploded on Jean-Claude Duvalier's departure from Haiti), one which sought to expurgate and banish the ghost of Trujillo from national terrain, it drew on an image of great popular resonance. As in Haiti, the devil has more often than not been adopted into a counterhegemonic narrative, whereby Dominicans embrace this trickster figure as a critique of the wealthy and powerful, one which affords an underclass paradigm of upward mobility through cunning and thievery, rather than through hard work. It thus speaks to the aversion to manual labor that is common to post-abolition slave societies. To champion the devil is to embrace the *tíguere* (tiger), the hustler who "works without sweating" and scales the social ladder by sleeping his way into the corridors of power, through his pretty face, his charming wiles, and his dance skills.[47] It should thus be no surprise that the devil can appear with a "Tony the Tiger" mask, since the tíguere is the quintessential popular antihero.[48] And the goat, another trickster, plays the savvy role in Dominican tales, often being paired with a dimwitted creature such as an ant.[49]

Part of the popularity of the demonic stems from a long history of the Catholic Church's vilification of popular forms of religiosity, such as vodú, as "Creole, homegrown, unorthodox, diverse and by extension, illegitimate, impure and evil," even though spirit possession and other signature pagan practices became widely diffused given the dearth of clergy in the interior into the twentieth century.[50] From the colonial period through the U.S. occupation (1916–24) and beyond, sorcery accusations branded African and creole religious practices as satanic.[51] Over time, the devil, alongside a range of lesser demonic figures, was embraced by Dominicans for its oppositional power against the Catholic elite. And, indeed, evidence that this ur-trickster is now a national culture hero par excellence is the fact that he not only comes out to play in carnival on independence day, but that he is often draped with the national flag.[52]

To excavate the prehistory of Dominican goat associations, however, one must take a detour into the colonial past.[53] Goat husbandry arrived with the Canary Island population, which comprised a large part of the Spanish immigration to the colony, probably due to the islands' role in provisioning ships, as well as to official colonization efforts to deter a creeping French island presence especially in the eighteenth century. The strong Canary Island influence resulted in distinctive pronunciation patterns, particularly in the central Cibao region.[54] But the goat arrived as a visitor to an island which by the sixteenth century was already crowded by wild boars. The highland *montero* peasants who lived in scattered settlements in the mountainous forested interior lived primarily off the wild porcine population, which had arrived with Columbus's ships and multiplied rapidly over time. These mountain peasants hunted feral boar and cattle, and provisioned the piratic community based at neighboring La Tortuga Island. Given the abundance of wild game, certain hunters, known as *bucaniers*, practiced an extravagantly wasteful style of hunting in which they would fell hundreds of boars in a single day's outing.[55] Indeed, hunters were classed into two groups: those who hunted bulls for hides (the *bucaniers*, who were considered far more skilled), and those who chased wild boar for the sale of meat and lard to planters.[56] The latter were simply called hunters and were famed for their technique of forming the meat into long strips, the distinctive form of smoking they used, and the resultant products' delicious taste. The two types of hunters appear to have been distinct social groups: some were European former bondsmen or ship crewmen; others were runaway slaves and their descendants who, as itinerant hunters, ate great quantities of meat (even raw, it was said) and would spend a year or two at a time in the mountains, only occasionally descending to La Tortuga for muskets, shot, *kleren* (homemade rum), and womanizing. While there were also cimarrón communities blending into this mountain peasantry—such as that of Maniel, which retained its autonomy from the sixteenth century through the nineteenth—the term *cimarrón* was applied to the free black peasantry due to its defiant nomadism and the fact that it gathered its primary starchy tuber crops—*ñame, batata, plátano*—and only secondarily cultivated *conucos*, or garden plots.[57]

Sidney Mintz has located the birth of creole culture in the interstices of the plantation, where slave women grew food crops to feed their families and for trade, yet the Dominican "reconstituted peasantry" was forged in the mountainous highlands of the heavily forested interior.[58] Colonial authorities blamed all manner of social ills on the monteros, whose resistance to settled agriculture the authorities considered a marker of savagery. As one lamented of these "lazy vagrants [*vagos, osiosos y vagabundos*]," "These kind of men don't have subsistence plots, farms, or other honest work to maintain themselves, nevertheless, they eat, drink, get drunk and succeed. Where will this lead them except for theft and evil?"[59] The fall of the montero, however, coincides with the decline of the thick interior woods, which lasted far later in this Spanish colony than elsewhere, due to plantation agriculture's late arrival. The woods were finally felled by the late nineteenth-century sugar boom driven by U.S. capital investment; at the peak of the sugar boom, in the late eighteenth century, the numbers of the "small farms of the poor that live from *monteria*, who spend the whole year without seeing the capital like the first Indians," reached twenty-seven thousand, an estimated one-fifth of the total population.[60]

Since agriculture was an emblem of civilization, this monteria contraband economy, no matter how lucrative, was seen as a national scourge, and the ranchers and their unruly cattle and pigs were the bane of nineteenth-century liberal reformers.[61] Critics railed against this "original affluent society," whose members apparently never worked and, worse yet, were proud of it.[62] In 1810 the traveler William Walton also took note of the abundant wild turtles, goats, fowl, and fish that fed the maroons around Neyba Bay, who "have retired into the recesses of the mountains, living on game and the spontaneous roots of the earth which they store in their hamlets with provident care."[63] He continued that they "live[d] in a kind of republican manner, intent only for their own safety, and governed by their own regulations," descending into town only when they had tortoiseshell, gold, or a "superabundance of cured game," which they exchanged for powder and clothing. They trapped their prey—principally the wild hog—with a snare fashioned from a sapling, which they placed beside a watering hole.[64] From the late eighteenth century onward, efforts were made to curb the rights of ranchers, and the feral boar, cattle, and goat which ran freely to the detriment of agriculture. The British traveler Robert Schomburgk noted that horticulture was a rarity and that only in certain areas did one see small patches of plantain, and fruits intercropped with cacao for domestic consumption.[65] And such was the case not only in the countryside. The freedmen and their pigs and cattle in the Los Minas neighborhood on the outskirts of the capital city of Santo Domingo were a constant menace to their neighbors' conucos.[66] But the threat posed by this subculture went beyond the purely economic. As the Englishman Jonathan Brown, who visited the island in 1837, put it indignantly:

> These blacks assume the manner of Spanish Hidalgos, forever smoking and forever lounging in their hammocks. A few of them are herdsmen or

mahogany cutters. In the vast solitudes of the interior, the former reside continually on horseback, with a lasso at their side, and a small case of cigars and a bottle of aguardiente [rum] at their saddle-bow. The flesh of their cattle serves them for food, and the hides are sent to the different towns upon the coast to be exchanged for their favorite luxuries. The whole country is not many removes from the tribes upon the Niger in point of civilization.[67]

Yet if swine became the key emblem of the montero economy, as the feral boar died off the goat became its substitute. Both beasts were notoriously unruly, roaming far and wide; both were sacrificial animals of choice, yet as creole symbols in a mestizo or mixed-race context, the cimarrón pig represented the interior black montero, while the goat was associated with Canary Islanders, and thus Spain. Nevertheless, both were essential components of the creole free-range economy, which forged this nation of mixed-race mestizos. So, like the European potato, first the pig and later the goat became key signs of the peasantry, of the "autochthonous body . . . a presocial state of isolation in which the poor were cut off from civilization and undifferentiated from each other and from nature" in the eyes of elites.[68] Like other emblems of *lo criollo*, these figures of nature were not vindicated until the twentieth century, when pigs and goats became symbols of freedom invoking the open-range economy, which was finally closed under the Trujillato (see fig. 10.3). Goats, boars, and cattle stood as a foil to the crony capitalist model of development under Trujillo, the foundation of a bucolic prehistory before Dominicans were captured by the state and the market. In contrast with other pastoralisms, however, these beasts were valued not due to the investment of labor time, but rather due to the fact that they provided subsistence with no labor investment.[69]

Since it is a creature which bridges the wild and the domestic, the goat conveys some of the wonder of the bush, and thus brings some of the magic of the monte into the patio. Maybe this is why the *chiva blanca* stands in for the sexually rapacious woman in several merengues, representing the appeal of the domesticated wild.[70] Edmund Leach's concept of anomalous animals that move across categorical divides helps explain the peculiar reverence for the goat, as does his notion of "tabooed ambiguity," since the goat (and the pig) also figure prominently in narratives of demonic animals and hybrid beasts, such as the spectral witch-animal, the bacá.[71] As signs of the wilderness, it should be no surprise, then, that bacás are associated with woods; they are said to sleep at night inside the Mapou tree.[72]

10.3 "Hey, hey! The ham is not just for you! I only wanted you to smell it!" The man with the cane represents Trujillo's Dominican Party, and the man running off with the "ham of liberty" is identified as the people. SOURCE: *UNIÓN CÍVICA*, 4 JUNE 1961. ARCHIVO GENERAL DE LA NACIÓN, SANTO DOMINGO.

Certainly Trujillo's goat apodo had much to do with a long prehistory in Europe of imaging Satan as a demonic goat or worse yet: a monstrous hybrid half-man, half-goat.[73] Trujillo thus bore an uncanny resemblance to the satyr, the Greek demonic hybrid with a voracious appetite for womanizing, wine, and pleasure of all sorts, a figure associated with mountains and woods. Goats have long been associated with the sacred, as living in a world apart, since they often appear in forests and thickets.[74] Popular tales convey the sense that the woods were also harbingers of supernatural forces; that the *monte* or bush is a feared as well as sacred space.[75] Tales of the enchantment of el monte have remained constant, from the eighteenth-century *negro incognito*, who was said to be able to metamorphose, to Trujillo's enemy Enrique Blanco, who took refuge in the mountains and could appear as an animal, to the *ciguapa*, a furry feminine creature, spied in thickets and mountain passes, which walks with feet askew and whose kidnapping antics form a genre of Dominican captivity tale.[76] Speckled goats—which Dominicans read as creole goats—are also favored in the bible (fig. 10.4).[77] Moreover, goats and pigs are animals of sacrifice for the *lwa* or misterios (gods) in Dominican vodú, and female goat is the favored food for Yemaja and Oshun in Santería as well as Brazilian Candomblé; goat meat is used in curative and sorcery rituals; it is also the festive food of choice for special occasions such as patron saint festivals.[78] A goat's blood is even requisite to seal secrecy oaths in Haiti.[79] The Haitian president Antoine Simon's life was protected by a series of buried charms and amulets and especially by a

10.4 Creole goats in the patio, Bánica, Elias Piña, Dominican Republic. PHOTO BY LAUREN DERBY.

beloved goat named Simalo, who was said to have been buried in a coffin like a man.[80]

Trujillo, the Goat

Yet another important context within which Trujillo's nickname must be located is the popular practice among Dominican men of totemic nicknaming. Nicknaming is a ubiquitous feature of Dominican popular culture, an offshoot of the popular rural belief that the name is a component of the person and can be used in witchcraft by one's enemies. Proper names are kept hidden, so that they might not be sold in a devil's pact, or to prevent rivals from using them to make a fatal spell called a *guanguá*.[81]

While nicknaming may serve a functional protective end, it is also a component of the elaborate verbal art of popular street banter which is characteristically male and which any self-respecting Dominican man must be proficient in. Dominican men express their respect for one another through a range of male epithets which are frequently military or bestial and through which they demonstrate proper demeanor by characterizing the other as a man of *dignidad* (dignity and seriousness), an *hombre serio*, a man of respect. Indeed, nicknames are an integral part of reputation, and Clif ford Geertz termed them "symbolic orders of person definition" which individuate and suggest something culturally significant about them.[82] As Antonio Lauria has put it, these "concrete deferential acts of *respeto* communicate many kinds of regard in which a person must be held—awe, trust, esteem for technical capacity, recognition of superior rank, and affection," characteristics which stick due to the social importance of nicknaming practices, which often completely efface real names as "Juan" and "José" are virtually forgotten in favor of *flaco* (skinny), *negro* (black), *el ranchero* (cowboy), or made up names such as *corrumbito*.[83]

While these greeting rituals are deferential, they ironically emerge most extravagantly among social equals. By indicating a posture of inferiority in a jocular manner, they flatter the recipient and establish a common male fraternal bond between speaking subjects; such rituals thus fall within Roberto DaMatta's characterization of "hierarchical social formula."[84] Yet to speak of these naming practices merely sociologically is to miss the expressive dimension of the labels themselves. As tokens of "exaggerated masculine prowess," these nicknames all express potential violence, and thus carry an edge.[85] Valentina Peguero has proposed that epithets such as *capitán* are a byproduct of the unusual extent of warfare on the island in the nineteenth century, but martial terms do not comprise the majority.[86] The majority of nicknames refer to animals—*tíguere* (tiger), *perrón* (large dog), *caballo* (horse), *león* (lion), *tiburón* (shark), *gorila* (gorilla)—and suggest size, fearlessness, and aggression, thus carrying a veiled threat and emblematizing predatory masculinity; that the nicknames have a violent cast, as Kevin Yelvington notes, is part and parcel of this style of black masculinity.[87] Adopting such ferocious imaginary animal masks is akin to Amerindian colonial rites in which Aztecs impersonated the deities by adorning themselves with feathers and in which, as Marcy Norton characterizes it, "the warrior's prowess was manifested and articulated through identification with fearsome predators."[88] In this sense, Dominican animal nicknaming enables men to playfully adopt the impressive characteristics of these powerful beasts in a ritual rebellion against their everyday plebeian status.[89]

The *tíguere* has become the central figure of Dominican underclass *hombria* (manhood), one which connotes cunning and illicit paths to upward mobility, from hustling to sleeping with women of higher status (although possessing enough sex appeal and thus outside interest to do so without becoming a *mantenido* or a "sugar baby"). Tigueraje represents a subset of nicknaming practices which are universal among Dominican men. As Antonio de Moya puts it, "*Tigueraje* is a life style and an attitude that combines the extreme traits of masculinity according to the street culture: slyness, courage, aggressiveness, indiscriminate sexual relations."[90] The tíguere "busca su chelito en la calle" (looks for money in the street) either through stealing or drug trafficking; alternatively, if the tíguere is sufficiently *listo* (ready), he can also be *fino* (high class), that is, he doesn't have to resort to violence, but can hustle *con corbata* (with a tie). The tíguere is a rogue and a trickster who wins through a mixture of chance, allure, and *habilidad* (skill), and with a hefty dose of sex appeal.[91] As James Scott puts it, the tíguere charms as the quintessential "trickster figure who manages to outwit his adversary and escape unscathed."[92] In a context in which high value is placed on being *decente* (decent) in public, these labels indicate that a Dominican man must also be a rule-breaking *hombre hombre* (manly man) and thus must in private accumulate sexual conquests as well; he must thus have a reputation among his male peers, even as he maintains respectability in the community at large.[93]

Like the Haitian *vagabòn* (vagabond), the tíguere is also attractive due to his very informality, as a form of rebellion against wage labor that is characteristic of all former slave plantation cultures, but one with particular appeal in the Dominican Republic due to its strong history of nonmarket monteria.[94] The aggressive edge of these terms also may relate to a form of underclass posturing which seeks to challenge U.S. and European men who come to the island as tourists to sexually consume Dominican women, which is especially galling given the fact that the United States invaded and occupied the Dominican Republic twice, in 1916 and 1965. Steven Gregory calls the attitude and comportment of these U.S. *conquistadores* (conquerors) imperial masculinity, "a power-laden social and semiotic architecture within which men fashion, interpret, and negotiate their relations with each other and the social world"—a framework that is rarely available to poor brown Dominican men.[95]

The gap between expectations and realities for young men in the Dominican Republic has widened significantly due to the economic vicissitudes of neoliberalism, and has in all likelihood changed the stakes for these everyday rites of masculinity. The economic downturn since the 1980s has eroded the *pequeña burguesia*, or lower middle class, and has augmented the rural and urban poor and swelled the ranks of the informal sector. It has also lessened the ability of young men to secure employment sufficient to maintain a family, which is requisite to respectable adult status. Notwithstanding the fact that the Dominican Republic has recently boasted high rates of economic growth, nearly a third of young people age fifteen to twenty-four are unemployed, and more than one-fifth of Dominicans live below the official poverty line of $2 per day.[96] After the 1980s there was a shift to tourism and free-trade assembly (now dispatched to Asia), but the fact that over 60 percent of those hired are women has caused social strain, since men are virtually excluded from job growth; the social fabric has been further challenged by the rising levels of unemployment and underemployment, which now stand at 15 percent and 40 percent respectively.[97] In this context, totemic labels help forge ties of manhood by conveying the image of big-man status. They enable working-class men to posture as men of agency with the money to conquer anyone, like the tourists who in a heartbeat can acquire virtually any Dominican woman of their choosing. And they do so by adopting the demeanor of power, since the speaking subjects lack the material trappings of social class.[98]

Dominican men use these predatory animal nicknames as emblems of street masculinity that also serve to keep the ever-present threat of homosexuality at bay by helping them to avoid a posture of subordination; as such the nicknames only appear to be deferential compliments or tokens. At times, these epithets even have a jocular tone in the form of "playful profanation," as in perrón, or large dog.[99] Subordinate masculinity might land one a dishonorable totemic identification, such as *cochón* (queer), *gallina* (hen), or *pájaro* (bird), which expresses the passive, feminized condition of being what Antonio de Moya has termed the "prostrate other."[100] These beastly labels, even if they are emblems of aggression, can somewhat surprisingly serve rituals of solidarity between men—as in other segmentary societies, for which, as Stanley Tambiah has said, exogamy and alliance, fighting and marriage, are conjoined.[101] Serving both as presentational rituals and as compliments, these male epithets amount to badges of male honor between and among men which enable a kind of pantomime for working-class men, who can thus perform a style of aggressive sexuality which is in reality available only to the wealthy.[102] As such these "tíguere leones" (tiger lions) and "tíguere gallos" (tiger roosters) resist an everyday social order in which men have little agency, by celebrating trickster wiles for scaling the social order in improper ways, such as sleeping with the boss's wife or with the American church volunteer.[103] This form of speech play is an important component of *relajarse*, a kind of banter, typically among men and accompanied by alcohol and laughter, that produces solidarity or *confianza* (trust) among men, by drawing on idioms of deference so as to paradoxically invoke an imagined brotherhood of studs.[104] Relajarse is thus a particular genre of the "oratorical eloquence" that is part and parcel of being male in the West Indies.[105]

Animal epithets first entered the formal political realm in the nineteenth century through the *gallera*, or cockfighting ring, as liberal and conservative parties were called *bolos* and *rabuses*—fighting cocks with and without barbed tails. The goat entered the fray when the goatee became popular among liberal politicians in the 1920s, as the country was opened to global commerce by the free-trade regime that accompanied the U.S. occupation, which brought a modern, less-hirsute look to urban Dominican men, as they took advantage of imported razors to trim their beards.[106]

To call Trujillo the goat invoked a popular culture of male animal speech play, but with a twist. Goats, of course, are patently not tigers; if the cunning and noble tiger has become the ultimate badge of Dominican manhood, the goat is its status obverse, a scavenger who eats anything and who is not hunter but prey. Goats are also mediators between bush and home, mortals and gods, as well as, today, being the preferred sacrificial animals to offer the misterios (spirits). As symbols of manhood, however, they underscore its most bestial aspect: pure indiscriminate sexual avarice without the art of romance or pretense of organized mating behavior.

As A. R. Radcliffe Brown has said, animals have a ritual value that enables them to articulate social worlds and, in this case, uproot an invasive regime of domination—the Trujillato—which exercised absolute power for far too long.[107] The feast of the goat turned the infrapolitics of the powerless into a gastropolitics of degradation, as Dominican men who had been shamed for decades by this ruthless dictator restored their dominance by eating him, in a rite which, drawing on Mikhail Bakhtin, destroyed the old order while it sowed and regenerated the body politic.[108] As Edmund Leach has said, "Men do not have to cook their food; they do so for symbolic reasons to show they are men and not beasts."[109] Eating is a metaphor for raw power, and it is not incidental that the tiburón (shark) appeals as a male emblem since it's a man-eater, and a *chivito* (little goat) is a term for a meat-eater.

Calling Trujillo a goat, then eating him as a form of resistance, was also an act of grotesque realism, since there could not have been anything more repulsive than actually consuming this most reviled of national icons; it was a form of deep play which had to be highly disguised indeed to actually become a pleasurable act (which may help account for the fact that today most Dominicans do not have a ready explanation for why Trujillo was called the goat, notwithstanding the title of Mario Vargas Llosa's fine novel).

Yet, as Clifford Geertz would say, the fiesta del chivo was certainly a story Dominicans desperately needed to tell themselves about themselves—that they had finally rectified the period of national humiliation, for once they were the ones eating the goat, rather than the reverse, and that they had thereby become a nation of hombre hombres, of tigers rather than of goats, once again.[110] And best of all, this was accomplished in encrypted form, drawing on deeply vernacular peasant practices related to billy goats and carnivalesque devils to disguise the oppositional politics of this dechoukaj, thus rendering it the perfect weapon of the weak.[111]

Notes

The epigraph is from Paul Austerlitz, *Merengue: Dominican Music and Dominican Identity* (Philadelphia: Temple University Press, 1997), 83. Special thanks to Martha Ellen Davis and Luis Cuesta for researching the history of this song for me.

This chapter is based on research conducted with funding from the University of California, Los Angeles (UCLA) International Institute, the UCLA Faculty Council on Research (COR) Grant, and the UCLA Center for the Study of Women. My research included interviews with anti-Trujillistas and others in Santo Domingo, as well as with goat farmers in the central frontier town of Bánica, in 2010. It also draws on research collected in the David Nicholls Archive, Regents Park College, Oxford University. It was revised during my tenure as an American Council of Learned Societies (ACLS) Burkhardt fellow at the Huntington Library. I presented a version at the conference "The Spanish Caribbean: Towards a Field of Its Own," in July 2011, where it benefited from comments from Frank Moya Pons among others; and at the Latin American Studies Association meeting in Rio de Janeiro, May 2009, where it was improved greatly by comments from Kevin Yelvington, Zeb Tortorici, Marion Traub-Werner, and the audience. Thanks also go to Lipe Collado for his eyewitness account of the fiesta del chivo (personal communication, July 2010), as well as to Abercio Alcántara and Irma Mora, who assisted my rural interviewing in Bánica.

1. "Colocan tarja conmemorativa del caída del dictador," *El Caribe*, 31 May 1962, 1–2.
2. In 1962 these festivities were organized by the Unión Cívica and the Movimiento 14 de Junio during the short window of civic liberty between the departure of Trujillo's family and the commencement of the repressive Balaguer regime (Lipe Collado, interview by author, 21 July 2009).
3. Allan Stoekl, "Introduction," in Georges Bataille, *Visions of Excess: Selected Writings, 1927–1939* (Minneapolis: University of Minnesota Press, 1985).
4. Frank Moya Pons, interview by author, 26 July 2001. Some people told me that Trujillo was called the goat because of how he bleated when he was killed; others said it was because he took over the national space the way a goat marks its territory.
5. Jorge Mañach, *Indagación del choteo* (Havana: La Verónica, 1940).

6. Peter Stallybrass and Allon White, *The Poetics and Politics of Transgression* (New York: Cornell University Press, 1986), 46–47.

7. The importance of goats increased after the U.S.-enforced slaughter of creole pigs in 1979 due to fears of a swine-flu epidemic spreading to the United States, which exterminated this race of pigs. See Manuel J. Andrade, *Folk-Lore from the Dominican Republic* (New York: American Folk-Lore Society, 1930), 46. For more on the colonial peasantry, see Raymundo González, *De esclavos a campesinos: Vida rural en Santo Domingo colonial* (Santo Domingo: Archivo General de la Nación, 2011).

8. Raymundo González, interview by author, July 2007.

9. Here I am drawing on Pierre Nora's "sites of memory," but in a context in which property was collective and not spatialized (*terrenos comuneros*), and was based on shares of access and usufruct not on private rights to particular bounded locations. See Nora's "Between Memory and History: Les Lieux de Memoire," *Representations* 26 (spring 1989): 7–25.

10. Claude Lévi-Strauss, *The Raw and the Cooked* (New York: Harper and Row, 1969); and Mary J. Weismantel, *Food, Gender and Poverty in the Ecuadorian Andes* (Philadelphia: University of Pennsylvania Press, 1998), 14.

11. The notion of "embodied history" (or habitus) is from Pierre Bourdieu, *The Logic of Practice* (Stanford: Stanford University Press, 1990); A. L. Beier, *Masterless Men: The Vagrancy Problem in England 1560–1640* (London: Methuen, 1985).

12. For more on this in Puerto Rico, see Francisco Scarano, "The Jíbaro Masquerade and the Subaltern Politics of Creole Identity Formation in Puerto Rico, 1745–1823," *American Historical Review* 101.5 (December 1996): 1398–431.

13. Annette Weiner, "Cultural Difference and the Density of Objects," *American Ethnologist* 21.2 (1994): 391–403; Fred R. Myers, "Introduction: The Empire of Things," in *The Empire of Things: Regimes of Value and Material Culture*, ed. Fred R. Myers (Santa Fe: School of American Research, 2001), 3–64. The term *proto-peasantry* is from Sidney W. Mintz and is intended to convey the way in which peasant formations in the Caribbean formed in the interstices of and as a mode of resistance to slavery. See Mintz's *Caribbean Transformations* (Baltimore: Johns Hopkins, 1984). Chicken is also used in sacrifice, but a goat is preferred.

14. Eduardo Archetti, *Guinea-Pigs: Food, Symbol and Conflict of Knowledge in Ecuador* (New York: Berg, 1997), 54.

15. Manuel Vicente Hernández González, *El sur dominicano (1680–1795): Cambios sociales y transformaciones económicos* (Santo Domingo: Archivo General de la Nación, 2008), 314. I am speaking of the period before the 1979 USAID-induced pig slaughter, which eradicated almost all the creole pigs on the island.

16. Javier Malagón Barceló, *Código Negro Carolino* (Santo Domingo: Editora Taller, 1974). For more on fencing legislation, see Richard Turits, *Foundations of Despotism: Peasants, the Trujillo Regime and Modernity in Dominican History* (Stanford: Stanford University Press, 2003), 57–59.

17. Lauren Derby, "Gringo Chickens with Worms: Food and Nationalism in the

Dominican Republic," in *Close Encounters of Empire: Writing the Cultural History of U.S.–Latin American Relations*, ed. Gilbert M. Joseph, Catherine C. LeGrand, and Ricardo D. Salvatore (Durham: Duke University Press, 1998), 451–96.

18. Weismantel, *Food, Gender and Poverty in the Ecuadorian Andes*, 6, 152; and Archetti, *Guinea-Pigs*. As Weismantel notes, while *comida criolla* is prized, it lacks status in relation to "white" foods.

19. Anton Blok, "Mediterranean Totemism: Rams and Billy Goats," in *Honour and Violence* (Malden, Mass.: Blackwell, 2001), 176.

20. The depravity of Abbes and his role in the repressive apparatus is well described in Mario Vargas Llosa, *La Fiesta del Chivo* (New York: Farrar, Straus and Giroux, 2000).

21. "Pueblo deplora el pleno asesinato del Benefactor," *La Nación*, 5 June 1961; "La Era de Trujillo no puede desaparecer," *La Nación*, 12 June 1961.

22. "El terror irrefrenado del gobierno se desata ametrallando al pueblo: Tanques de guerra AMD causan pánico en calles," *Unión Cívica*, 14 September 1961.

23. "Sigue el terror en La Vega patrocinado por 'los paleros,'" *Unión Cívica*, 14 September 1961.

24. Lipe Collado, interview by author, July 2009. Collado attended the original fiesta del chivo when he was fifteen and a youth member of the 14 de Julio movement.

25. Interview by Abercio Alcántara, Bánica, May 2009.

26. Thanks to the Duke anonymous reader for suggesting this interpretation.

27. These observations about goat behavior were culled from interviews, conducted from 2008 to 2011, with goat farmers and Dominicans who had grown up in rural areas.

28. Andrade, *Folk-Lore from the Dominican Republic*, 16; José Labourt, *Sana, sana, culito de rana* (Santo Domingo: Editora Taller, 1982), 177–79. More Dominican goat sayings can be found in Juan Carlos García, "Breve que te quiero breve," *El Nacional*, 16 July 2010, 17; and at "El chivo en República Dominicana," Boque Chivo Blog, 31 July 2007, http://boquechivo.diariolibre.com/blog/?p=3.

29. Blok, "Mediterranean Totemism," 174. See Antonio de Moya, "Power Games and Totalitarian Masculinity in the Dominican Republic," in *Caribbean Masculinities: Working Papers*, ed. Rafael L. Ramírez, Victor I. García-Toro, Ineke Cunningham (San Juan: HIV/AIDS Research and Education Center, University of Puerto Rico, 2002), 105–7, for a discussion of public cuckolding rituals in the Dominican Republic. For a tale about a goat being "scapegoated" for a man named Pedro Animal who beats his wife to death for her jealousy (and covers her with goat tripe to hide his crime), see "Juan Sonso y su compadre Pedro," in Andrade, *Folk-Lore from the Dominican Republic*, 45.

30. Notably, however, the term for this behavior is *cabrón*, not *chivo*.

31. Andrade, *Folk-Lore from the Dominican Republic*, 23, 77.

32. Michael T. Taussig, *Defacement: Public Secrecy and the Labor of the Negative* (Stanford: Stanford University Press, 1999), 5.

33. William H. Beezley, *Judas at the Jockey Club and Other Episodes of Porfirian Mexico* (Lincoln: University of Nebraska Press, 2004), 89–90.

34. Fernando Cervantes, *The Devil in the New World: The Impact of Diabolism in New Spain* (New Haven: Yale University Press, 1994).

35. Elizabeth McAlister, *Rara! Power and Performance in Haiti and Its Diaspora* (Berkeley: University of California Press, 2002), 3.

36. Dagoberto Tejeda Ortiz, *El carnaval dominicano: Antecedentes, tendencias y perspectivas* (Santo Domingo: Editora Amigo del Hogar, 2008), 207.

37. Elaine Breslaw, *Witches of the Atlantic World: A Historical Reader and Primary Sourcebook* (New York: New York University Press, 2000), 243.

38. José Limón, *Dancing with the Devil: Society and Cultural Poetics in Mexican-American South Texas* (Madison: University of Wisconsin Press, 1994), 174.

39. Tejeda Ortiz, *El carnaval dominicano*, 211; Limón, *Dancing with the Devil*, 174.

40. From the town of La Vega; see Tejeda Ortiz, *El carnaval dominicano*, 210.

41. Norman Cohn, quoted in Breslaw, *Witches of the Atlantic World*, 51; Labourt, *Sana, sana, culito de rana*, 163.

42. James A. Arnold, *Monsters, Tricksters, and Sacred Cows: Animal Tales and American Identities* (Charlottesville: University of Virginia Press, 1996), 9–10.

43. Gabriel García Márquez, Nobel Lecture, 8 December 1982. Thanks to Raul Moreno who brought this text to my attention. For more on European werewolf lore, see Montague Summers, *The Werewolf* (London: K. Paul, Trench, Trubner, 1933).

44. David Pratten, *The Man-Leopard Murders: History and Society in Colonial Nigeria* (Bloomington: Indiana University Press, 2007); Allen F. Roberts, "Perfect Lions, Perfect Leaders: A Metaphor for Tabwa Leadership," *Journal des Africanistes* 53.1 (1983): 93–105; Malcolm Ruel, "Were-Animals and the Inverted Witch," in *Witchcraft Confessions and Accusations*, ed. Mary Douglas (London: Tavistock, 1970), 333–50; and León Garcia Garagarza's essay in this volume.

45. Harold Courlander, *The Drum and the Hoe: Life and Lore of the Haitian People* (Berkeley: University of California Press, 1973), 98; Katherine Smith, "*Lansetcòd*: Market, Memory and Mimicry in Haitian Carnival," in *Kanaval: Vodou, Politics and Revolution on the Streets of Haiti*, by Don Constantino, Richard Fleming, and Leah Gordon, ed. Stuart Baker (London: Soul Jazz Publishing, 2010), 71–78.

46. U. A. Casal, "The Goblin Fox and Badger and Other Witch Animals of Japan," *Folklore Studies* 18 (1959): 1–2; Dan Boneh, "Mystical Powers of Hyenas: Interpreting a Beduin Belief," *Folklore* 98.1 (1987): 57–58; Bill Ellis, "Why Is a Lucky Rabbit's Foot Lucky? Body Parts as Fetishes," *Journal of Folklore Research* 39.1 (January–April 2002): 67–68. For more on the bacá, see Lauren Derby, *The Dictator's Seduction: Politics and the Popular Imagination in the Era of Trujillo* (Durham: Duke University Press, 2009), chap. 6; and Christian Krohn-Hansen, *Political Authoritarianism in the Dominican Republic* (New York: Palgrave Macmillan, 2008), chap. 7.

47. The quote emerged from a focus group I conducted on *tigueraje*, Bánica, Elias

Piña, October 2008. For more on the tíguere, see Lipe Collado, *El tíguere domin-icano* (Santo Domingo: Talleres Gráficos de la UASD, 1981); and Christian Krohn-Hansen, "Masculinity and the Political among Dominicans," in *Machos, Mistresses, Madonnas: Contesting the Power of Latin American Gender Imagery*, ed. Marit Melhuus and Kristi Anne Stølen (New York: Verso, 1996), 108–33.

48. I observed the devil depicted with a Tony the Tiger mask at Los Diablos del Sur, Good Friday, Bánica, Elias Piña, April 2009.

49. Andrade, *Folk-Lore from the Dominican Republic*, 294.

50. McAlister, *Rara!*, 122. Although McAlister discusses Haiti, the Catholic Church in the Dominican Republic has traditionally undergone similar policies.

51. In 1961 the state slaughtered religious pilgrims at Palma Sola due to their hetero-dox beliefs. See Lusitania Martínez, *Palma Sola: Su geografía mística y social* (Santo Domingo: Ediciones Centro de Desarrollo del Espíritu Empresarial, 1991); and Juan Manuel Garcia, *La massacre de Palma Sola: Partidos, lucha política y el asesi-nato del general, 1961–63* (Santo Domingo: Editora Alfa y Omega, 1986). For a discussion of similar practices of diabolization in Haiti, see Kate Ramsey, *The Spirits and the Law: Vodou and Power in Haiti* (Chicago: University of Chicago Press, 2011).

52. Such as los Diablos de Barahona (see Tejeda Ortiz, *El carnaval dominicano*, 208).

53. See Igor Kopytoff's call for biographies of things, in his essay "The Cultural Biography of Things: Commoditization as Process," in *The Social Life of Things*, ed. Arjun Appadurai (New York: Cambridge University Press, 1986), 64–94.

54. John M. Lipski, "The Spanish of the Canary Islands," Penn State Personal Web Server, http://www.personal.psu.edu/jml34/Canary.htm; and Irene Pérez Guerra, ed., *Estado actual de los estudios lingüisticos y filológicos en la República Dominicana* (Santo Domingo: Patronato de la Ciudad Colonial, 2000).

55. While Oviedo had said these hunters could fell five hundred heads in a morning, later accounts report numbers closer to one hundred. See Gonzalo Fernández de Oviedo y Valdés, *L'histoire naturelle et general des Indes, isles, et terre ferme de la grande mer Oceane*, trans. Jean Poleur (Paris: Michel de Vascosan, 1556); M. L. E. Moreau de St. Mery, *Descripción de la parte española de Santo Domingo* (Ciudad Trujillo: Editora Montalvo, 1944); Peter R. Galvin, *Patterns of Pillage: A Geogra-phy of Caribbean-Based Piracy in Spanish America, 1536–1718* (New York: Peter Lang, 1999), 114.

56. Alexandre Olivier Esquemeling, *Buccaneers of America: A True Account of the Most Remarkable Assaults Committed of Late Years upon the Coast of the West Indies by the Buccaneers of Jamaica and Tortuga* (New York: Dover, 1967), 54; Jean Baptiste Labat, *The Memoirs of Pere Labat, 1693–1705* (New York: Routledge, 1970).

57. Yvan Debbasch, "Le Maniel: Further Notes," in *Maroon Societies: Rebel Slave Communities in the Americas*, ed. Richard Price (Garden City, N.Y.: Anchor, 1973), 143–48.

58. Mintz, *Caribbean Transformations*.

59. Raymundo González, "El comegente," in *Homenaje a Emilio Cordero Michel*

(Santo Domingo: Academia Dominicana de la Historia, 2004), 199–200: "Esta clase de hombres no tienen conucos, ni labranzas, ni otro oficio honesto con que mantenerse, sin embargo, comen, beven, se emborrachan y triunfan. De donde ha de salir esto sino del robo y maldad?"

60. "Rancherias de gentes pobres que viven de la monteria . . . los quales pasan al año sin ver las capitales al modo de los primeros indios" (Antonio Sánchez Valverde, *La idea del valor de la isla española* [Ciudad Trujillo: Editora Montalvo, 1947], 148). See González, "El comegente," 203, for a quote in which someone curses the woods for providing a cover-up for these escaped slaves. Moreno Fraginals notes that the rise of sugar was the end of the Caribbean forest, although in the Dominican Republic this process started later than in Cuba, with the late nineteenth-century U.S.-led sugar invasion (*Sugarmill: The Socioeconomic Complex of Sugar in Cuba, 1760–1860* [New York: Monthly Review Press, 1976]).

61. See Raymundo González, "Ideologia del progreso y campesinado en el siglo XIX," *Ecos* 1.2 (1993): 25–44; and Cornelius Pauw, *A General History of the Americans* (Rochdale: Printed by and for T. Wood, 1806), 7, who states, "It is agriculture that has led man by the hand from a savage state to a politic constitution: the more cultivated the soil, the more abundant the harvest, the sooner will the cultivators humanize," a position which Sánchez Valverde debates at length in *La idea del valor en la isla española*.

62. Marshall Sahlins, *Stone Age Economics* (Chicago: Aldine-Atherton, 1972).

63. William Walton, *Present State of the Spanish Colonies* (London: Longman, Hurst, Orme and Brown, 1810), 32.

64. Walton, *Present State of the Spanish Colonies*, 33–35. Juan B. Pérez says he caught wild pig in the 1930s (*Geografia y sociedad* [Santo Domingo: Editora del Caribe, 1972], 269) and Lipe Collado described to me how he hunted feral goats in Azua in his childhood in the 1950s.

65. Cited in Bernardo Vega and Cordero Michel, *Asuntos Dominicanos en archivos ingleses* (Santo Domingo: Fundación Cultural Dominicana, 1993), 15.

66. Sánchez Valverde, *La idea del valor en la isla española*, 146fn200.

67. J. Brown, *The History and Present State of Sto. Domingo* (Philadelphia: W. Marshall, 1837), 288.

68. Catherine Gallagher and Stephen Greenblatt, "The Potato in the Materialist Imagination," in *Practicing New Historicism* (Chicago: University of Chicago Press, 2000), 111, 114.

69. Contrast this with Sharon Hutchinson, "The Cattle of Money and the Cattle of Girls among the Nuer," *American Ethnologist* 19.2 (May 1992): 294–316; and Jean Comaroff and John Comaroff, "Goodly Beasts and Beastly Goods: Cattle and Commodities in a South African Context," *American Ethnologist* 17.2 (May 1990): 195–216.

70. As in the *chiva blanca* (comp. Pedro Mendoza). Interestingly, creole pigs are also eroticized, as in the drinking story "La puerquita con las nalgitas doraditas," which is about a man who goes to collect a piglet for Christmas dinner, becomes sexually

aroused, and finally consummates his lust with the beast (after which he cannot bring himself to eat it). This story emerged in a focus group on tigueraje, Santo Domingo, October 2008.

71. Edmund R. Leach, "Anthropological Aspects of Language: Animal Categories and Verbal Abuse," in *New Directions in the Study of Language*, ed. Eric H. Lenenberg (Cambridge: Massachusetts Institute of Technology Press, 1964), 39.

72. Courlander, *The Drum and the Hoe*, 98. For the anomalous animal as a mediation of structural opposition, see Claude Lévi-Strauss's masterful study of Amerindian mythology, *Mythologiques*, 3 vols. (New York: Harper and Row, 1969). Mary Douglas also discusses how anomalous animals are taboo because they are betwixt and between, in "The Abominations of Leviticus" and "The System Shattered and Renewed," both in *Purity and Danger* (Harmondsworth: Penguin, 1970), 51–71, 196–220, respectively.

73. Norman Cohn cited in Breslaw, *Witches of the Atlantic World*, 51.

74. Émile Durkheim, *Elementary Forms of Religious Life* (New York: Free Press, 1995), 322.

75. Lydia Cabrera, *El Monte*, quoted in Roberto González Echeverria, "Biografia de un cimarrón," in *The Voice of the Masters: Writing and Authority in Modern Latin American Literature* (Austin: University of Texas Press, 1988), 119.

76. For a story of a ciguapa that kidnapped a man and never returned his children, see Juan B. Perez, *Geografia y sociedad* (Santo Domingo: Editora del Caribe, 1972), 243. On the negro incognito, see González, "El comegente." The *caco* chieftain Charlemagne Peralte was also said to be able to assume other forms, as was the 1930s Dominican outlaw Enrique Blanco. See Roger Gaillard, *Charlemagne Péralte le Caco* (Port-au-Prince: R. Gaillard, 1982), 333.

77. This was noted to me by small ranchers in the central frontier town of Bánica.

78. The meanings of goat sacrifice in Brazil are treated in Brian Brazeal, "A Goat's Tale: Diabolical Economies of the Bahian Interior," in *Activating the Past: History and Memory in the Black Atlantic World*, ed. Andrew Apter and Lauren Derby (Newcastle upon Tyne: Cambridge Scholars Publishing, 2010), 267–94.

79. George Eaton Simpson, *Religious Cults of the Caribbean: Trinidad, Jamaica and Haiti* (Rio Piedras: Institute of Caribbean Studies, Puerto Rico, 1980), 19–20; Lillian Nérette Louis, *When Night Falls: Kric Krac! Haitian Folktales* (Englewood: Libraries Unlimited, 1999), 50, 55; Alfred Metraux, *Voodoo in Haiti*, trans. Hugo Charteris (New York: Oxford University Press, 1959), 36.

80. Metraux, *Voodoo in Haiti*, 54.

81. Labourt, *Sana, sana, culito de rana*, 19–21.

82. Peter Wilson and Clifford Geertz, cited in Frank Manning, "Nicknames and Number Plates in the British West Indies," *Journal of American Folklore* 87.344 (April–June 1974): 130, 132. For more on Caribbean masculinity, see Roger Abrahams, *The Man-o-Words in the West Indies* (Baltimore: Johns Hopkins University Press, 1983).

83. Anthony Lauria, "'Respeto,' 'Relajo' and Inter-personal Relations in Puerto Rico,"

Anthropological Quarterly 37.2 (April 1964): 56; Manning, "Nicknames and Number Plates in the British West Indies," 130. Thanks to Kevin Yelvington for suggesting this citation to me. Elijah Anderson notes the importance of greeting among African Americans, in *Street Wise: Race, Class, and Change in an Urban Community* (Chicago: University of Chicago Press, 1990), 168.

84. Roberto DaMatta, *Carnival, Rogues and Heroes: An Interpretation of the Brazilian Dilemma* (Notre Dame: University of Notre Dame Press, 1991), 198.

85. Manning, "Nicknames and Number Plates in the British West Indies," 130.

86. Valentina Peguero, *Militarization of Culture in the Dominican Republic, from the Captains General to General Trujillo* (Lincoln: University of Nebraska Press, 2004).

87. Kevin Yelvington, "Flirting in the Factory," *Journal of the Royal Anthropological Institute* 2 (1996): 325–28. A celebration of Dominican animal banter called "Dominican Tigueraje Animal Allusion 'Street Talk,'" appears in the DR1 online message board, http://www.dr1.com/forums/general-stuff/111509-dominican-tigueraje-animal-allusion-street-talk.html.

88. Marcy Norton, "Going to the Birds: Birds as Things and Beings in Early Modernity," in *Early Modern Things: Objects and Their Histories, 1500–1800*, ed. Paula Findlen (London: Routledge, 2013), 53–83.

89. Max Gluckman, "Rituals of Rebellion in South-East Africa," in *Order and Rebellion in Tribal Africa*, ed. Max Gluckman (London: Cohen and West, 1963), 1–37.

90. De Moya, "Power Games and Totalitarian Masculinity in the Dominican Republic," 114.

91. For more on the tíguere, see Derby, *The Dictator's Seduction*, chap. 5; and Christian Krohn-Hansen, "Masculinity and the Political among Dominicans: The Dominican Tiger," in *Machos, Mistresses, Madonnas: Contesting the Power of Latin American Gender Imagery*, ed. Marit Melhuus and Kristi Anne Stølen (New York: Verso, 1996), 108–33.

92. James Scott, "Domination, Acting, Fantasy," in *The Paths to Domination, Resistance, and Terror*, ed. Carolyn Nordstom and Joann Martin (Berkeley: University of California Press, 1992), 66. See also Lewis Hyde, *Trickster Makes This World: Mischief, Myth, and Art* (New York: Farrar, Straus and Giroux, 1998).

93. Abrahams, *The Man-o-Words in the West Indies*. For more on circum-Caribbean masculinity, see Roger N. Lancaster, *Life Is Hard: Machismo, Danger, and the Intimacy of Power in Nicaragua* (Berkeley: University of California Press, 1992).

94. Steven Gregory discusses informality as a means of resistance to wage labor in *The Devil behind the Mirror: Globalization and Politics in the Dominican Republic* (Berkeley: University of California Press, 2007), 191.

95. Gregory, *The Devil behind the Mirror*, 135. See also José E. Limón, "Carne, Carnales and the Carnivalesque," in *Dancing with the Devil: Society and Cultural Poetics in Mexican-American South Texas* (Madison: University of Wisconsin Press, 1994), 123–40.

96. Marie Michael, "The Dominican Republic: Latin America's Latest Economic

'Miracle'?" *Dollars and Sense Magazine*, March–April 2001, http://www.dollars andsense.org/archives/2001/0301toc.html.

97. Thomas K. Morrison and Richard Sinkin, "International Migration in the Dominican Republic: Implications for Development Planning," *International Migration Review* 16.4, special issue, "International Migration and Development" (winter 1982): 819–36. I thank Carel Alé for this reference.

98. DaMatta, *Carnival, Rogues and Heroes*, 234.

99. Erving Goffman, *Interaction Ritual: Essays in Face-to-Face Behavior* (Chicago: Aldine, 1967), 87.

100. De Moya, "Power Games and Totalitarian Masculinity in the Dominican Republic," 132; Lancaster, *Life Is Hard*, 273.

101. Stanley Tambiah, "Animals Are Good to Think and Good to Prohibit," *Ethnology* 8.4 (October 1969): 423.

102. Goffman, *Interaction Ritual*, 72.

103. These double-barrel composite categories were suggested to me in an interview with César Augusto Zapata, 2009.

104. Mikhail Bakhtin, cited in Limón, "Carne, Carnales and the Carnivalesque."

105. Yelvington, "Flirting in the Factory"; Manning, "Nicknames and Number Plates in the British West Indies"; Peter J. Wilson, *Crab Antics: The Social Anthropology of English-Speaking Negro Societies of the Caribbean* (New Haven: Yale University Press, 1973).

106. Derby, *The Dictator's Seduction*, chap. 1.

107. Cited in Boneh, "Mystical Powers of Hyenas."

108. Mikhail Bakhtin, *Rabelais and His World* (Bloomington: University of Indiana Press, 1984).

109. Edmund Leach, *Claude Lévi-Strauss* (New York: Penguin, 1976), 102.

110. Clifford Geertz, "Notes on the Balinese Cockfight," *Interpreting Cultures: Selected Essays* (New York: Basic, 1973). On the hombre hombre, see Lancaster, *Life Is Hard*, 239.

111. James Scott, *Weapons of the Weak: Everyday Forms of Peasant Resistance* (New Haven: Yale University Press, 1985).

Loving, Being, Killing Animals

NEIL L. WHITEHEAD

Since the appearance of Peter Singer's *Animal Liberation* (1975), followed by Tom Regan and Singer's *Animal Rights and Human Obligations* (1976), an incremental but clearly visible shift in the public view of human-animal relations has occurred, inspired by a growing output of books, articles, and films, the appearance of organizations and grassroots movements, and lifestyle changes.[1] Previously obscured from critical inquiry, nonhuman nature became the object of philosophical discourse, mostly confined to universities in Europe and the United States.[2]

The result has been a series of reforms leading to more humane treatment of animals, the spread of direct-action politics around such issues as hunting, trapping, lab testing, and animal farming, and greater public readiness to take animal interests seriously, leading, for example, to stiffer prison sentences in cases of animal cruelty. There is a general heightened awareness, thanks partly to the Darwinian legacy, that humans and animals occupy the same temporal space, their fates organically bound together within the same planetary ecology. However, this liberal cultural framework fails to escape the logic of capitalism and colonialism, since the universalization of "human" rights and the extension of those rights to "animals" begs many questions as to animality and humanity as well as about the emancipatory potential of the human rights discourse itself. This issue is important since, as Regina Horta Duarte suggests in this volume, "during the first decades of the twentieth century, the processes of constructing national identities in various Latin American countries were decisively linked to the sciences of the natural world."

Moreover, the logic of domination is inherent in our attempts to write animals in, just as with the category of "children," the perceived lack of opportunity or ability to "speak for oneself" invites the rescuing discourse of inherent "rights" to supplant this silence. Thus there are both political and theoretical issues at stake here: on the one hand the advocacy of inherent value and moral rights which are part of our progressive liberal modernity, and on the other a scientifically and medically inflected perception of the human as an integral category of ontology. If we are to move forward from the arguments made over thirty years ago by Singer and Regan, then, as do authors in this volume, we must begin to not just center the "animal" but simultaneously to decenter the "human."[3] John Soluri writes in his chapter for this volume, "I am hesitant to reject liberal notions of 'human rights' or to affirm 'animal rights' where animals are implicitly understood to be individuals. The historical and ecological meanings of animals are always contextual—conditioned by when and where they live." In the same way that the "human rights" discourse has been welded to the exercise of global domination by Western governments, so too does restricting animals to the role of historical "victims" threaten to reproduce the way in which liberal discourse on human social justice is apt to confine political resistance to symptoms of trauma or to historical victimization and so erase the eruptive political meanings of such resistance.[4] In this way the uncritical call for "animal rights" can become a disguised form of neocolonial control, as in contemporary Asia, Africa, or Latin America.[5] The legal and moral necessity for Western governments, NGOs, and liberal activists to intervene in local contexts is then advanced under the guise of a "deep ecology" and rhetorically driven by a variety of preservationist and conservationist discourses.[6] Particularly prominent in this elision of the conservationist and neocolonialist agenda is the "free-market environmentalist" movement, whose guiding light, Terry L. Anderson, has produced a series of works over the last decade promoting the idea that free-market approaches can also be an effective means to conserve wildlife by giving local peoples, "tribals," a financial stake in wildlife and ecological management programs.[7] Notably such programs often have at their core some form of wildlife conservation that fits well with the tourist and recreational priorities of rich Europeans and Americans. Overall this "free-market" strategy, by calling for the creation of politically stable market conditions and the use of local labor in servicing those markets, not only marginalizes existing forms of local livelihood but also recreates local peoples as market-oriented consumers. The wildlife and animals likewise are commodified as tourist and recreational objects, in turn producing a hierarchy of value that prioritizes iconic species such as tigers, jaguars, and lions, or elephants, rhinos, and spectacular bird-life. The insertion of such highly capitalized entrepreneurial projects into marginal or even contested regions of postcolonial states can also produce a convergence of "ecological" and "state security" interests.[8] Moreover, the subtle conjoining of capitalist economic development, wildlife conserva-tion, and "human" rights means that even such liberal and progressive NGOs as Cultural Survival become complicit in forms of conservancy that func-tion as a means of socially engineering local receptivity to capitalist forms of economy.[9]

So even as we need to reject anthropomorphism as a means to construct an animal historiography, and even though the authors in this volume show that writing a specific history of animals is itself problematic, the purposes of writing histories must at the least be redirected to excavating the human-animal dyad as a historically dynamic, and largely ignored, feature of those "histories of humans." Just as the idea of "landscape" began to permit new kinds of histories in which human agency, even if not abandoned, was seen as only one component of a set of complex relationships among humans and their landscapes, which necessitated closer attention to ecological process and nonhuman behaviors, so the challenge going forward is to, in turn, breach the ecological-animal/human divide. This is partly achieved in this volume through the examination of what Neel Ahuja terms elsewhere trans-species intimacies, connectivities, and formations created through the ex-ercise of imperial biopower, that is, the way in which power over humans is often via power over animals, or the animalization of humans.[10] Equally, as Duarte suggests for Brazil, in both colonial and postcolonial contexts "the search for institutions which would serve the country's entry into the mod-ern world" were also serviced by a "preservationist discourse" such that "the denunciation of the extermination of birds was a recurring theme that was certainly in accord with international movements then on the rise." Like-wise, as Heather McCrea cogently demonstrates with regard to Mexico, not just the animal but the microbial was part of this emergent modernity, and the eradication of cholera or malaria could become a means for the spread of governmentality to the wild and untamed regions beyond the rule of law. Indeed, the establishment of the modern state in North America and Eu-rope in the late nineteenth century also hinged in part on analogous cam-paigns of public health, as in the iconic case of Typhoid Mary in the United States.[11] Whether through the construction of scientific knowledge or the practice of medical health campaigns, the eradication of insects as vectors for viruses, pathogens, bacteria, microbes, and so forth—all part of living systems if not of animal life—has been at the core of attempts to establish modernity. McCrea notes the central role played by the Rockefeller Foundation's International Health Board in encouraging and aiding the campaigns against yellow fever, malaria, and cholera in Mexico, and this imperial interest in "public health" was no less evident inside the United States itself. As one of the organizations that led to the foundation of the contemporary Centers for Disease Control (CDC), the Rockefeller Foundation and its campaigns abroad illustrate a linear connection between the construction of bioregimes of health and sanitation in South America and the biopower of the contemporary U.S. state. As McCrea argues, "Animals, insects, and frequently the indigenous Maya were all alike cast as purveyors of filth and illness," and so we are still battling "Mexican" or "Asian" flu viruses emanating from pigs and chickens that would invade the national space. This in turn gives rise to what McCrea describes in this volume as "the ideal of the healthy state [that] required the regulation and control of animals, insects, and indigenous populations according to popularized notions associated with 'good' and 'bad' species and racial hierarchies." The advent of the "healthy state" is also the moment for the invention of the modern, racially biologized human.[12] Analogous campaigns against the disease of drug use (or even "jihadism") are part of this formulation of modern state power as well. It is therefore not surprising to learn that the CDC hosted, often in conjunction with the CIA and the military, a variety of bioexperiments on U.S. citizens, especially on the relatively powerless populations of African Americans, mental patients, and orphaned children.[13] Likewise the U.S. National Park Service and departments of natural resources at the state level are continually engaged in battling "invasive" species—zebra mussels, lampreys, Asian carp, Japanese beetles, ash borers, elm leaf beetles, and snakehead fish, alongside a host of plants such as garlic mustard, millefoil, and knotweed—which perpetuates the idea that the agencies of the state are defenders of our biointegrity and political freedoms. Just as the *garrapata* (tick), discussed in McCrea's chapter, dramatizes the voracious and endemic threat from the foreign that modernity must eradicate, in a dystopian vision of biomutation "killer bees" now iconically track the advance of the threatening biology of the other. As Martha Few also notes in her chapter, pathogens and insects seem not to count as "animals," but our zoological attitudes are certainly informed and shaped by ideas about pathogens and insects. Neatly and saliently conjoining those ideas is *A Plague of Sheep*, a classic work on the explosion of sheep populations in Mexico and how they overgrazed and displaced the indigenous humans.[14] As Alfred Crosby documented more widely of this "Columbian Exchange" in the Americas, and as the chapters in this volume more precisely demonstrate, the changing landscapes and biotic taxa of the Americas were perhaps the first evidence of the stirrings of that modern biopower which has now moved from the exterior body into the interior body of neurological mind and somatic chemistry, mediated through medical science.[15]

To begin an escape from this discourse of power and domination over a threatening and foreign "nature," and following many of the chapters in this volume, we might direct our attention to that aspect of non-Western thinking about animals which has been called its "perspectival quality."[16] The perspectival character of classificatory and cosmological systems, common to many indigenous peoples of the Americas and indeed elsewhere, results in viewing the world as inhabited by different sorts of subjects or persons, human and nonhuman, who apprehend reality from distinct points of view. This is not simply a relativism, but in fact challenges an opposition between relativism and universalism.[17]

As many anthropologists have already concluded, the classic distinction between Nature and Culture cannot be assumed to describe domains internal to non-Western cosmologies. Indeed, the results of ethnographic research over the last two decades implies a redistribution of the predicates subsumed within the paradigmatic sets that traditionally oppose one another under the headings of Nature and Culture: universal and particular, objective and subjective, physical and social, fact and value, the given and the instituted, necessity and spontaneity, immanence and transcendence, body and mind, animality and humanity, and many more.[18]

This has led to the suggestion that "multinaturalism," as opposed to Western "multiculturalism" is a key and revealing difference here. Multiculturalism, the plenitude of culture, is founded for Western thought on an assumption of the unity of nature but the plurality of cultures. The Natural proceeds from a supposedly objective universality of body and substance; the Cultural is generated by the subjective particularity of spirit and meaning. In this scheme indigenous conceptions suppose a spiritual unity and a corporeal diversity in which culture or the subject would be the form of the universal, while nature or the object would be the form of the particular. A critique of the distinction between Nature and Culture is therefore relevant here for the light it may shed on perspectivist cosmologies and how they can inform the attempt to "center animals." Eduardo Viveiros de Castro has neatly summarized the ethnographic materials as follows: "Typically, in normal conditions, humans see humans as humans, animals as animals and spirits (if they see them) as spirits; however animals (predators) and spirits see humans as animals (as prey) to the same extent that animals (as prey) see humans as spirits or as animals (predators)."[19] In this way the houses or villages of peccaries or jaguars, their places, give rise to their own particular habits and characteristics, their own forms of culture. Fur, feathers, claws, beaks are indeed seen as body decorations and, as such, as cultural instruments for loving and killing. So, too, in a direct inversion of sociobiology, they see their social system as human institutions of chiefs, shamans, rituals, and marriages.[20] As the editors of this volume point out, "The vast number of works dealing with indigenous peoples in the Americas attests to the astrological and cosmological significance animals had in many communities." However, indigenous perspectivism is not just a latent, ahistorical, symbolic category, but a theory of cultural practice. For example, in hunting and war, we encounter perhaps the most fundamental way of loving, being, and killing animals (others) and humans (ourselves). In this way hunting and war are the primordial relationship through which humanity and animality are created. Nor is it a given as to who occupies what position across this putative divide. As Few writes in this volume, "In other ways locusts acted like humans—swarming and flying together across the land, acting as a kind of community, and eating human, not insect, food." This humanity of the animals is usual, not exceptional, in indigenous Amazonia, where, as the Huaorani say, "We blow hunt and spear kill." The material partners in these violent exchanges are monkeys and peccaries in our eyes, but for the Huaorani, the peccaries are human for the way in which they attack villages in groups, destroying gardens, houses, and property, and even taking life if they find someone in the village. The Huaorani therefore fight peccaries with the weapons of war, not of hunting, that is, with spears, not blowpipes. Blowpipes are used to hunt monkeys for food. Moreover, because the gift of food is not to be taken lightly, the violence of killing is supplanted by love and identification: it is the weapon, the dart, which kills the beloved monkey, not the human person. *Kinship with Monkeys*, a recent ethnography of the Guajá, takes this metaphor to a logical outcome in which the relations of love and identity yield to the ultimate intimacy of cannibalism.[21] For the Guajá and the Patamuna, with whom I am familiar, all forms of plant and animal life, and especially monkeys, have *ekati* (animation) and are woven into a comprehensive kinship system or webs of affinity or love.[22] Therefore, for the Guajá, all consumption of animate life can be considered a form of incestuous cannibalism, an act of loving and killing through the intimacy of eating. More widely in South America the idiom of cannibalism is adjoined to funerary practices for close kin, and grief at the loss of another is "consumed" metaphorically and literally in acts of cannibalism. Indeed, for the Euro-American, the Eucharist of Christianity is also a moment of love and killing expressed through the consumption of the blood and flesh of the sacrificed Christ, metaphorically for some, but literally for others.[23]

Monkeys, as well as dogs, may be breastfed by a human mother and achieve a social status that recalls but exceeds that of Western "pets," as discussed in Zeb Tortorici's chapter in this volume. For captives of war are also understood as "pets" and, like monkeys, may be finally (if terminally) fully absorbed into the sociocultural matrix through cannibalism. The sexuality of love is displaced onto bodily proximity as the rules of incest track those of cannibalism. However, the intimacy of killing and eating is at least as strong as that of copulation, especially in an un-Freudian world. Loving, being, and killing animals thus dissolves the binaries of nature-culture, human-animal, lover-pet, and through a deictic perspectivism opens the possibility for corporeal transformation as spirits (subjectivities, identities) occupy variant material forms. The logics of killing and loving are therefore guided through reference to a spiritual terrain in which appropriate and inappropriate actions are articulated in a theory of practice as ritual. In turn, shamanism, the mastery of ritual as practice and performance, legitimately can be understood as the "high-science" of indigenous society.[24] Thus the Guajá monkeys are very different "persons" than the monkeys of Ahuja's chapter, who have a "contradictory status as, on the one hand, figures of progress aligned with modern biomedical technology, and, on the other, as 'invasive species' that symbolize the multiple violences of U.S. imperialism and neoliberal development policy." As the pioneering work of Emiko Ohnuki-Tierney has already made clear, the monkey is a mirror that reflects a shifting, not stable, boundary to the human that may be crossed in either direction.[25] For this reason, even mainstream primatologists, whose commitment to zoological science might suggest otherwise, now question not just the culturally dependent character of scientific knowledge, but also the ultimate feasibility of its purposes as a mode of inventing the human.[26] As Ahuja writes in this volume, "A history of imported monkeys in Puerto Rico might help scholars theorize knowledge production in the humanities, as well as in . . . biopolitical theory, critical species studies, and science studies." Furthermore, writing on the Caribbean and the popular identification of the former dictator of the Dominican Republic, Rafael Trujillo, with a goat,

Derby reminds us that not just monkeys but also goats might come to play such a role, since what is important is not just the analogical character of the bioanatomical properties of monkeys or goats when viewed alongside humans, but also how this phenomenon reveals the way in which culture, power, and symbolization unfold through time.

The commercial seal hunters discussed by Soluri, if they can be considered hunters at all, clearly do not occupy such subject positions and are arguably themselves so alienated from their labor as to make the possibility of loving and being all but unattainable. Although this did not preclude a kind of mawkishness which might have affected sealers, Soluri writes that "seal hunters' descriptions of Patagonian animals—be they fur seals, penguins, or guanacos—are unsentimental. On the rare occasion when a sealer expressed a sentiment toward an animal, the object of affection was invariably a mascot." Here the relations of production are those of killing as an industrial process, a ghastly portent of holocausts to come, just as animal husbandry foreshadows human slavery. What is seen as cultural transcendence or universalism from the perspective of multiculturalism becomes shared experience in a multinatural viewpoint. This might therefore also be the starting point for centering animals in histories informed not by a politics of identity, but by that of experience. The hunter's experience is shared with the prey, as hunting itself is always a mimesis of prey—what they do you must also do in order to find and kill them. Killing by being an animal gives rise to affinity. Even in the must-have commodity- and gadget-driven world of U.S. sport hunting, this mimesis of prey is evident, even to nonhunters, in the endless reproduction of images of prey, a hyperreality of baleful but grateful prey.[27] Nevertheless, most hunters will certainly assent to the idea that animals are all individuals who fail to match expectations generated by scientific zoologies in endless ways.

However, the fact that indigenous cosmologies centered animals in ways different than did European cosmologies should not obscure how, both in Europe or America, local or "folk" ideas and practices were under heavy pressure from the burgeoning modernity of science and associated systems of classification. Just as Keith Thomas has shown for Europe that there were fundamental changes in attitudes to the "natural world" from the sixteenth century on, so too does Adam Warren in this volume reveal how a particular native healing system was melded with European medical taxonomy: "By the eighteenth century, in fact, such local practices involving animal-based treatments had spread beyond indigenous communities and had come to constitute a key feature of popular home medical guides known as *recetarios* in Peru."[28] Such a situation might also be interpreted as one in which the indigenous was modernized through appropriation of indigenous healing practices that involved animals and the "translation" of those practices into pharmacy. Given the conjunction of humoral ideas in both Peruvian and European magical healing practices, this process may appear almost invisible, or even as a valuable "rescuing" of vanishing cultures. It may be that, as Warren notes of the particular recetario he discusses, such a work "clearly reflects the complex concepts and theories of animal- and plant-based healing central to southern Andean indigenous medical thought, furthering their transfer into popular colonial medical practice." But such a process of translation, however benign, necessarily subtly alters the frameworks of thought that had previously informed Peruvian "humoral" healing. This is certainly the case today with the explosion of neoshamanism.[29] Culturally unappealing aspects of indigenous practices, such as sacrifice (human and animal), dark shamanism, and the use of ecstatic narcotics, are apt to be erased, while biopirates from international corporations traverse the Amazon searching for pharmaceutical elements derived from indigenous manipulations of animals and plants. This means that we still need to assess the overall significance of animal manipulations in indigenous systems before we can better evaluate the place of animals in both Peruvian and European healing practices.

Another potential form of history—informed not by a politics of identity but by that of experience—emerges from the way in which "zoology" fails to properly represent animals. Queering the nonhuman and noting the biological "exuberance" of animals which our human identity otherwise acts to constrain could be the basis for histories in which interpreting animal actions become much more central.[30] Animals are not passive recipients of human agency, but are active in shaping our culture, as both McCrea and Soluri emphasize. However, Soluri rightly concludes in his discussion of sealers that there are major difficulties left to resolve in how this agency might be read from "human" documents: "Finally, we remain with the vexing question of whether fur seals played a role in this history beyond that of passive victim. Can historical sources be read 'against the current' to reveal the agency of seals?"

Even if history is apparently only a human cultural practice, it is not universally so in any one sense of the idea of history.[31] But in fact the work of ethnohistorians and debates on memory and history in anthropology over the last two decades strongly suggest that actually we have yet to recognize how animals are historical agents and how their historicities are constituted, in the same way that previously non-Western histories among small-scale societies were obscured, or thought not to exist at all, because of the absence of textual forms of recording. A war with the jaguars or the peccaries, as narrated in the histories of Huaorani or Patamuna hunters, is most certainly centered on the agency of animals. The chiefs and warriors of the jaguars are known as individuals and recognized if encountered. In the case of the Patamuna, the whole of the Siparuni Mountain is the jaguar's kingdom and for that very reason no one goes there. Should a Siparuni jaguar trouble the Patamuna, then this is indeed a reason for war, as it would be with their human neighbors, although sometimes jaguars are more agreeable neighbors than men. Those rapacious goats, rats, and cats, as much as the animals held in closer bondage (horses, sheep, cattle, and pigs), were all involved in the "conquest" of the Americas, as has often been noted. Perhaps we should therefore picture the feral animals in particular as historical agents in their own right, for their social projects were necessarily no less colonial than those of the humans who enabled their passages to the "new" world. As Tortorici writes in his chapter on marrying dogs: "Part of the project of 'centering animals' in Latin American history is to shift the focus from the discourses on and about animals to the actual histories of those animals in order to better understand their mutable relationships with the humans around them."

This also entails acknowledging the dark side of animal behaviors. So even as we may note the power relations inherent in such acts as the turkey rape committed by Pedro Na, as discussed by the editors in their introduction, there is no reason to suppose that animals are always or merely human victims. For example, several cases of elephants raping and attacking rhinos have been reported, with up to sixty-three rhinos being killed in just a short space of time. Video footage of such rape and other transspecies forms of aggression are widely available.[32] Just as the newly "human" peoples of indigenous America came to occupy the noble savage cultural slot in the eighteenth century, in similar fashion today the animals have come to express the ethically purest, even if robustly Darwinian, behavioral capacities, since they are untrammeled by the contamination of culture, uncomplicated by religions and politics, and uninflected by the moral imperatives of society and law. "Without faith, law, and reason" was how the early missionaries famously characterized the native peoples of South America, and as such they were precisely then "natural" men.[33] Today it is the naturalness of animals that makes them worthy and apposite emblems of a condition of moral integrity that our dystopian narratives of humanity have undermined.

If historical meaning does not emanate from the human alone, the key issue is to write histories which are meaningful *from the animal's point of view*. Certainly, as Few and Tortorici suggest, even without this ideal outcome, centering animals leads to a more complex historiography in the conventional sense. As Crosby first showed through the notion of a "Columbian exchange," histories need to take account of ecological and biological processes, and "centering animals" reminds us of the way in which human history is inseparable from and profoundly defined by ideas of the animal.[34] In intellectually reaching the "species" boundary, we have anyway vastly improved the writing and thinking about the past, but whether that boundary can in some sense be crossed requires further thought, not just about animals but also about how history itself is constituted.

Beyond the animal as historical agent is the context of action through which human-animal relationships are established. The interpretation of landscape or habitat is no less important than issues of agency in the effort to center animals, since landscapes are a form of history and memory that is material in the way that human texts are. The changing course of a river, the falling of an ancient tree, a stormy excess of snow, rain, wind, sunlight, or shade all dynamically alter the field of animal-human interaction. Too exclusive a concentration on particular moments or forms of interaction as historiographically significant or prominent may be to miss the significance evinced in behavior which can only be considered in this broader temporal and spatial plane. In this way shared experience of place, not just experience of the other as other, can be relevant here.[35]

The direct experience of the deep Amazonian forest may produce epistemic possibilities that cannot be imagined from the urban contexts of most academic practice. Here a general lack of visual sightlines, the relative absence of mammals, and the prevalence of insects over even the birds is experientially an inherent challenge to normative ideas of dominion over the natural. This is why the "green hell" of Amazonia is such a standard trope in travel and fictional writing.[36] Indeed the order of natural things is changed dramatically in the depths of the forest, for here insects, often the least considered of animals, dominate all. As Few also shows in her chapter on locusts, "Sources, written by European travelers, include descriptions of locust plagues in ways that work to further exoticize the New World and its animal species for its European audiences." Leading to that classic Western trope through which "it can be said that while swarming, locusts acted as a kind of conquering 'army,' though the goal was simply to eat and reproduce, not to establish a colony." This exactly echoes more recent fictive accounts, such as the famous short story by Carl Stephenson, "Leiningen versus the Ants" (1938).[37] It is therefore relevant to understand that for many Amazonian dwellers the "invasion" of such ants is a welcome opportunity, as the ants eagerly devour roaches, mice, and other household "pests." The short ethnographic narrative that follows might serve also as an exemplar to begin a history of animals as agents:

> So the ants are known as *o coração*, but Brasilino said that it was really two words run together: *corre chão* (running floor). I have not yet found out what species they are, but there are black ones and red ones and they do sting. They arrive (seemingly at random) and invade usually new houses at night. They all come in cleaning the walls and roof and floor of anything in their way. One must stem the tide to avoid complete invasion with some kind of deterrent such as buckets of water. However, when one ant begins the exodus back into the forest, they all follow. We were invaded twice while we were there, but no one else's house was targeted by the ants. On another note, Brasilino and Dona Luiza were driven from their fine *sítio* inside of Lago Grande by *saúva* (leaf-cutter ants). They had lived there a number of years already when the ants moved in. They had several dwellings and a flour house and Dona Luiza planted many fruit trees. But the ants were determined and resisted all kinds of attempts to extinguish them including fire and water. Finally Brasilino moved his family to the mouth of the lake where they live now. We visited the old homestead site in 2005 and the ants were still living there, thirty-five years later.[38]

Even the "great white hunter" may experience animality in surprising ways. The famed missionary to Africa in the nineteenth century, David Livingstone, writes the following of a lion hunt, a vignette which incidentally, as it occurs in the opening passages of the book, inaugurates an image of Livingstone in Christ-like sacrifice:

> When in the act of ramming down the bullets, I heard a shout. Starting, and looking half round, I saw the lion just in the act of springing upon me. I was upon a little height; he caught my shoulder as he sprang, and we both came to the ground below together. Growling horribly close to my ear, he shook me as a terrier dog does a rat. The shock produced a stupor similar to that which seems to be felt by a mouse after the first shake of the cat. It caused a sort of dreaminess, in which there was no sense of pain nor feeling of terror, though quite conscious of all that was

happening. It was like what patients partially under the influence of chloroform describe, who see all the operation, but feel not the knife. This singular condition was not the result of any mental process. The shake annihilated fear, and allowed no sense of horror in looking round at the beast. This peculiar state is probably produced in all animals killed by the *carnivora*; and if so, is a merciful provision by our benevolent Creator for lessening the pain of death.[39]

Sustained engagement with other ontologies and epistemologies through ethnography, hunting (or fishing), or even domesticity and husbandry thus opens new channels for centering the animal in history, while paying attention to the meaning-laden contexts of action and behavior that provide an extended hermeneutic for the interpretation of ultimately ineffable others, whatever their forms of speciation. So the editors Few and Tortorici are certainly right in their observation that "another strand of the work on animals in colonial Latin America evokes considerable sophistication in the analyses of hybrid human-animal deities, shape-shifting and diabolical animals, animal metaphors, and animal symbolism as represented in archival documentation and codices. The animals themselves, however, have rarely been accorded a standpoint that would make them the primary focus of such works."

Which prompts the question: What, then, is an animal? The perennial instability in that category, as much as in that of the "human," suggests that "real animals" will forever elude us. For it is ultimately an impossible desire, familiar from ethnography, of trying to occupy the subject position of the other. By definition this cannot be achieved, for to do so would to be that other whose ineffable nature cannot be explained in its own terms, for that is where the project of knowledge of others begins, with uninterpreted utterance and behavior. Therefore, knowing others, including the animals, must always be a matter of trying to make exotic experience accessible, rather than explicable. Explanation may well function as mode of access, but in itself can never exhaust the meanings in and of the actions of others. Anthropology, in the field of the human, has long realized that totalizing "holistic" explanation is impossible. No one can say why others do as they do, since we cannot even do that for ourselves. It is the purpose of knowledge, rather than hope for the completeness of knowledge, that needs to drive our explanatory projects of both humans and animals. This volume has done much to advance that project.

Notes

1. Peter Singer, *Animal Liberation* (New York: Random House, 1975); and Tom Regan and Peter Singer, eds., *Animal Rights and Human Obligations* (Englewood Cliffs: Prentice-Hall, 1976).

2. Cary Wolfe, *Animal Rites: American Culture, the Discourse of Species, and Posthumanist Theory* (Chicago: University of Chicago Press, 2003).

3. Erica Fudge, Ruth Gilbert, and Susan Wiseman, eds., *At the Borders of the Human: Beasts, Bodies and Natural Philosophy in the Early Modern Period* (Houndmills, U.K.: Macmillan, 1999).

4. The notion of "trauma" is deployed to suggest that, rather than making political or cultural choices, enactors of "suicide bombing" and other forms of violent resistance are in effect responding to the condition of PTSD (post-traumatic stress disorder). In fact the analysis in Didier Fassin and Richard Rechtman's *The Empire of Trauma: An Inquiry into the Condition of Victimhood* (Princeton: Princeton University Press, 2009) is deeply flawed since it ironically fails to recognize that, in the Palestinian case, which is extensively discussed, the very notion of "martyrdom" has explicitly changed to avert this inference. Thus, while the notion of *shahid* (martyr) implies victimization, the more recent and relevant idea of *istishhadi* (martyrous one) is a proactive notion that emphasizes the heroism in the act of sacrifice over the victimization that is also part of the act. See Neil L. Whitehead and Nasser Abufarha, "Suicide, Violence, and Cultural Conceptions of Martyrdom in Palestine," *Social Research* 75.2 (2008): 395–415.

5. For example, see Charles C. Geisler, "Endangered Humans: How Global Land Conservation Efforts Are Creating a Growing Class of Invisible Refugees," *Foreign Policy* 130 (2002): 80–81; John Knight, ed., *Natural Enemies: People-Wildlife Conflicts in Anthropological Perspective* (London: Routledge, 2001); François Lamarque, *Human-Wildlife Conflict in Africa: Causes, Consequences and Management Strategies* (Rome: Food and Agriculture Organization of the United Nations, 2009); and John Terborgh, ed., *Making Parks Work: Strategies for Preserving Tropical Nature* (Washington, D.C.: Island Press, 2002).

6. See Wolfe, *Animal Rites*, 26–27.

7. Anderson is the executive director of the Property and Environment Research Center, senior fellow at the Hoover Institute, and adjunct professor at the Stanford Graduate School of Business. His major works include (with Donald R. Leal) *Free Market Environmentalism* (New York: Palgrave, 2001 [1991]) and *Enviro-capitalists: Doing Good while Doing Well* (Lanham, Md.: Rowman and Littlefield, 1997).

8. A good illustration of these intertwined relationships of security, consumerism, and conservation is evident in a recent news story that reported that "heavily armed men have kidnapped six volunteers from WWF-India who were counting the tiger population at a reserve in India's remote northeast, an official from the

conservation group said Monday" (Wasbir Hussain, "6 WWF Volunteers Kidnapped at Manas Tiger Reserve in India," *Huffington Post*, 7 February 2011, http://www.huff ingtonpost.com/2011/02/07/wwfvolunteerskidnapped-_n_819461.html). Simi- larly, the tragic death of Diane Fossey, who was found murdered in the bedroom of her cabin in Virunga Mountains, Rwanda, on 26 December 1985, underscores the ways in which advancing animal rights can lead to the occlusion of human rights.

9. This has been the case in Zimbabwe, where the colonial farming class retained a high degree of control over the most productive farmlands even after indepen- dence from Britain. In this context local participation in wildlife conservation, through the World Wildlife Fund's sponsorship of the Communal Areas Manage- ment Program for Indigenous Resources (CAMPFIRE) program, became a means to preserve those colonial landholdings and to offer alternative sources of income to historically dispossessed black farmers. "By engaging in activities such as safari hunting and nature tourism, CAMPFIRE projects earned US$1,384,083 (Z$20 mil- lion) in 1997, according to the NGO Zimbabwe Trust" (Kristin B. Gunther, "Can Local Communities Conserve Wildlife? Visions of the Future: The Prospect for Reconciliation," *Cultural Survival Quarterly* 23.4 (winter 1999), http://www.cul turalsurvival.org/publications/cultural-survival-quarterly/zimbabwe/can-local -communities-conserve-wildlife.

10. Neel Ahuja, "Abu Zubaydah and the Caterpillar," *Social Text* 29.1 106 (spring 2011): 127–49.

11. Mary Mallon (1869–1938), known as Typhoid Mary, was the first person in the United States to be identified as a healthy carrier of typhoid fever. As a cook, she infected more than fifty people, three of whom died from the disease. Because of her denial that she was responsible for spreading the disease, she was forcibly quarantined twice and finally died in quarantine.

12. Just as the Christian missionaries from the sixteenth century to the nineteenth pursued the proper and moral arrangement of domestic space to guard against the sin-disease of unregulated sexuality, campaigns against malaria mandated that mosquito nets be hung around hammocks or beds, thus dividing the home life into smaller "safe" zones; those who did not abide by these new arrangements became the new "primitive" in a regime of modern medical science and state public-health committees.

13. The MK-Ultra experiments with LSD and the medical experiments in which syph- ilis was purposely left untreated are just two of the more egregious examples of this hidden history. See Neil L. Whitehead, "Post-human Anthropology," *Identi- ties: Global Studies in Culture and Power* 16.1 (2009): 1–32.

14. Elinor G. K. Melville, *A Plague of Sheep: Environmental Consequences of the Con- quest of Mexico* (Cambridge: Cambridge University Press, 1994).

15. Alfred W. Crosby, *The Columbian Exchange: Biological and Cultural Consequences of 1492* (Westport: Greenwood, 1972). See also Whitehead, "Post-human Anthro- pology."

16. Kaj Århem, "Ecosofía makina," in *La selva humanizada: Ecología alternativa en el*

trópico húmedo Colombiano, ed. François Correa (Bogotá: Fondo Editorial CEREC, 1993), 109–26.

17. See Eduardo Viveiros de Castro, "Cosmological Deixis and Amerindian Perspectivism," *Journal of the Royal Anthropological Society* 4.3 (2000): 469–88.

18. Jacques Derrida, *The Animal That Therefore I Am,* trans. David Willis (New York: Fordham University Press, 2008).

19. Castro, "Cosmological Deixis and Amerindian Perspectivism," 470.

20. Edward O. Wilson, *Sociobiology: The New Synthesis* (Cambridge: Harvard University Press, 1975).

21. Loretta A. Cormier, *Kinship with Monkeys: The Guajá Foragers of Eastern Amazon* (New York: Columbia University Press, 2003).

22. Fernando Santos-Granero, *The Power of Love: The Moral Use of Knowledge amongst the Amuesha of Central Peru* (London: Athlone, 1991).

23. For more discussion of the themes of intimacy and cannibalism, see Beth Conklin's *Consuming Grief: Compassionate Cannibalism in an Amazonian Society* (Austin: University of Texas Press, 2001); and Neil L. Whitehead's discussion of cannibal cosmology and its relation to Christian idioms, in Hans Staden, *Hans Staden's True History: An Account of Cannibal Captivity in Brazil,* ed. and trans. Neil L. Whitehead and Michael Harbsmeier (Durham: Duke University Press, 2008).

24. A term derived from a Patamuna gloss of the term *piaii,* more often translated as "shamanism."

25. Emiko Ohnuki-Tierney, *The Monkey as Mirror: Symbolic Transformations in Japanese History and Ritual* (Princeton: Princeton University Press, 1989).

26. Henri Atlan and Frans B. M. de Waal, *Les frontières de l'humain* (Paris: Editions le Pommier, 2007).

27. Robert Brightman, *Grateful Prey: Rock Cree Human-Animal Relationships* (Berkeley: University of California Press, 1993).

28. Keith Thomas, *Man and the Natural World: Changing Attitudes in England, 1500–1800* (New York: Penguin, 1983).

29. In the sixteenth century Sir Walter Raleigh returned to England from the Orinoco River with three shamans: Ragapo, "Harry," and Cayoworaco. They took lodgings near the Tower of London and frequently visited Raleigh, who busied himself with teaching them English and various experiments in medicinal preparations. See Neil L. Whitehead, ed. and trans., *The Discoverie of the Large, Rich and Bewtiful Empire of Guiana by Sir Walter Ralegh,* Exploring Travel Series, vol. 1 (Manchester: Manchester University Press, 1997), 3, and American Exploration and Travel Series, vol. 71 (Norman: Oklahoma University Press, 1997), 30–31.

30. See Bruce Bagemihl, *Biological Exuberance: Animal Homosexuality and Natural Diversity* (New York: St. Martin's, 1999); and Noreen Giffney and Myra J. Hird, *Queering the Non/Human* (London: Ashgate, 2008).

31. Neil L. Whitehead, ed., *Histories and Historicities in Amazonia* (Lincoln: University of Nebraska Press, 2003).

32. On YouTube, for example, see "Horny Elephant Raped Rhino," http://www.you tube.com/watch?v=QJ7lruKlzK8. Also see Charles Siebert, "An Elephant Crack-up?," *New York Times*, 8 October 2006, http://www.nytimes.com/2006/10/08/magazine/08elephant.html?_r=1.

33. Anthony Pagden, *The Fall of Natural Man: The American Indian and the Origins of Comparative Ethnology* (Cambridge: Cambridge University Press, 1987).

34. Crosby, *The Columbian Exchange*.

35. Steven Feld and Keith H. Basso, eds., *Senses of Place* (Santa Fe: School of American Research Press, 1996).

36. Neil L. Whitehead, "South America/The Amazon: The Forest of Marvels," in *The Cambridge Companion to Travel Writing*, ed. Peter Hulme and Tim Youngs (Cambridge University Press, 2002), 122–39.

37. Carl Stephenson, "Leiningen versus the Ants," *Esquire* 10.6 (December 1938).

38. Kent Wisniewski, personal communication, 11 January 2010.

39. David Livingstone, *Missionary Travels and Researches in South Africa: Including a Sketch of Sixteen Years' Residence in the Interior of Africa* (New York: Harper, 1858), 12.

RECOMMENDED BIBLIOGRAPHY

Achim, Miruna. *Lagartijas medicinales: Remedios americanos y debates científicos en la Ilustración*. Mexico City: Conaculta / Universidad Autónoma Metropolitana, Cuajimalpa, 2008.

Aguilar, Sandra. "Nutrition and Modernity: Milk Consumption in 1940s and 1950s Mexico." *Radical History Review* 110 (2011): 36–58.

Alves, Abel A. *The Animals of Spain: An Introduction to Imperial Perceptions and Human Interaction with Other Animals, 1492–1826*. Leiden: Brill, 2011.

Anderson, Virginia DeJohn. *Creatures of Empire: How Domestic Animals Transformed Early America*. Oxford: Oxford University Press, 2004.

Andrews, Thomas G. "Contemplating Animal Histories: Pedagogy and Politics across Borders." *Radical History Review* 107 (2010): 139–65.

Animal Studies Group, ed. *Killing Animals*. Champaign: University of Illinois Press, 2006.

Archetti, Eduardo P. *Guinea-Pigs: Food, Symbol and Conflict of Knowledge in Ecuador*. New York: Berg, 1997.

Århem, Kaj. "Ecosofía Makuna." *La selva humanizada: Ecología alternativa en el trópico húmedo colombiano*, ed. François Correa, 109–26. Bogotá: Fondo Editorial CEREC, 1993.

Ariel de Vidas, Anath. "A Dog's Life among the Teenek Indians (Mexico): Animals' Participation in the Classification of Self and Other." *Journal of the Royal Anthropological Institute* 8.3 (2002): 531–50.

Armstrong, Philip. "The Postcolonial Animal." *Society and Animals* 10.4 (2002): 413–19.

Arnold, James A., ed. *Monsters, Tricksters, and Sacred Cows: Animal Tales and American Identities*. Charlottesville: University of Virginia Press, 1996.

Asúa, Miguel de, and Roger French. *A New World of Animals: Early Modern Europeans on the Creatures of Iberian America*. Burlington: Ashgate, 2005.

Bagemihl, Bruce. *Biological Exuberance: Animal Homosexuality and Natural Diversity*. New York: St. Martin's, 1999.

Bazant, Milada. "Bestialismo, el delito nefando, 1800–1856." *Documentos de Investigación* 66 (2002): 1–22.

Beirne, Piers. *Confronting Animal Abuse: Law, Criminology, and Human-Animal Relationships.* Lanham: Rowman and Littlefield, 2009.

Benes, Peter. *New England's Creatures: 1400–1900.* Boston: Boston University Press, 1995.

Benson, Elizabeth P. *Birds and Beasts of Ancient Latin America.* Gainesville: University of Florida Press, 1997.

Benson, Etienne. "Animal Writes: Historiography, Disciplinarity, and the Animal Trace." *Making Animal Meaning,* ed. Linda Kalof and Georgina M. Montgomery, 3–16. Ann Arbor: Michigan State University Press, 2011.

Berger, John. *About Looking.* New York: Vintage, 1992.

Berlin, Brent. *Ethnobiological Classification: Principles of Categorization of Plants and Animals in Traditional Societies.* Princeton: Princeton University Press, 1992.

Brandes, Stanley. "Torophiles and Torophobes: The Politics of Bulls and Bullfights in Contemporary Spain." *Anthropological Quarterly* 82.3 (2009): 779–94.

Brantz, Dorothee, ed. *Beastly Natures: Animals, Humans, and the Study of History.* Charlottesville: University of Virginia Press, 2010.

Brazeal, Brian. "A Goat's Tale: Diabolical Economies of the Bahian Interior." *Activating the Past: History and Memory in the Black Atlantic World,* ed. Andrew Apter and Lauren Derby, 267–94. Newcastle upon Tyne: Cambridge Scholars, 2010.

Brienen, Rebecca. "From Brazil to Europe: The Zoological Drawings of Albert Eckhout and Georg Marcgraf." *Early Modern Zoology: The Construction of Animals in Science, Literature and the Visual Arts,* ed. Karl A. E. Enenkel and Paul J. Smith, 273–317. Leiden: Brill, 2007.

Brower, Matthew. *Developing Animals: Wildlife and Early American Photography.* Minneapolis: University of Minnesota Press, 2010.

Brown, Eric C. "Insects, Colonies, and Idealization in the Early Americas." *Utopian Studies* 13.2 (2002): 20–37.

——, ed. *Insect Poetics.* Minneapolis: University of Minnesota Press, 2006.

Bulliet, Richard W. *Hunters, Herders, and Hamburgers: The Past and Future of Human-Animal Relationships.* New York: Columbia University Press, 2005.

Burt, Jonathan. *Animals in Film.* London: Reaktion, 2002.

Cabezas Carcache, Horacio. "Producción agropecuaria." *Historia general de Guatemala,* ed. Jorge Luján Muñoz, 3:294–303. Guatemala City: Asociación de Amigos del País, Fundación para la Cultura y el Desarrollo, 1995.

Christian, William, Jr. "Sobrenaturales, humanos, animales: Exploración de los límites en las fiestas españolas a través de las fotografías de Cristina García Rodero." *La Fiesta en el mundo hispánico,* 13–32. Toledo: Universidad de Castilla–La Mancha, 2004.

Coe, Sophie D. *America's First Cuisines.* Austin: University of Texas Press, 1994.

Coleman, Jon T. *Vicious: Wolves and Men in America.* New Haven: Yale University Press, 2004.

Cook, Alexandra Parma, and Noble David Cook. *The Plague Files: Crisis Management in Sixteenth-Century Seville.* Baton Rouge: Louisiana State University Press, 2009.

Cormier, Loretta A. *Kinship with Monkeys: The Guajá Foragers of Eastern Amazon.* New York: Columbia University Press, 2003.

Corona M., Eduardo, and Joaquín Arroyo Cabrales, eds. *Relaciones hombre-fauna: Una zona interdisciplinaria de estudio.* Mexico City: Instituto Nacional de Antropología e Historia, 2002.

Costlow, Jane, and Amy Nelson, eds. *Other Animals: Beyond the Human in Russian Culture and History.* Pittsburgh: University of Pittsburgh Press, 2010.

Crandon-Malamud, Libbet. *From the Fat of Our Souls: Social Change, Political Process, and Medical Pluralism in Bolivia.* Berkeley: University of California Press, 1991.

Creager, Angela N. H., and William Chester Jordan, eds. *The Animal/Human Boundary: Historical Perspectives.* Rochester: University of Rochester Press, 2003.

Cronon, William. *Changes in the Land: Indians, Colonists, and the Ecology of New England.* New York: Hill and Wang, 1983.

——, ed. *Uncommon Ground: Toward Reinventing Nature.* New York: W. W. Norton, 1995.

Crosby, Alfred W. *The Columbian Exchange: Biological and Cultural Consequences of 1492.* Westport: Greenwood Press, 1972.

——. *Ecological Imperialism: The Biological Expansion of Europe, 900–1900.* New York: Cambridge University Press, 1986.

Cueto, Marcos. "Appropriation and Resistance: Local Responses to Malaria Eradication in Mexico, 1955–1970." *Journal of Latin American Studies* 37 (2005): 533–59.

——. "Sanitation from Above: Yellow Fever and Foreign Intervention in Peru, 1919–1922." *Hispanic American Historical Review* 72.1 (2002): 1–22.

Cushman, Gregory. "'The Most Valuable Birds in the World': International Conservation Science and the Revival of Peru's Guano Industry, 1909–1965." *Environmental History* 10 (July 2005): 477–509.

Derby, Lauren. "Bringing the Animals Back In: Writing Quadrupeds into the Environmental History of Latin America and the Caribbean." *History Compass* 9.8 (2011): 602–21.

——. "Gringo Chickens with Worms: Food and Nationalism in the Dominican Republic." *Close Encounters of Empire: Writing the Cultural History of U.S.–Latin American Relations,* ed. Gilbert M. Joseph, Catherine C. LeGrand, and Ricardo D. Salvatore, 451–96. Durham: Duke University Press, 1998.

Derrida, Jacques. *The Animal That Therefore I Am.* Trans. David Willis. New York: Fordham University Press, 2008.

Descola, Philippe. *In the Society of Nature: A Native Ecology in Amazonia.* Cambridge: Cambridge University Press, 1996.

Dinzelbacher, Peter. "Animal Trials: A Multidisciplinary Approach." *Journal of Interdisciplinary History* 32.3 (2002): 405–21.

Dopico Black, Georgina. "The Ban and the Bull: Cultural Studies, Animal Studies, and Spain." *Journal of Spanish Cultural Studies* 11.3–4 (2010): 235–49.

Duarte, Regina Horta. "Pássaros e cientistas no Brasil: Em busca de proteção, 1894–1938." *Latin American Research Review* 41.1 (2006): 3–26.

Dunlap, Thomas R. "American Wildlife Policy and Environmental Ideology: Poisoning Coyotes, 1939–1972." *Pacific Historical Review* 55.3 (August 1986): 345–69.

Earle, Rebecca. "'If You Eat Their Food . . . ': Diets and Bodies in Early Colonial Spanish America." *American Historical Review* 115.3 (2010): 688–713.

Edwards, Peter. *Horse and Man in Early Modern England.* London: Hambledon Continuum, 2007.

Evans, E. P. *The Criminal Prosecution and Capital Punishment of Animals: The Lost History of Europe's Animal Trials.* New York: E. P. Dutton, 1906.

Fausto, Carlos. "A Blend of Blood and Tobacco: Shamans and Jaguars among the Parakanã of Eastern Amazonia." *In Darkness and Secrecy: The Anthropology of Assault Sorcery and Witchcraft in Amazonia,* ed. Neil L. Whitehead and Robin Wright, 157–78. Durham: Duke University Press, 2004.

Few, Martha. "'El daño que padece el bien común': Casta revendedoras y los conflictos por la venta de carne en Guatemala Colonial, 1650–1730." *Mesoamérica* 49 (2007): 1–24.

———. *Women Who Live Evil Lives: Gender, Religion, and the Politics of Power in Colonial Guatemala.* Austin: University of Texas Press, 2002.

Fissell, Mary. "Imagining Vermin in Early Modern England." *History Workshop Journal* 47 (1999): 1–29.

Flórez Malagón, Alberto G., ed. *El poder de la carne: Historias de ganaderías en la primera mitad del siglo XX en Colombia.* Bogotá: Editorial Pontificia Universidad Javeriana, 2008.

Fudge, Erica. "A Left-Handed Blow: Writing the History of Animals." *Representing Animals,* ed. Nigel Rothfels, 3–18. Bloomington: University of Indiana Press, 2002.

———. *Perceiving Animals: Humans and Beasts in Early Modern English Culture.* Chicago: University of Illinois Press, 2002.

Fudge, Erica, Ruth Gilbert, and Susan Wiseman, eds. *At the Borders of the Human: Beasts, Bodies and Natural Philosophy in the Early Modern Period.* Houndmills, U.K.: Macmillan, 1999.

Funes Monzote, Reinaldo. "Facetas de la interacción con los animales en Cuba durante el siglo XIX: Los bueyes en la plantación esclavista y la Sociedad Protectora de Animales y Plantas." *Signos Históricos* 16 (2006): 80–110.

———. "Los orígenes del asociacionismo ambientalista en Cuba: La Sociedad Protectora de Animales y Plantas, 1882–1890." *Naturaleza en declive: Miradas a la historia ambiental de América Latina y el Caribe,* ed. Reinaldo Funes Monzote, 267–309. Valencia: Centro Francisco Tomás y Valiente UNED Alzira-Valencia / Fundación Instituto de Historia Social, 2008.

Garfield, Seth. "A Nationalist Environment: Indians, Nature, and the Construction of the Xingu National Park in Brazil." *Luso-Brazilian Review* 41.1 (2004): 139–67.

Gerbi, Antonello. *Nature in the New World: From Christopher Columbus to Gonzalo Fernández de Oviedo.* Trans. Jeremy Moyle. Pittsburgh: University of Pittsburgh Press, 1975.

Giffney, Noreen, and Myra J. Hird, eds. *Queering the Non/Human*. London: Ashgate, 2008.

González-Montagut, Renée. "Factors That Contributed to the Expansion of Cattle Ranching in Veracruz, Mexico." *Mexican Studies / Estudios Mexicanos* 15.1 (1999): 101–30.

Graham, Robert Cunninghame. *Horses of the Conquest: A Study of the Steeds of the Spanish Conquest*. Norman: University of Oklahoma Press, 1949.

Gruzinski, Serge. *Man-Gods of the Mexican Highlands: Indian Power and Colonial Society, 1520–1800*. Stanford: Stanford University Press, 1989.

Guerrini, Anita. *Experimenting with Humans and Animals: From Galen to Animal Rights*. Baltimore: Johns Hopkins University Press, 2003.

Haraway, Donna. *Primate Visions: Gender, Race, and Nature in the World of Modern Science*. London: Routledge, 1989.

———. *When Species Meet*. Minneapolis: University of Minnesota Press, 2008.

Hardouin-Fugier, Elisabeth. *Bullfighting: A Troubled History*. Chicago: University of Chicago Press, 2010.

Hermitte, Esther. "El concepto de nahual en Pinola, México." *Ensayos antropológicos en los Altos de Chiapas*, ed. Norman McQuown and Julian Pitt-Rivers. Mexico City: Instituto Indigenista Interamericano, 1989.

Hirschkind, Lynn. "Sal / Manteca / Panela: Ethnoveterinary Practice in Highland Ecuador." *American Anthropologist* 102.2 (2000): 290–302.

Hribal, Jason. "'Animals Are Part of the Working Class': A Challenge to Labor History." *Labor History* 44.4 (2003): 435–53.

Iriarte, J. Agustín, and Fabian M. Jaksic. "The Fur Trade in Chile: An Overview of Seventy-Five Years of Export Data (1910–1984)." *Biological Conservation* 38 (1986): 243–53.

Johnson, Sara E. "'You Should Give Them Blacks to Eat': Waging Inter-American Wars of Torture and Terror." *American Quarterly* 61.1 (2009): 65–92.

Kalof, Linda, and Georgina M. Montgomery, eds. *Making Animal Meaning*. Detroit: Michigan State University Press, 2011.

Kay, Sarah. "Legible Skins: Animals and the Ethics of Medieval Reading." *Postmedieval: A Journal of Medieval Cultural Studies* 2.1 (2011): 13–32.

Kohn, Eduardo. "How Dogs Dream: Amazonian Natures and the Politics of Transspecies Engagement." *American Ethnologist* 34.1 (2007): 3–24.

Kosek, Jake. "Ecologies of Empire: On the New Uses of the Honeybee." *Cultural Anthropology* 25.4 (2010): 650–78.

Landry, Donna. *Noble Brutes: How Eastern Horses Transformed English Culture*. Baltimore: Johns Hopkins University Press, 2009.

Lee, Raymond. "Cochineal Production and Trade in New Spain to 1600." *The Americas* 4.4 (1948): 449–73.

Livingston, Julie, and Jasbir K. Puar. "Interspecies." *Social Text* 29.1 106 (2011): 3–14.

Lockhart, James. *The Nahuas after the Conquest: A Social and Cultural History of the*

Indians of Central Mexico, Sixteenth through Eighteenth Centuries. Stanford: Stanford University Press, 1992.

Lockwood, Jeffrey A. *Grasshopper Dreaming: Reflections on Killing and Loving.* Boston: Skinner House Books, 2002.

———. *Locust: The Devastating Rise and Mysterious Disappearance of the Insect that Shaped the American Frontier.* New York: Basic Books, 2004.

López Austin, Alfredo. *Cuerpo humano e ideología: Las concepciones de los antiguos nahuas.* 2 vols. Mexico City: Universidad Nacional Autónoma de México, 1980.

MacGregor-Loaeza, Raúl. "Los insectos y las antiguas culturas mexicanas: Un ensayo etnoentomológico." *Revista de la Universidad de México* 29.6–7 (1975): 8–13.

Marichal, Carlos. "Mexican Cochineal and the European Demand for American Dyes, 1550–1850." *From Silver to Cocaine: Latin American Commodity Chains and the Building of the World Economy, 1500–2000,* ed. Steven Topik, Carlos Marichal, and Zephyr Frank, 76–92. Durham: Duke University Press, 2006.

McCrea, Heather. *Diseased Relations: Epidemics, Public Health, and State-Building in Yucatán, Mexico, 1847–1924.* Albuquerque: University of New Mexico Press, 2011.

———. "On Sacred Ground: The Church and Burial Rites in Nineteenth-Century Yucatán, Mexico." *Mexican Studies / Estudios Mexicanos* 23.1 (winter 2007): 33–62.

McNeill, J. R. *Mosquito Empires: Ecology and War in the Greater Caribbean, 1620–1914.* Cambridge: Cambridge University Press, 2010.

Melville, Elinor G. K. *A Plague of Sheep: Environmental Consequences of the Conquest of Mexico.* Cambridge: Cambridge University Press, 1994.

Millones, Luis, and Renata Mayer. *La fauna sagrada de Huarochirí.* Lima: Instituto de Estudios Peruanos, 2012.

Monaghan, John. "The Person, Destiny, and the Construction of Difference in Meso-america." RES: *Anthropology and Aesthetics* 33 (spring 1998): 137–46.

Montgomery, Georgina M., and Linda Kalof. "History from Below: Animals as Historical Subjects." *Teaching the Animal: Human-Animal Studies across the Disciplines,* ed. Margo DeMello, 35–47. Brooklyn: Lantern Books, 2010.

Morales, Edmundo. *The Guinea Pig: Healing, Food, and Ritual in the Andes.* Tucson: University of Arizona Press, 1995.

Murrin, John M. "'Things Fearful to Name': Bestiality in Early America." *The Animal / Human Boundary: Historical Perspectives,* ed. Angela N. H. Creager and William Chester Jordan, 115–56. New York: University of Rochester Press, 2002.

Myers, Kathleen Ann. *Fernández de Oviedo's Chronicle of America: A New History for a New World.* Trans. Nina M. Scott. Austin: University of Texas Press, 2007.

Norton, Marcy. "Animals in Spain and Spanish America." *Lexikon of the Hispanic Baroque: Transatlantic Exchange and Transformation,* ed. Kenneth Mills and Evonne Levy. Austin: University of Texas Press, forthcoming 2013.

———. "Going to the Birds: Birds as Things and Beings in Early Modernity." *Early Modern Things: Objects and Their Histories, 1500–1800,* ed. Paula Findlen, 53–83. London: Routledge, forthcoming 2013.

Oliver, Lilia V. *Un verano mortal: Análisis demográfico y social de una epidémica de cólera: Guadalajara, 1833.* Guadalajara: Gobierno de Jalisco Secretaria General Unidad Editorial, 1986.

Ortiz de Montellano, Bernard R. *Aztec Medicine, Health, and Nutrition.* New Brunswick: Rutgers University Press, 1990.

Parrish, Susan Scott. "The Female Opossum and the Nature of the New World." *William and Mary Quarterly* 54.3 (1997): 475–514.

Pearson, Susan J., and Mary Weismantel. "Does 'the Animal' Exist? Toward a Theory of Social Life with Animals." *Beastly Natures: Animals, Humans, and the Study of History,* ed. Dorothee Brantz, 17–37. Charlottesville: University of Virginia Press, 2010.

Penyak, Lee. "Criminal Sexuality in Central Mexico, 1750–1850." PhD diss., University of Connecticut, 1993.

Pérez, Louis A., Jr. "Between Baseball and Bullfighting: The Quest for Nationality in Cuba, 1868–1898." *Journal of American History* 81.2 (1994): 493–517.

Phillips, Carla Rahn, and William D. Phillips Jr. *Spain's Golden Fleece: Wool Production and the Wool Trade from the Middle Ages to the Nineteenth Century.* Baltimore: Johns Hopkins University Press, 1997.

Pilcher, Jeffrey M. *The Sausage Rebellion: Public Health, Private Enterprise, and Meat in Mexico City, 1890–1917.* Albuquerque: University of New Mexico Press, 2006.

Pineo, Ronn F., and James A. Baer, eds. *Cities of Hope: People, Protests, and Progress in Urbanizing Latin America, 1870–1930.* Boulder: Westview, 2000.

Pitt-Rivers, Julian. "Spiritual Power in Central America: The Naguals of Chiapas." *Witchcraft Confessions and Accusations,* ed. Mary Douglas, 183–206. London: Routledge, 2004.

Ramos-Elorduy, Julieta. "Insectos comestibles." *Arqueología Mexicana* 6.35 (1999): 68–73.

Regan, Tom. *The Case for Animal Rights.* Berkeley: University of California Press, 2004.

Regan, Tom, and Peter Singer, eds. *Animal Rights and Human Obligations.* Englewood Cliffs, N.J.: Prentice-Hall, 1976.

Reyes García, Luis. *Anales de Juan Bautista.* Mexico City: Centro de Investigaciones y Estudios Superiores en Antropología Social, 2001.

Riaño, Pablo. *Gallos y toros en Cuba.* Havana: Fundación Fernando Ortiz, 2002.

Ríos, Arcadio, and Félix Ponce. "Tracción animal, mecanización y agricultura sostenible." *Transformando el campo cubano: Avances de la agricultura sostenible,* ed. Fernando Funes et al., 159–66. Havana: Asociación Cubana de Técnicos Agrícolas y Forestales, 2001.

Ritvo, Harriet. *The Animal Estate: The English and Other Creatures in the Victorian Age.* Cambridge: Harvard University Press, 1987.

———. "Animal Planet." *Environmental History* 9.2 (2004): 204–20.

Rothfels, Nigel. *Savages and Beasts: The Birth of the Modern Zoo.* Baltimore: Johns Hopkins University Press, 2002.

Rydström, Jens. *Sinners and Citizens: Bestiality and Homosexuality in Sweden, 1880–1950.* Chicago: University of Chicago Press, 2003.

Sahagún, Bernardino de. *Florentine Codex: General History of the Things of New Spain.* Trans. Arthur J. O. Anderson and Charles E. Dibble. Santa Fe: School of American Research; Salt Lake City: University of Utah Press, 1950–82.

Salisbury, Joyce. *The Beast Within: Animals in the Middle Ages.* New York: Routledge, 1994.

Santos-Granero, Fernando. *The Power of Love: The Moral Use of Knowledge amongst the Amuesha of Central Peru.* London: Athlone, 1991.

Sarmiento Ramírez, Ismael. *Cuba entre la opulencia y la pobreza: Población, economía y cultura material en los primeros 68 años del siglo XIX.* Madrid: Agualarga Editores, 2004.

Sarreal, Julia. "Disorder, Wild Cattle, and a New Role for the Missions: The Banda Oriental, 1776–1786." *The Americas* 67.4 (April 2011): 517–45.

Schiavini, Adrian. "Los lobos marinos como recurso para cazadores-recolectores: El caso de Tierra del Fuego." *Latin American Antiquity* 4 (1993): 346–66.

Schiebinger, Londa L. *Nature's Body: Gender in the Making of Modern Science.* New Brunswick: Rutgers University Press, 2004.

Schwartz, Marion. *A History of Dogs in the Early Americas.* New Haven: Yale University Press, 1997.

Shapiro, Kenneth. "Human-Animal Studies: Growing the Field, Applying the Field." Ann Arbor: Animals and Society Institute, 2008.

Shubert, Adrian. *Death and Money in the Afternoon: A History of the Spanish Bullfight.* New York: Oxford University Press, 1999.

Singer, Peter. *Animal Liberation.* New York: Random House, 1975.

Skabelund, Aaron. "Can the Subaltern Bark? Imperialism, Civilization, and Canine Cultures in Nineteenth-Century Japan." *JAPANimals: History and Culture in Japan's Animal Life,* ed. Gregory M. Pflugfelder and Brett L. Walker, 195–243. Ann Arbor: Center for Japanese Studies, University of Michigan, 2005.

Slater, Candace. *Entangled Edens: Visions of the Amazon.* Berkeley: University of California Press, 2002.

S. Lopes, Maria Aparecida de. "Los patrones de la criminalidad en el estado de Chihuahua: El caso del abigeato en las últimas décadas del siglo XIX." *Historia Mexicana* 50.3 (2001): 513–53.

Sluyter, Andrew. "The Ecological Origins and Consequences of Cattle Ranching in Sixteenth-Century New Spain." *Geographical Review* 86.2 (1996): 161–77.

Sousa, Lisa. "The Devil and Deviance in Native Criminal Narratives from Early Mexico." *The Americas* 59.2 (2002): 161–79.

Souza, Laura de Mello e. *The Devil and the Land of the Holy Cross: Witchcraft, Slavery, and Popular Religion in Colonial Brazil.* Austin: University of Texas Press, 2004.

Suárez, José, Arcadio Ríos, and Pedro Soto. "El tractor y la tracción animal." *Revista de Ciencias Técnicas Agropecuarias* 14.2 (2005): 40–43.

Swart, Sandra. *Riding High: Horses, Humans and History in South Africa.* Johannesburg: Witswatersrand University Press, 2011.

Taunay, Afonso de Escragnolle, and Odilon Nogueira de Matos. *Zoologia fantástica do*

Brasil (séculos XVI e XVII). São Paulo: Edusp, Museu Paulista Universidade de São Paulo, 1999.

Teixeira-Pinto, Márnio. "Being Alone amid Others: Sorcery and Morality among the Arara, Carib, Brazil." In *Darkness and Secrecy: The Anthropology of Assault Sorcery and Witchcraft in Amazonia*, ed. Neil L. Whitehead and Robin Wright, 215–43. Durham: Duke University Press, 2004.

Thomas, Keith. *Man and the Natural World: Changing Attitudes in England, 1500–1800*. New York: Penguin, 1983.

Tortorici, Zeb. "Against Nature: Sodomy and Homosexuality in Colonial Latin America." *History Compass* 10.2 (2012): 161–78.

——. "Animals and Archives: Making Sense of the *Discurso Filosófico Sobre el Lenguage de los Animales*." *e-misférica* 10.1. http://hemisphericinstitute.org/hemi/en/e-misferica-101.

——. "Contra Natura: Sin, Crime, and 'Unnatural' Sexuality in Colonial Mexico, 1530–1821." PhD diss., University of California, Los Angeles, 2010.

Ulloa, Astrid, and Luis Guillermo Baptiste-Ballera, eds. *Rostros culturales de la fauna: Las relaciones entre los humanos y los animales en el contexto colombiano*. Bogotá: Instituto Colombiano de Antropología e Historia, 2002.

Urton, Gary. "Animals and Astronomy in the Quechua Universe." *Proceedings of the American Philosophical Society* 125.2 (1981): 110–27.

Valdés Aguilar, Rafael. *El cólera: Enfermedad de la pobreza*. Culiacán, Sinaloa: Universidad Autónoma de Sinaloa, 1993.

Vassberg, David E. "Concerning Pigs, the Pizarros, and the Agro-pastoral Background of the Conquerors of Peru." *Latin American Research Review* 13.3 (1978): 47–61.

Viveiros de Castro, Eduardo. *A inconstância da alma selvagem e outros ensaios de antropologia*. São Paulo: Cosac and Naify, 2002.

——. "Cosmological Deixis and Amerindian Perspectivism." *Journal of the Royal Anthropological Institute* 4.3 (1998): 469–88.

Weil, Kari. *Thinking Animals: Why Animal Studies Now?* New York: Columbia University Press, 2012.

Whitehead, Neil L. *Dark Shamans: Kanaimà and the Poetics of Violent Death*. Durham: Duke University Press, 2002.

——. *Of Cannibals and Kings: Primal Anthropology in the Americas*. University Park: Pennsylvania State University Press, 2011.

——. "Post-human Anthropology." *Identities: Global Studies in Culture and Power* 16.1 (2009): 1–32.

Whitehead, Neil L., and Michael Wesch, eds. *Human No More: Digital Subjectivities, Unhuman Subjects and the End of Anthropology*. Boulder: University of Colorado Press, 2012.

Whitehead, Neil L., and Robin Wright, eds. *In Darkness and Secrecy: The Anthropology of Assault Sorcery and Witchcraft in Amazonia*. Durham: Duke University Press, 2004.

Wilcox, Robert W. "Zebu's Elbows: Cattle Breeding and the Environment in Central

Brazil, 1890–1960." *Territories, Commodities and Knowledge: Latin American Environmental Histories in the Nineteenth and Twentieth Centuries,* ed. Christian Brannstrom, 218–46. London: Institute for Latin American Studies, 2003.

Wolfe, Cary. *Animal Rites: American Culture, the Discourse of Species, and Posthumanist Theory.* Chicago: University of Chicago Press, 2003.

Yarrington, Doug. "Cattle, Corruption, and Venezuelan State Formation during the Regime of Juan Vicente Gómez, 1908–35." *Latin American Research Review* 38.2 (2003): 9–33.

CONTRIBUTORS

Neel Ahuja is assistant professor of postcolonial studies in the Department of English and Comparative Literature at the University of North Carolina, Chapel Hill. His work on species has appeared in *Social Text*, PMLA, and the *Journal of Literary and Cultural Disability Studies* as well as in the edited volume *Postcolonial Green: Environmental Politics and World Narratives* (2010). He is currently working on a book manuscript titled *Bioinsecurities: Embodiment, Disease Interventions, and the Politics of U.S. National Security.*

Lauren Derby is associate professor of Latin American history at the University of California, Los Angeles. Her publications include *The Dictator's Seduction: Politics and the Popular Imagination in the Era of Trujillo* (Duke University Press, 2009), which won the Gordon K. and Sybil Lewis award from the Caribbean Studies Association and the Bolton-Johnson Prize from the Council on Latin American History, American Historical Association, as well as receiving honorable mention for the Bryce Wood Book Award from the Latin American Studies Association. She was coeditor, with Andrew Apter, of *Activating the Past: History and Memory in the Black Atlantic World* (2010). She is currently writing a book on demonic animal narratives in Haiti and the Dominican Republic and is coediting a reader on the Dominican Republic.

Regina Horta Duarte is professor of Brazilian history at Universidade Federal de Minas Gerais and researcher level 1B of CNPq (Brazil's National Council of Scientific and Technological Development). She is the editor of the journal *Historia Ambiental Latinoamericana y Caribeña* (HALAC), published by the Sociedad Latinoamericana y Caribeña de Historia Ambiental (SOLCHA). She is the author of *A imagem rebelde* (1991), *Noites circences: Espectáculos de circo e teatro em Minas Gerais no século XIX* (1995), *Historia e natureza* (2002), *A Biologia militante: O Museu Nacional, especialização científica, divulgação do conhecimento e práticas políticas no Brasil: 1926–1945* (2010), and several journal articles. She is currently researching the rise of gated communities in Brazil in the 1960s and 1970s, focusing on their political and environmental aspects.

Martha Few is associate professor of Latin American history and director of graduate studies at the University of Arizona in Tucson. She is the author of *Women Who Live*

Evil Lives: Gender, Religion, and the Politics of Power in Colonial Guatemala (2002), and is completing the book *All of Humanity: Colonial Medicine, Indigenous Healing, and Public Health in Enlightenment Central America.* Her research has also been published in the journals *Ethnohistory, British Journal for the History of Science,* and *Mesoamérica.* She will turn next to a history of locusts and locust plagues in Mesoamerica before the introduction of pesticides, from the late postclassic era to the end of the nineteenth century.

Erica Fudge is professor of English studies at the University of Strathclyde, Glasgow, where she is also the director of the British Animal Studies Network. She is the author of four books: *Perceiving Animals: Humans and Beasts in Early Modern English Culture* (2000), *Animal* (2002), *Brutal Reasoning: Animals, Rationality, and Humanity in Early Modern England* (2006), and *Pets* (2008). She edited *Renaissance Beasts: Of Animals, Humans, and Other Wonderful Creatures* (2004) and coedited *At the Borders of the Human: Beasts, Bodies and Natural Philosophy in the Early Modern Period* (1999) and, with the Animal Studies Group, *Killing Animals* (2006). She has also written articles that have appeared in journals including *Textual Practice, Oxford Literary Review, Angelaki,* and *New Formations.*

Reinaldo Funes Monzote is professor of history at the Universidad de La Habana and director of the Geohistorical Research Program in the Fundación Antonio Núñez Jiménez de la Naturaleza y el Hombre. He is the author of *El despertar del asociacionismo científico en Cuba, 1876–1920* (2004) and *De bosque a sabana: Azúcar, deforestación y medio ambiente en Cuba: 1492–1926* (2004), revised and published in English as *From Rainforest to Cane Field in Cuba: An Environmental History since 1492* (2008). He also recently edited *Naturaleza en declive: Miradas a la historia ambiental de América Latina y el Caribe* (2008).

León García Garagarza was born and raised in Mexico City and received his doctorate in colonial Latin American history from the University of California, Los Angeles, in 2011. He was awarded the 2012 Audrey Lumsden-Kouvel Fellowship by the Newberry Library, and in 2011–12 he was a postdoctoral fellow at the Smithsonian Institution, where he conducted research on Mesoamerican religions. He is currently working on a book manuscript, *New Perspectives on the Mexican Apostolic Inquisition Trials against Idolater Indians (1536–40).*

Heather McCrea is associate professor of history at Kansas State University in Manhattan, Kansas. Her book *Diseased Relations: Epidemics, Public Health, and State Building in Yucatán, Mexico, 1847–1924* (2011) explores the intersections between Western and Maya medical practices during nineteenth- and early twentieth-century epidemics of smallpox, cholera, yellow fever, and malaria. She has published essays in *Mexican Studies / Estudios Mexicanos* and in the edited volume *Converging Worlds: Communities and Cultures in Colonial America* (2011). McCrea's new work, *Gulf of Disease,* investigates the creation of trans-Caribbean identity concurrent to public-health campaigns to combat tropical maladies.

John Soluri is associate professor of history and director of global studies at Carnegie Mellon University. He is the author of *Banana Cultures: Agriculture, Consumption, and Environmental Change in Honduras and the United States* (2005). His articles have appeared in *Environmental History, Hispanic American Historical Review,* and *Latin American Research Review.* He is a founding member of the Sociedad Latinoamericana y Caribeña de Historia Ambiental (SOLCHA). His current research is centered on animals, commodity markets, borders, and environmental change in Patagonia and Tierra del Fuego.

Zeb Tortorici is assistant professor in the Department of Spanish and Portuguese Languages and Literatures at New York University and previously taught history at Stanford University as an American Council of Learned Societies (ACLS) New Faculty Fellow and at Tulane University as a visiting assistant professor. He is working on a book manuscript titled *Contra Natura: Desire, Colonialism, and the Unnatural.* His essays have been published in the journals *e-misférica, History Compass, Journal of the History of Sexuality,* and *Ethnohistory* as well as in the edited collections *Queer Youth Cultures* (2008), *Death and Dying in Colonial Spanish America* (2011), and *The Human Tradition in Colonial Latin America* (2013). With Daniel Marshall and Kevin Murphy he is currently coediting an issue of *Radical History Review* on the topic of "Queering Archives."

Adam Warren is associate professor at the University of Washington, where he holds the Howard and Frances Keller Endowed Professorship in History. He is a specialist in Peruvian history, colonial Latin American history, and the history of medicine and is the author of *Medicine and Politics in Colonial Peru: Population Growth and the Bourbon Reforms* (2010). He has published numerous articles and book chapters, including "An Operation for Evangelization: Friar Francisco González Laguna, the Cesarean Section, and Fetal Baptism in Late Colonial Peru," *Bulletin of the History of Medicine,* and "Medicine and the Dead in Lima: Conflicts over Burial Reforms and the Meaning of Catholic Piety, 1808–1850," in *Death and Dying in Colonial Spanish America* (2011).

Neil L. Whitehead was professor of anthropology at the University of Wisconsin, Madison. He is the author of *Dark Shamans: Kanaimà and the Poetics of Violent Death* (Duke University Press, 2002). His recent edited and coedited volumes include *Virtual War and Magical Death: Technologies and Imaginaries for Terror and Killing* (Duke University Press, 2013), *Human No More: Digital Subjectivities, Unhuman Subjects, and the End of Anthropology* (2012), *Of Cannibals and Kings: Primal Anthropology in the Americas* (2011), *Anthropologies of Guyana* (2009), and *Hans Staden's True History: An Account of Cannibal Captivity in Brazil* (Duke University Press, 2008), among numerous other books and journal publications.

Note: *Italic* page numbers indicate figures.

Animal Liberation (Singer), 9, 329

animal research. *See* vivisection

animal rights: neocolonialism in call for, 330; philosophical and legal trend in favor of, 221; rise of animal-protection societies and, 221, 222

Animal Rights and Human Obligations (Regan and Singer), 329

animal-rights movement: animal studies linked with, 9; insects in, 64, 85n14; on monkeys of Puerto Rico, 194, 197

Animals of Spain, The (Alves), 7

animal studies, 5–10; animal-rights movement and, 9; benefits of, to understanding of human history, x, 3–4, 339; definition of, 9; future of, 10; growth of field, ix–x; history of field, 9; insects as topic in, 17, 63, 64, 85n5; methodological challenges of, 3, 245; recent developments in, 5–10; review of literature in, 5–10

Animal Studies Group, 16, 85n14

animal testing. *See* vivisection

Anjoso, Padre, 99

Anselm, Saint, 98

ants: agency of, 340; birds' consumption of, 287, 294; insecticides used against, 62

Antarctica, "discovery" of, 264n9

Antennae (journal), 63

anthropocentrism: and animal agency, 14; in colonial Mexico, 105, 109–13; in Cuban animal-protection society, 226–28; of historical sources, 113

anthropomorphism: of birds in Brazil, 291; of dogs in colonial Mexico, 109–11; problems with term, 109

Anthrozoos, 9

antibiotics, resistance to, 170

antihumanism, 9

apodos (nicknames), in Dominican culture, 315–19. *See also* Trujillo, Rafael

Apollo Parnopios (deity), 63

Apostolic Inquisition, 55n18

Appleman, William Henry, 253

Archive of the Indies (AGI), 144n3

archives, colonial: limitations of, 4; visibility of animals in, 19

Arctocephalus australis. See fur seal, South American

Arctocephalus philippii, 264n7

Argentina: bird-protection societies of, 277; conservation policies of, 258, 259–60; fur seals in (*see* fur seal hunting, in Patagonia)

armadillos, medicinal use of, 141, 148n62

Armas, Joseph, 100–102, 106

Armstrong, Philip, 11

Arnold, James, 309

Arozarena, Ramón de, 218

Arrevillaga, Tomás de, 79, 82

Arroyo Cabrales, Joaquín, *Relaciones hombre-fauna*, 7

Ashur (deity), 63

Asociación Cubana para la Protección de Animales y Plantas, 235

Asociación Ornitológica del Plata (Argentina), 277

ASPCA, 225

Assyrians, on locusts, 63

Asúa, Miguel de, *A New World of Animals*, 7

asylum for beggars, in Cuba, 232, 241n95

Athearn, George F., 255

Atkins, Edwin, 218–19

Atlapolco (Mexico): Catholicism in, 53n1; Teton's influence in, 31, 36, 40, 41

Audiencia of Guatemala, 70, 78–80, 82, 84n3, 88n45, 91n91

Audubon, John James, 276

Augustine, Saint, 73, 98

Aurora del Yumurí (periodical), 223

Austria, animal-protection laws in, 222

authoritarianism, in Brazil, 271

Aves do Brasil, As (Goeldi), 271

Avetine, Saint, 98

Aymara language: Cobo's study of, 127, 128; Kallawaya use of, 146n22

Aztec: diet of, 60n68; ritual lustral baths of, 57n36

Babcock, Charles Almanzo, 276–77; *Bird Day: How to Prepare for It*, 277

bacás, 309–10, 313

Bachiller y Morales, Antonio, 214

Bachman, George, 184, 190–91

bad air, 71, 158

Bakhtin, Mikhail, 319

Balaguer regime (Dominican Republic), 320n2

Balbuena, Antonio, 99–100, 106, 117n37

Balmaceda, José Manuel, 243, 259

Baltimore (Maryland), animal-protection societies in, 223

baptisms: of animals, 96, 97 (*see also* baptisms, dog, in colonial Mexico); of dolls, 100, 115n28; salt in, 100, 101, 116n32; Teton's opposition to, 31, 36, 40, 51

baptisms, dog, in colonial Mexico, 18, 93–113; anthropocentrism and, 109–13; descriptions of, 99–102; Inquisition on, 97, 99–102, 108–9, 111, 117n37; social meanings of, 105–9

Baptiste-Ballera, Luis Guillermo, *Rostros culturales de la fauna*, 7

"barbarians," Maya depicted as, 153, 162, 169

"barbarism," vs. civilization, in Yucatán Peninsula, 151, 156–57, 160

Barcelona (Spain), animal-protection societies in, 220, 226

barn owl, 229, 240n77

Barrios, Francisco de, 119n61

Basterrechea, Thoribio, 93–96, 100, 102, 111, 115n25

Bastien, Joseph, 131, 133, 134, 135, 137, 139, 147n43

Batman, 308

bats, 229

Bauduy, Pedro, 218

Bautista, Juan. See *Anales de Juan Bautista*

Beach, Harlan Page, *A Geography and Atlas of Protestant Missions*, 165

bears, laws protecting, 222

bedbugs, 62

bees, 62, 332

Beezley, Bill, 307

beggars, asylum for, in Cuba, 232, 241n95

behavioral field studies, on monkeys, 184, 189, 190, 191, 198

"Beneficencia para con los animals" (De Weiss), 223, 239n49

Benes, Peter, *New England's Creatures*, 6

Benítez, Jaime, 198–99

Bergh, Peter, 225, 240n65

Berlin, Brent, 157

Bermúdez, Atiliano, 230

bestiality: animal agency in, 13–14; Bible on, 2, 21n5; common victims of, 14; evidence used in cases of, 14; historical information revealed in cases of, 2–3; punishments for, 1–2, 21n5, 111

Bible: on animal worship and idolatry, 98, 115n18; on bestiality, 2, 21n5; cosmovision based on, 51–52; European animals in, 51–52; goats in, 314; locusts in, 63, 66, 67, 85n10, 86n18, 89n70

Bierhorst, John, 59n56, 59n62

Bilac, Olavo, 287

Bioculture (company), 197

bioinsecticides, 90n89

biology, in U.S. and Brazilian national identity, 270

biomedical research: on animals (*see* vivisection); on humans, 332, 343n13

biopolitics, 154, 171–72nn18–19

Bird Day (Brazil), 286–87, 291

Bird Day (U.S.), 276–77, 286

Bird Day: How to Prepare for It (Babcock), 277

BirdLife International, 295

birds: animal-protection societies on, in Cuba, 228, 229, 240n77; extinction of, 278, 295; international conservation movements for, 276–78, 283–84; international migration of, 277–78; laws protecting, 222, 276, 277, 278; medicinal use of, in Andes, 137, 141–42, 148n49; watching of, 288–89. *See also* birds, in Brazil; birds, in United States; *specific types*

birds, in Brazil, 15–16, 270–95; in agriculture, utility of, 283, 284, 287, 288–90, 294; change in sensibilities and, 277, 285–86, 295; demand for feathers of, 272–77, 280, 281; early scientific writings on, 271; ethical arguments against killing of, 282–83, 285–86; indigenous peoples and, 279; intellectuals' opposition to killing of, 278,

Candomblé, goats in, 314

cannibalism, 50, 334–35

canoes, 248

Canuc, Juan, 1

capitalism, in free-market environmentalism, 330–31

capital punishment: of animals, 96, 107, 111;
for bestiality, 21n5, 111

Cárdenas (Cuba): horses in, 237n16; oxen
in, 215

Caribbean: importation of monkeys to, 180;
lack of native species of monkeys in, 180;
shape-shifting in, 309–10. *See also* Cuba;
Dominican Republic; Puerto Rico

Caribbean Primate Research Center
(CPRC), 194, 196–97

carnivalesque: in Lent celebrations, 307–8;
in sacraments for dogs, 102, 106, 108, 111

Carpenter, Clarence Ray: behavioral field
studies of, 184, 191, 198; in breeding of
monkeys, 183, 197; establishment of Cayo
Santiago colony by, 180–81, 183, 184–85,
187; history of Cayo Santiago written by,
198–200

carros ballena, 219

Case for Animal Rights, The (Regan), 9

Caste War of Yucatán (1847–1902), 156–57;
casualties of, 157, 174n36; cholera pandemic in origins of, 151, 156; elite perceptions of Maya in, 156–57, 162–63; start of,
156

castration, for bestiality, 1

Castro, Fidel, 306

cataloging practices: humoral systems in,
131–32, 136–42; for plants vs. animals,
136–42

catfish, medicinal use of, 139

Cathars, 97–98

Catholic Church: on ancestral Mexican
deities as devils, 40, 57n35; blessings for
animals in, 98; on burials during cholera
outbreaks, 159; cosmovision of, 51–52; on
creole religious practices, 310, 324n51; on
humanity of Indians, 119n63; Jesuit missions of, in Andes, 127–28, 133, 146n19; in
locust-extermination campaigns, 65, 72–

74, 77; on sacramental desecration, animals in, 96–97; on souls, 97–98, 108–9;
theological conceptions of animals in,
96–97, 98; on tlaciuhqueh (diviners), 34.
See also inquisitions

cats: big, 46, 59nn58–59; domestic, mock
trials of, 107

cattle. *See* cows

Cavanilles, Antonio José, 147n30

caves, in Yucatán Peninsula, 152

Cayo Santiago monkey colony (Puerto
Rico), 180–202; administration of, turnover in, 188–92, 196; behavioral field studies on, 189, *190*, 198; caretakers of, 200;
community outreach on behalf of, 196–
97; disease among, 189; escape of monkeys from, 186, 189; establishment of,
180–81, 183–88; funding for, 189, 191–92,
196, 197; habitat redesign for, 185–86; histories written by researchers at, 198–200;
killed for use in vaccine production, 183;
location of, 185; media coverage of, 181,
186–87, *187*, *188*, 197; photos of island, *185*,
187, *188*; population of, 189; public
resistance to, 187; Puerto Rican takeover
of, 181, 196–97; spinoff institutions of, 191;
tourism at, 197

CDC, U.S., 332

cemeteries, in cholera outbreaks, 159–60,
175n43

cencoccopi, 47

cenotes (sinkholes), 152, 154

Centers for Disease Control (CDC), U.S.,
332

Central Intelligence Agency (CIA), U.S., 332

Cervantes, Miguel (witness), 101

Chakrabarty, Dipesh, 201

chapulínes, use of term, 65–66. *See also*
locusts

Chapultepec (Mexico), 52, 66

Charcas (Bolivia), Delgar in, 129

Charles River Laboratories, 197

Charles Shearer (vessel), 256

Charnay, Claude-Joseph Déserié M., 150

chemical insecticides: birds and, 294;
environmental impact of, 64; ethics of,

chemical insecticides (*cont.*)
64, 86n15; in government-directed extermination campaigns, 62; human warfare
in development of, 90n89; insect resistance to, 170; modern popularity of, 62
chemical warfare, 90n89
chestnut-bellied euphonia, demand for
feathers of, 276
Chichimeca, 48–49, 60n74
chickens: in cockfighting, 222, 226, 227, 228,
230; indigenous consumption of, 44;
indigenous names for, 24n32; interspecies
grazing with turkeys, 1, 2–3; sacrifice of,
in Dominican Republic, 321n13; transformation of humans into, 39
chicome malinalli, 46
Chilam Balam de Nah, 158
Chile: civil war of 1891 in, 259; conservation
policies of, 258–60; fish farming in, 261;
fur seals in (*see* fur seal, South American;
fur seal hunting, in Patagonia)
China: fur seal exports to, 244, 261; locusts
in, 73
chinchilla, medicinal use of, 123, 124, 139, 141
cholera: symptoms of, 154; transmission of,
154. *See also* cholera, in Yucatán Peninsula
cholera, in Yucatán Peninsula, 154–64;
burial practices during outbreaks, 159–60,
173n32; in Caste War period, 151, 156; consumption of animal byproducts and, 160–
61; first pandemic of 1830s, 151, 154–56,
171n7; intemperance linked with vulnerability to, 155; livestock slaughter and,
160–61; Maya treatments for, 153; prevention strategies for, 156; untended animals
and, 155–56, 157, 158
choteo, 304
Chrisostomo, Francisco, 80
Christianity: on compassion toward animals,
98; on killing of animals, 98; locusts in, 63–
64, 67, 85n10, 86n18, 89n70; spread of, as
justification for colonialism, 71; theological
conceptions of animals in, 96–97, 98. *See
also* Bible; Catholic Church
Chucuito (Peru), Cobo in, 128
chupacabras legend, 194–96

chuqui chuqui, medicinal use of, 139
CIA, U.S., 332
cicadas, vs. locusts, 86n17
cigarra, use of term, 66, 86n17. *See also*
locusts
cimarróns (runaway slaves), in Dominican
Republic, 305, 311
circuses, animal cruelty in, 230–31
civilization, vs. "barbarism," in Yucatán Peninsula, 151, 156–57, 160
civil rights, historical expansion of, 222
class, social: in international movement for
protection of animals, 221; in nicknames
of Dominican Republic, 316–17; in pet
ownership in colonial Mexico, 106
cleanliness, of body and soul, in vulnerability to disease, 155
climate, of Yucatán Peninsula, 151–52
clubs, fur seal hunting with, 253–54, 257
Cobo, Bernabé: life of, 127–28; sources of
indigenous knowledge of, 127–28, 131. *See
also* Historia del Nuevo Mundo
cochineal, 27n60
cockfighting: in Cuba, 226, 227, 228, 230; in
England, 222
coconepipilpipil, 47
Codex Borbonicus, 56n27
Codex Telleriano-Remensis, 37
Codice Chimalpopoca, 46
Código de Caça (Brazil, 1943), 293
cognitive abilities, of fur seals, 247, 263
Cohuatépec (Mexico): Catholicism in, 53n1;
Teton's influence in, 31, 36, 40
Coleman, Jon T., 112; *Vicious: Wolves and
Men in America*, 6
Collado, Lipe, 322n24, 325n64
Collao (Bolivia), Cobo in, 128
Colón (Cuba), oxen in, 215
colonial archives: limitations of, 4; visibility
of animals in, 19
colonialism: locusts' impact on, 64–65, 83–
84; review of literature on history of, 5–
7; spread of Christianity as justification
for, 71
Columbian Exchange, The (Crosby), 6, 31,
333, 339

Columbia University: in Cayo Santiago monkey colony, 180, 183, 190–91; School of Tropical Medicine, 180, 183–85, *184*, 190–91

Columbus, Christopher: on exotic species, 8; transportation of animals by, 6, 311

commodification of animals, ubiquity of, 12. *See also specific forms*

commoners. *See* macehuales

Communal Areas Management Program for Indigenous Resources (CAMPFIRE), 343n9

compassion toward animals, Christianity on, 98

condor, medicinal use of, 141, 143

Conill, Oscar, 233

Conklin, Harold, 157

Consejo del Estado (Dominican Republic), 306

conservation movements: for birds, 276–78, 283–84; free-market environmentalist approach to, 330–31

conservation policies: on birds, 276; on fur seals, 258–60, 261

constitutional federalism, of Brazil, 271

constitutional monarchy, of Brazil, 271

constitutions, Brazilian, 271, 294

consumption of animals, 15–16; as cannibalism, 334–35; during cholera outbreaks, 160–61; human conquest of animals through, 161–62. *See also* diet, human; *specific animals*

Cook, Alexandra Parma, 89n67, 89n71, 91n97

Cook, James, 250

Cook, Noble David, 89n67, 89n71, 91n97

copalli, 61n78

Cormier, Loretta A., *Kinship with Monkeys*, 334

corn cultivation: in creation stories, 49; European livestock and, 41, 42, 43; indigenous reliance on, 44; locusts and, 69, 70

Corona M., Eduardo, *Relaciones hombre-fauna*, 7

Correa, Mariano, 94

Cortés, Hernán, 6, 38

Cosio, Toribio de, 79

cosmogony, Mesoamerican, 45–49

cosmovision: Catholic, 51–52; definition of, 54n5; human-animal transformation in, 34, 50, 54n9; Mesoamerican, 32, 34, 44–45, 50, 54n9

cows: agency of, x; agricultural use of manure of, x, 218; animal-protection societies on milk of, 231; indigenous names for, 9. *See also* cows, in colonial Mexico; oxen, in Cuba

cows, in colonial Mexico: exportation of hides of, 44; feral populations of, 44; indigenous consumption of, opposition to, 12, 39; indigenous consumption of, origins of, 44; indigenous handling of, ban on, 43; introduction and spread of, 41–44; population of, 43, 44; reproduction rates of, 41; transformation of humans into, 31, 39, 41

coyo guanca, medicinal use of, 141

CPRC (Caribbean Primate Research Center), 194, 196–97

creation stories, Mesoamerican, 45–49

creole, definition of, 88n46

creole culture, origins of, 311–12

creole goats. *See* goats, in Dominican Republic

creole identity, goat as sign of, 304

creole pigs. *See* pigs, in Dominican Republic

creole religious practices: Catholic vilification of, 310, 324n51; goats in, 314

Cressy, David, 96, 102, 108

Criminal Prosecution and Capital Punishment of Animals, The (Evans), 96

Cristóbal Gundlach, Juan, 229, 240nn77–78

Cronon, William, 69

Crosby, Alfred W., *The Columbian Exchange*, 6, 31, 333, 339

cruelty to animals, in Cuba, 220–34; in bull-fighting, 226, 227, 228, 230, 232–33; in cockfighting, 226, 227, 228, 230; media coverage of, 230–31; prevention of (*see* protection of animals, in Cuba)

Cruzada de Amor aos Pássaros, 291

Cuba: animal labor in (*see* labor, animal, in

Cuba (*cont.*)

Cuba); economic crisis of 1990s in, 209–10; establishment of republic of (1902), 233; Maya rebels exported to, 163; railroad system of, 212; sugarcane cultivation in, 210–20, 215, 234–35; Ten Years War in, 220; treatment of animals in (*see* cruelty to animals, in Cuba; protection of animals, in Cuba)

Cuban Association for the Protection of Animals and Plants, 235

Cuban Humane Society for the Protection of Children and Against the Cruelty of Animals, 233–34

Cuban Society for the Protection of Animals and Plants. *See* Sociedad Cubana Protectora de Animales y Plantas

cuckolds, 307

Cueto, Marcos, 146n23

Cueva (Puerto Rico), free-ranging monkey populations on, 191, 192

culture, human: vs. nature, problems with classic distinction between, 333; recognition of role of animals in, ix; "so-called," ix

Curcio-Nagy, Linda A., 112

cuy, medicinal use of, 141

Dalrymple, Alexander, 250

DaMatta, Roberto, 316

Darnton, Robert, "Workers Revolt," 106–7

darts, 334

Darwin, Charles, 248

Darwinism, social, 162

DDT, 62

death penalty. *See* capital punishment

deforestation: in Cuba, 211, 212; in Dominican Republic, 312, 325n60

deities: ancestral Mexican, as devils, 40, 57n35; animal, prevalence of, 7–8; Mesoamerican views on boundaries of category of, 34, 44–45; tlaciuhqueh (diviners) as, 33

Delgado de Najera, Tomás, 72–73, 74

Delgar, Martín, 12–13, 123–43; audience of, 136, 142; cataloging of plants vs. animals by, 136–42; embracing of indigenous

medicine, vs. his contemporaries, 124–25, 130; goal of, 136; *Historia del Nuevo Mundo* (Cobo) used by, 125, 126, 128, 129, 130, 137; humoral framework of, 131–42; "Libro de medicina y cirugía para el uso de los pobres," 131, 144n8; life of, 127, 128–29; on mediums for treatments, 137, 148n49; published lists of Kallawaya medicinal plants and, 130–31; *Recetario eficáz para las familias*, 123, 124, 130; reputation of, 129, 146n27; scholarship on, lack of, 146n23; sources of indigenous knowledge of, 128–31; Spanish and Andean practices merged by, 126–27; travels of, 128, 129; writings attributed to, 124, 130, 144n8

demons, in Dominican culture, 307–10, 313–14

dengue fever, 170

dengue shock syndrome, 170

Derby, Lauren, 4, 11, 18–19, 195, 336

Derrida, Jacques, 205n52

Descola, Philippe, 155, 162; *Nature and Society*, 126–27

Desecheo Island (Puerto Rico), free-ranging monkey populations on, 191, 192

devils: in Dominican culture, 307–10, 308, 313–14; Mexican ancestral deities recast as, 40, 57n35; shape-shifting by, 309

De Weiss, "Beneficencia para con los animals," 223, 239n49

diablo cojuelo, 308

Diario de la Marina, 230–31

Dias, Gonçalves, 287

Díaz, Porfirio, 166

Dibble, Charles E., 54n8

Dickinson, Edward Ross, 171n18

Diego Ramírez Islands (Chile), fur seal hunting on, 256, 257

diet, fur seal, 26n20

diet, human: animal-protection societies on, 231; in creation stories, 46–49; of elite/nobles, 44, 47, 60nn67–68; European livestock in, opposition to, 12, 31, 32, 39–41, 50–53; European livestock in, origins of, 44; fat in, 249, 262; fish in, 47,

248; fur seals in, 249, 262; locusts in, 66, 83; protein sources in, 44. *See also specific foods*

disease: bad air and, 71, 158; germ theory of, 158–59, 164; among horses, 233; intemperance linked with vulnerability to, 155; locusts linked to, 65, 70–72; miasma theories of, 156, 159, 160; among monkeys, 189, 196; transmission of, advances in understanding of, 164. *See also* disease, in Yucatán Peninsula; vectors of disease; *specific diseases*

disease, in Yucatán Peninsula, 17, 149–70; during Caste War, 156–57; cholera, 151, 154–64, 171n7; dengue fever, 170; evil eye and, 157–58; insects linked to, 151; livestock blamed for spread of, 155–56, 157–58; Maya beliefs about origins of, 158; Maya blamed for, 153, 154, 155, 163; mosquitoes blamed for, 164–69; Rockefeller campaigns to eradicate, 151, 165–68; state-building concurrent with, 152; yellow fever, 151, 153, 164–69

disease moments, 152, 154, 171n11

disease research, monkeys in, 180, 182–83

dissection of animals, in anatomy classes, 123

divination, in indigenous religion, 33

diviners. *See* tlaciuhqueh

divine wrath, locusts as symbol of, 64, 67

"Does 'the Animal' Exist?" (Pearson and Weismantel), 10

dog-ancestors, 48–49, 60n74

dogs: change in sensibilities toward, 98–99; in creation stories, 47, 48–49, 60nn73–74; dissection of, in anatomy classes, 123; feral, in Puerto Rico, slaughter of, 194; hairless, 49; purebred, 105–6; sacramental desecration by, 96, 114n12; stray, in Cuba, 233; working, 99, 106. *See also* dogs, in colonial Mexico

dogs, in colonial Mexico, 18, 93–113; anthropocentrism and, 105, 109–13; anthropomorphism of, 109–11; change in sensibilities toward, 98–99, 109–13; church dogma challenged by, 96, 102,

108–9; funerals for, 101, 108, 112; indigenous breeding of, 49; indigenous consumption of, 49; reproduction of, 107–8; satirical commemoration of life of, 102–5, 106–7, 112; social meanings of, 105–9, 112, 118n53. *See also* baptisms, dog, in colonial Mexico; weddings, dog, in colonial Mexico

Dogs of the Conquest (Varner and Varner), 5–6

dolls, baptism of, 100, 115n28

Dolores de Cuebas, María, 100, 101, 106, 108, 110, 112

domestic animals. *See* livestock; pets

domestic-wild dualism: fur seals in, 258; goats in, 313; insects in, 64

Dominican Republic: agriculture in, late arrival of, 304, 311–12; creole culture in, origins of, 311–12; creole religious practices in, 310, 324n51; devils in culture of, 307–10, *308*; diet in, 305–6; economy of, 304–6, 311–13, 317–18; goats in (*see* goats, in Dominican Republic); nicknames in culture of, 315–19; U.S. invasions of, 317

Dopico Black, Georgina, 4

Douglas, Mary, 326n72

doves, white-tailed, hunting of, 274

drag, transspecies, 109

Dresser, Norine, 108

DRNA (Natural Resources and Environment Department), Puerto Rican, 194

droughts, in 1-Rabbit years, 37, 52, 56n24

drug use: campaigns against, 332; in Puerto Rico, 197

Duany, Jorge, 182

Duarte, Regina Horta: chapter by, 270–95; comments on, 11, 15–16, 329, 331

Dumont, Alejandro, 214

Duvalier, Jean-Claude, 310

eared seals. *See* otariids

Earle, Rebecca, 12

ear pain, treatment of, 141

Easter celebrations, in Dominican Republic, 307–8

Eastern Europe, collapse of socialism in, 209

ecology, European, transferred to colonies, 6–7
economic human-animal relationships, 12. *See also specific types*
economics: in free-market environmentalism, 330–31; of fur seal hunting, 244, 251–53, 257, 259, 267nn44–45
economy: of Cuba, crisis of 1990s in, 209–10; of Dominican Republic, 304–6, 311–13, 317–18
education: dissection of animals in, 123; U.S., birds in, 277
eggs, medicinal use of, 148n49
elephants, dark side of behavior of, 338
elite, Brazilian: demand for feathers among, 272–74; in Revolution of 1930, 271
elite, colonial: elimination of indigenous peoples sought by, 17–18; funerals of, 117n41; in locust-extermination campaigns, 78; satire on extravagance of, 102–5
elite, Cuban, on care and treatment of animals, 217, 220
elite, Guatemalan, in locust-extermination campaigns, 79, 80
elite, Mexican: on animals in spread of disease, 157; in Caste War, 156–57, 162–63; diet of indigenous, 44; ideology of, 162; oppression of Maya by, 166; in public health campaigns, 152–53, 169; on species and human race hierarchies, 152, 153, 162
Empire of Trauma, The (Fassin and Rechtman), 342n4
Emu (magazine), 277
encomiendas, 43–44
end times. *See* eschatology, indigenous
energy sources, in Cuba, 213, 235, 237n11
England: animal-protection laws in, 222; animal-protection societies in, 15, 210, 220, 222; bird feathers in, 275, 277; bird-protection laws in, 278; bird watching in, 289; change in sensibilities toward animals in, 99; manure in agriculture in, x
English colonists: animal husbandry promoted by, 58n44; livestock introduced by, 57n40

Enlightenment, in ideology of elite, 162
entertainment, sacraments for dogs as, 94, 105–9, 110–11
entomology, professionalization of, 62, 84n2
environmental movement: free-market, 330–31; insects in, 64
epidemics: cholera, 151, 154–56, 171n7; after fall of Tenochtitlan, 41; introduction of European livestock and, 41; locusts linked to, 65, 70–72; swine-flu, 321n7
erosion, from overgrazing, 44
eschatology, indigenous: after Spanish Conquest, 36–39; of Teton, 32, 33, 35, 36–39, 49–53
Esposito, Roberto, 180
Estado Novo (Brazil), 271, 292
estancias, 42, 43, 57n39
Esteyneffer, Juan de, *Florilegio medicinal*, 131–32
ethics and morality: birds associated with, 291–92; of killing of birds, 282–83, 285–86; of mass killing of insects, 64, 85–86nn14–15; vulnerability to disease linked to, 155
ethnic identity, precontact, based on lineage, 55n22
ethnography, transspecies, 7
Eucharist: animals eating, 96, 114n12; metaphorical vs. literal meaning of, 335
eugenics, in Puerto Rico, 182
Europe: ecology of, transferred to colonies, 6–7; laws on protection of animals in, 222, 277; livestock imported from, 41–44; locusts of, 67; personhood in, 34–35. *See also specific countries*
Evans, E. P., 85n12; *The Criminal Prosecution and Capital Punishment of Animals*, 96
evil eye, 157–58
excommunication, of locusts, 64, 85n12
Exodus, Book of, 63, 85n10, 115n18
exorcisms, of locusts, 73–74, 89n67
exotic species: of birds, in Brazil, 282; descriptions of, 8; names for, 8–9; reactions to, 8–9; as scientific curiosities to Europeans, 130, 146n29. *See also specific types*

experience, vs. identity, politics of, 336–37

experimentation, animal. *See* vivisection

exportation of fur sealskins. *See* fur seal hunting, in Patagonia)

exportation, from colonial Mexico, of live-stock products, 44

exportation, from Brazil: of feathers, 275–76; political power associated with, 271; wealth created by, 274

Express (vessel), 251–52

extermination, origins of term, 82. *See also* insects; locusts

extinctions, bird, 278, 295

Falkland Islands, fur seal hunting on, 250, 258, 259–60, 264n9, 268n70

famine, locusts linked to, 65, 66, 70, 72, 83

fandangos, 94, 114n4

Farfán, Agustín, *Tractado breve de anothomía y chirugía*, 147n34

fashion: bird feathers in, 272–77, 273, 280, 281; bird skins in, 276; fur in, 262

Fassin, Didier, *The Empire of Trauma*, 342n4

fasting, in locust-extermination campaigns, 72

fat, animal: in indigenous diets, 249, 262; medicinal use of, 137, 141

FDA, U.S., monkeys of Puerto Rico and, 191, 192, 193

feast of the goat, 302–3, 304, 307, 310, 319–20

feathers, in Brazil: amount exported, 275–76; consumer demand for, 272–77, 280, 281; decline in popularity of, 276; most valued types of, 275–76

feces, animal: agricultural use of, x, 218, 229; medicinal use of, 148n49

feces, human, in spread of cholera, 154

federalism, constitutional, of Brazil, 271

Federal Tariff Act of 1913 (U.S.), 276

Feliciano Velázquez, Primo, 46

female sexuality, in Puerto Rico, 182

feral populations: of cows, in colonial Mex-ico, 44; of dogs, in Puerto Rico, 194; of goats, in Dominican Republic, 304, 305; as historical agents, 338

Ferdinand II of Aragon, 97

Fernández de Lizardi, José Joaquín, 117n40

Fernández Díaz, Pedro, 233

fertilizers: manure as, x, 218; in sugarcane cultivation, 218

Few, Martha, 11, 17, 26n52, 182, 200, 332, 334, 338, 339, 341

fiesta del chivo (feast of the goat), 302–3, 304, 307, 310, 319–20

Finlay, Carlos, 164

fire, in locust-extermination campaigns, 74, 75, 80–81

firearms: in bird hunting, 274–75, 275; in fur seal hunting, 254, 254–56

fish: in creation stories, 47; in indigenous diets, 47, 248; medicinal use of, 139

fishing industry, 261; in Puerto Rico, 196–97

Fiske green bagasse furnaces, 218–19

Fissell, Mary, 83, 92n116

Five Suns, myth of, 46–49, 51–53

fleas, 62

Fletcher, Richard, 153

Florence (schooner), 259

Florentine Codex (Sahagún), 8, 35, 56n27

Florilegio medicinal (Esteyneffer), 131–32

flying-pens, 218

Fon-Fon! (magazine), 272–73, 273, 291

Food and Drug Administration (FDA), U.S., monkeys of Puerto Rico and, 191, 192, 193

food shortages, locusts linked to, 65, 66, 70, 83

forests, decline of: in Cuba, 211, 212; in Dominican Republic, 312, 325n60

Fossey, Diane, 343n8

Foster, George (naturalist), 250

Foster, George M. (anthropologist), 131–34, 135, 147n43

Foucault, Michel, 172n19

France: animal-protection laws in, 222, 228; animal-protection societies in, 220, 223; bird feathers in, 275, 276; bird-protection societies of, 277; dog weddings in, 99; Paris Exposition of 1889, 131; pet-keeping practices in, 105, 107

Francis of Assisi, Saint, 98

freedom, pigs and goats as symbol of, 313, *314*

free-market environmentalism, 330–31

French, Roger, *A New World of Animals*, 7

Frente Unido Pro-Defensa del Valle de Lajas, El, 193, 195, 196

Freyre, Gilberto, 212

Frías y Jacott, José Francisco de. *See* Pozos Dulces, count of

Frontera, José Guillermo, 191, 198–99

Fudge, Erica, 3, 9, 19, 126, 143, 145n12, 201, 245–46; "A Left-Handed Blow," 10, 85n14

funding, for Cayo Santiago monkey colony, 189, 191–92, 196, 197

funeral ceremonies: colonial extravagance in, 102–3, 117n41; for dogs, 101, 108, 112; dogs' role in, 49

Funes Monzote, Reinaldo, 11, 15

fur: changes in consumer demand for, 261–62; medicinal use of, 148n49

fur seal, South American, 246–48; agency of, 254, 261, 263, 337; cognitive abilities of, 247, 263; conservation policies for, 258–60, 261; demand for fur of, 261–62; diet of, 26n20; fur of, 247, 255, 265n19; habitat of, 244, 246–47, 260–61; in indigenous diets, 249, 262; population of, decline in, 252–54, 258–59; population of, modern, 260; reactions of, to human attacks, 26n52, 254–55, 263; reproduction of, 247–48, 259–60, 265n16, 266n21; social behaviors of, 255; taxonomic classification of, 246–47. *See also* fur seal hunting, in Patagonia

fur seal hunting, in Patagonia, 15–16, 243–63; bans on, 259–60; characteristics of fur seals enabling, 247–48, 260; conservationist policies on, 258–60, 261; dangers to hunters of, 253, 257, 267n48; decline of, in nineteenth century, 244; economics of, 244, 251–53, 257, 259, 267nn44–45; historical sources on, lack of, 245; hunters' views on animals and, 257–58, 336; indigenous peoples and, 244–45, 248–50, 251, 262; by North Atlantic hunters, 244, 250–58, 251, 261–62; number of seals killed, 244, 251, 267n39; origins of, 243–44, 250;

in pelagic zone, 249, 259; reactions of seals to, 26n52, 254–55, 263; scholarship on, 244, 264n9; techniques of, 249, 253–57, 254, 259, 261

Gagá secret societies, 308

Gage, Thomas, 67, 68, 69, 74–75, 77, 87n28

Galen, 71

gallina de la tierra. *See* turkeys

gallinazo, 145n18

Gann, Thomas, *The Maya Indians of Southern Yucatan and Northern British Honduras*, 157

garbage collection, during cholera outbreaks, 156, 159, 160

García, Nicolás, 80

García Garagarza, León, 11, 12, 13

García Villarraza, Juan, 225, 226, 234, 240n65

García y García, Apolinar, 153

Garibay, Ángel María, 36

garrapatas (ticks), 149–50, 332

Gazeta de Guatemala (newspaper), 68

Geertz, Clifford, 315–16, 320

Geography and Atlas of Protestant Missions, A (Beach), 165

Gerhard, Peter, 54n6

Germany: animal-protection societies in, 220, 226; bird feathers in, 275

germ theory of disease, 158–59, 164

Gil y Lemos, Francisco, 144n3

glanders, 233

goatees, 304, 319

goats: indigenous consumption of, 60n67; indigenous names for, 24n32. *See also* goats, in Dominican Republic

goats, in Dominican Republic, 18–19, 302–20, *315*; arrival of, in colonial era, 310–11; in celebrations of anniversaries of Trujillo's death, 302–3, 306, 307–10, 320n2; characteristics of, 306–7; consumption of, 304, 305–6, 314; creole, 314, *315*; in economy, 304–6; feral populations of, 304, 305; fiesta del (feast of the), 302–3, 304, 307, 310, 319–20; pigs' decline in rise of, 313, 321n7; sacrifice of, 304, 314, 321n13; as symbol of freedom, 313; Trujillo as (*see*

Historia del Nuevo Mundo (Cobo) (*cont.*)
  reproduction of, in later works, 147n30;
  on turkey buzzard, 145n18; writing of, 128
Historia de los mexicanos por sus pinturas, 45, 46
historical sources: anthropocentric nature of,
  113; on fur seal hunting, 245; limitations of,
  4; on locusts, 63–64; on Mesoamerican
  religion, 32; by scientific researchers, on
  monkeys, 198–200; visibility of animals in,
  19. *See also specific sources*
history, animal: growth of field, ix–x; as
  human history of animals, 201. *See also*
  animal studies
history, human: animals as victims in, 330,
  338; benefits of centering animals in, x, 3–
  4, 339; recognition of role of animals in,
  x; visibility of animals in, 3, 4, 5, 19
Hogar, El, 233
Holland, animal-protection societies in, 220
Honduras, bird hunting in, 278
honey, medicinal use of, 137, 148n49
"Honras fúnebres a la perra Pamela" (anonymous), 102–5, 106–7, 112, 117n40
hookworms, 157, 165, 168, 182
hornos de bagazo verde, 218–19
horses: baptisms of, 100; indigenous names
  for, 9; indigenous use of, 43. *See also*
  horses, in Cuba
horses, in Cuba: animal-protection societies
  on, 228, 230, 231, 233; cruelty to, 210, 228,
  230, 231, 233, 241n87; disease among, 233;
  labor of, in sugar plantations, 212; number of, 237n16
Horses of the Conquest (Graham), 5–6
Huaorani, 334
Huastecs, on turkeys and nobles, 48
Hugo, Victor, 287
human(s): as category of difference vs. substance, 9; decentering of, 13, 14, 330;
  determination of humanity of, 119n63. *See
  also* human(s), boundaries of category of
human(s), boundaries of category of, 3–4;
  European views on, 34; Mesoamerican
  views on, 34–35, 44–46; religious sacraments for animals and, 108
human-animal studies. *See* animal studies

human-animal transformation: in creation
  stories, 45–49; after eating European livestock, 31, 39–41, 50–51; in Mesoamerican
  cosmovision, 34, 50, 54n9; by nahuallis,
  34; by tlaciuhqueh (diviners), 34, 54n9
human imperialism, 108
human nature, Brazilian views on, 293–94
human rights, historical expansion of, 222,
  329–30
human sacrifice, turkeys as substitute for, 47
Humboldt Current, 246
hummingbirds, demand for feathers of, 276
humor: of goat as Trujillo's nickname, 304;
  in sacraments for dogs, 107–8, 110–11
humoral systems: in Andean medicine, 125,
  132–36, 137–40; cataloging practices
  based on, 131–32, 136–42; in colonial
  Guatemalan medicine, 71; dissemination
  of beliefs about, 132–34; as framework for
  Delgar's writings, 131–42
hunger, locusts linked to, 65, 66, 70, 83
hunting, 15–16; of birds (*see* birds); of fur
  seals (*see* fur seal hunting, in Patagonia);
  of monkeys in Puerto Rico, 194; vs. war,
  weapons of, 334
hunting clubs, in Brazil, 274
Hurtarte, Juan Lucas de, 76
husbandry, animal: in Cuba, decline of, 213–
  15; in English colonies, x, 58n44; by indigenous people, 43, 44, 57n39, 58n44; interspecies grazing in, 1, 2–3; labor force
  needed for, 58n45; land destroyed by, 41–
  44; in Mexico, introduction and expansion of, 41–44; in Patagonia, land concessions for, 243. *See also specific animals*
hydraulic energy, in Cuba, 213, 237n11
Hystoire du Mechique, 46–48, 60n73, 61n78

identity: creole, goat as sign of, 304; ethnic
  vs. linguistic, 55n22; vs. experience, politics of, 336–37; national, biology in, 270
idolatry, Bible on, 98, 115n18
Iguaçu National Park (Brazil), 294
ihb (International Health Board). *See*
  Rockefeller Foundation, International
  Health Board (ihb) of

Ihering, Hermann von: on birds of Brazil, 280–86, 287, 289, 290; on nature as totality, 293

Ihering, Rodolpho von, 289, 290; "Protegei os passarinhos," 287

ihiyotl, 54n13, 59n50

ik (winds), in origins of disease, 158

Ilarione da Bergamo, 114n4

Illustrated London News, 186, 187

imperialism: British, 187; human, 108; U.S., in Puerto Rico, 181, 182–88, 195–96, 199. *See also* colonialism

Imperial Museum (Brazil), 280, 288

Incas, Kallawaya as healers of, 124, 127

incest, 335

India: animal-welfare regulations of, 187; conservation of tigers in, 342n8; importation of monkeys to Puerto Rico from, 180–81, 183–88, 192

Indian Protection Service (Brazil), 279

Indians. *See* indigenous peoples; North American Indians; *specific groups*

indigenous peoples: animal deities of, 7–8; animal husbandry by, 43, 44, 57n39, 58n44; Catholic Church on humanity of, 119n63; diet of (*see* diet, human); elite's wish to eliminate, 17–18; on exotic species, names of, 8–9; in free-market environmentalism, 330–31; languages of (*see* languages, indigenous); perspectivism of, 333–34; review of literature on, 7–8. *See also specific groups*

indigenous population declines: estimates of, 39; European livestock in, 42, 43–44; fur seal hunting and, 244–45; locusts in, 65

indigo cultivation, locusts and, 69, 81–82

industrial development: in Puerto Rico, 192; rise of animal-protection societies and, 221

Industrial Revolution, locust extermination in, 90n89

Ingenio, El (journal), 223

inquisitions: Apostolic Inquisition, 55n18; origins and goals of, 96–97; Portuguese Inquisition, 97; Spanish Inquisition, 97. *See also* Mexican Inquisition

Insect Poetics (Brown), 63, 85n5, 87n29

insects: in animal studies field, 17, 63, 64, 85n5; "beneficial," 62; bird consumption of, in agricultural lands, 283, 284, 287, 288–90, 294; chemicals used against (*see* chemical insecticides); disease transmission by, 17; eradication of, in establishment of modernity, 331–32; ethics of mass killing of, 64, 85–86nn14–15; government campaigns to exterminate, 62; as pests vs. vectors, 17, 151, 164; technological innovations used against, 77–78, 90n89; ubiquity of human desire to kill, 62, 64. *See also specific types*

Instituto Oswaldo Cruz, 282

intellectuals, on killing of birds in Brazil, 278, 284, 286, 294–95

intemperance, vulnerability to cholera linked to, 155

internal combustion engine, in Cuban agriculture, 219

International Health Board (IHB). *See* Rockefeller Foundation, International Health Board (IHB) of

International Ornithological Congress, 277

International Society for the Protection of Animals, 220

International Union for the Conservation of Nature (IUCN), 260

invasive species: monkeys in Puerto Rico as, 181, 192, 194; U.S. battles against, 332

Iraq, National Zoo of, 194

Isabella I of Castile, 97

Isidore, Saint, 98

Isla de los Estados (Argentina), fur seal hunting on, 250

Islamic cultures, locusts in, 63

Italy: animal-protection movement in, nineteenth-century lack of, 297n16; protection of birds in, 284–85

Itatiaia National Park (Brazil), 294

Itzá Maya, tributary labor system used in wars against, 76

Itzpapalotl (deity), 49

IUCN (International Union for the Conservation of Nature), 260 jaguars, 46, 59nn58–59, 338

Jamaica, sugar production in, 218

Jerome, Saint, 71, 98

Jesuits: Andean missions of, 127–28, 133, 146n19; Cobo as, 127–28, 133

Jesus Christ, 51–52

Jiménez Martínez, Eduardo, 303

John Paul II, Pope, 98

Jordan, Robert Michael, 196

Jornal do Comercio, 289–90

Juan Fernández fur seal, 264n7

Juan Fernández Island (Chile), fur seal hunting on, 250, 264n7

Judeo-Christian cultures, locusts in, 63–64, 85n10

Juli (Peru), Cobo in, 127

Kaiselbard, Edmundo, 196–97

Kallawaya, 12, 123–43; as healers of Incas, 124, 127; humoral system of, 133–36, 137–40; languages spoken by, 146n22; published lists of substances used by, 130–31; secret language of, 146n22; as source of Cobo's indigenous knowledge, 127–28, 131; as source of Delgar's indigenous knowledge, 130–31; spread of healing practices of, 131; travels of, 124, 127, 131

Kawéskar, fur seal hunting and, 244–45, 248

Kessler, Matt J., 186, 199

Kete, Kathleen, 105, 112

K'iche' Maya: creation story of, 45, 59n51; on locusts, 66

kidney stones, treatment of, 140–41

killing of animals: Christianity on, 98; for human medical benefit, 143; vs. insects, 64; ubiquity of, 16–17

Kinship with Monkeys (Cormier), 334

Knox, Thomas W., 170n1; *The Boy Travellers in Mexico*, 149–52, *150*, 165–66

Koford, Carl, 192

Kohn, Eduardo, 200

Kritsky, Gene, 86n17

Kuhtmann, J. F., 226

labor, animal, in Cuba, 15, 209–36; in animal-protection movement, 223–24,

228; growth of, after collapse of Soviet Union, 209–10; replacement of, with machines, 209, 210, 213–14, 218–19, 234–35; on sugar plantations, 210–20

labor, animal, in Guatemala, 81

labor, human: in animal husbandry vs. agriculture, 58n45; in encomiendas vs. repartimiento system, 43–44; in fur seal hunting, 252–54; in locust-extermination campaigns, 74–77, 79–81, 83; locusts' impact on, 65, 83

ladybugs, 62

Lajas Valley (Puerto Rico), free-ranging monkey populations in, 193–94

land, agricultural. *See* agriculture

Landa, Antonio de, *Guía del administrador de ingenio*, 217

land concessions, in Tierra del Fuego, 243

land grants, by Spanish Crown, 42

land rights, indigenous, colonial protections for, 43, 58n45

landscapes: as form of history, 339; human relationships with, 331

langosta, use of term, 65–66, 86n17. *See also* locusts

languages, indigenous: in creation stories, 45; identity based on, 55n22; names for exotic species in, 8–9; words for humans and animals in, 34–35, 54n12. *See also specific languages*

La Parguera (Puerto Rico), monkey colony of, 191, 192–93

Latin America, international migration of birds in, 277–78. *See also specific countries*

Latour, Bruno, x

Lauria, Antonio, 316

laws: in Brazil, on protection of birds, 278, 281, 292–93; in Cuba, on protection of animals, 210, 223, 224, 226, 228, 230–31, 233; in Europe, on protection of animals, 222, 277; international, on protection of birds, 222, 276, 277, 278; in Patagonia, on fur seal hunting, 258–60; in United States, on protection of birds, 276, 278, 283–84, 286; in Yucatán Peninsula, on livestock slaughter, 160–61; in Yucatán Peninsula, on untended animals, 155–56

Lazear, Jesse, 164
Lea, Henry Charles, 96
Leach, Edmund, 313, 319
leather: from bird skins, 276; from cow hides, 44
"Left-Handed Blow, A" (Fudge), 10, 85n14
legal precedence, importance of, in Mexican Inquisition, 117n37
legislation. *See* laws
Lei das Contravenções Penais (Brazil, 1941), 293
"Leiningen versus the Ants" (Stephenson), 340
Lent celebrations, 307–8, *308*
leones, 261
Leptospira icteroides, 179n89
Lévi-Strauss, Claude, 304
Leviticus, Book of, 2, 21n5, 66
Leyenda de los soles, 46–48, 59n56
liberalism: and birds of Brazil, 278, 280, 282, 283, 285; goats as symbol of, 304
"Libro de medicina y cirugía para el uso de los pobres" (Delgar), 131, 144n8
lice, 62, 168
Life magazine, 186, 187, *188*
Lima (Peru): Cobo in, 127, 145n18; Delgar in, 129
Lingue pour la Protection des Oiseaux (France), 277
livestock: animal-protection societies on, 222; in indigenous diets, opposition to, 12, 31, 32, 39–41, 50–53; in indigenous diets, origins of, 44; introduction and spread of European, 41–44; in locust-extermination campaigns, 81; regulation of slaughter of, during cholera outbreaks, 160–61; regulation of slaughter of, for protection of animals, 226, 228; reproduction rates of, 41, 42; spread of disease blamed on, 155–56, 157–58, 161; types of, 41; vaccination of, 160–61. *See also specific types*
Livingston, Julie, 5, 17
Livingstone, David, 340–41
llama, medicinal use of, 141, 142
lobos marinos ("sea wolves"), 261

Lockhart, James, 8, 24n32
Lockwood, Jeffrey A., 86n15, 90n89
locusts: vs. cicadas, 86n17; devouring associated with, 66, 86n18; duration of plagues of, 90n78; European vs. New World species of, 67; historical sources on, 63–64; history of cultural views of, 63–64; history of plagues of, 63–64, 66; human consumption of, 66, 83; lifecycle of, 67–68, 75; physical appearance of, 67–68; precontact, 66, 88n58; swarming behavior of, 68; technology in extermination of, 90n89. *See also* locusts, in colonial Guatemala,
locusts, in colonial Guatemala, 17, 62–84; agricultural production affected by, 65, 68–70, 83; bodies of, disposal of, 70–71, 77; colonialism influenced by, 64–65, 83–84; community-organized campaigns against, 65, 74–77; disease associated with, 65, 69–72; handbooks on, 81–82; as historical agents, 69; lack of response in, to human attacks, 26n52; physical appearance of, 67–68; prevention of, 78; religious strategies against, 65, 72–74, 77; Spanish experts on, 78–79; state-directed campaigns against, 62–63, 65, 78–84; terms used for, 65–67
López Austin, Alfredo, 45, 54n5, 133
López de Azpestia, Juan, 71, 76–77, 80, 81
López de la Paliza, Juan Antonio, 93, 94
LSD, 343n13
Luderwaldt, Hermann, 286
lustral baths, 40, 57n36
Luther, Martin, 109

macaw: demand for feathers of, 276; medicinal use of, 141
macehuales (commoners): in creation stories, 45–46; diet of, 47–48, 60nn67–68; Inquisition persecution of, 36, 55n21; Teton as, 35–36
Madrid (Spain), animal-protection societies in, 220, 226, 232
maize. *See* corn cultivation
majá de Santa María, 229, 240n78

Memoria sobre la industria pecuaria en la Isla de Cuba (Pozos Dulces), 222–23

Mencos, Martín Carlos de, 70, 88n45

Mendieta, Balthasar, 99

Mendoza, Antonio de, 43

mercury poisoning, remedies for, 129, 146n25

Mérida (Mexico): cholera in, 160; mosquito eradication in, 167

Mesoamerican(s): on boundaries between humans and animals, 34–35, 44–46; cosmogony of, 45–49; cosmovision of, 32, 34, 44–45, 50, 54n9; locusts in culture of, 66, 86n18; medical culture of, 71. *See also specific groups*

Mesoamerican religion: colonial sources on, 32; divination in, 33; human-animal transformation in, 54n9; lustral baths in, 57n36; multidisciplinary approach to, 32; tlaciuhqueh (diviners) in, 33–35

Mexica(s): eschatological beliefs of, 37–38; locust plagues and, 66, 89n58

Mexican Inquisition: baptism in jurisdiction of, 36, 55n18; dog baptisms in, 99–102, 108–9, 111, 117n37; dog weddings in, 18, 93–96, 108–9, 111; legal precedence in, importance of, 117n37; macehuales (commoners) in, 36, 55n21; prisons of, 94, 114n7; tlaciuhqueh (diviners) tried by, 36–37, 55n21

Mexican Revolution, 166

Mexico: bird-protection laws in, 278; building "healthy" state of, 152, 164, 168, 332; chupacabras legend in, 195; colonial (*see* New Spain); disease in (*see* disease; public health); Revolution of, 166; Rockefeller Foundation's relationship with, 151, 165, 166, 178n79; species and human race hierarchies in, 152, 153; U.S.-led invasion of (1914), 165. *See also specific cities*

Mexico City, recetarios published in, 131

Meza, José de, 81

miasma theories of disease, 156, 159, 160

Michmaloyan (Mexico): animal husbandry in, 42, 57n38; location of, 54n6; Teton in, 32, 36

Mieth, Hansel, 186, *188*

migration, bird, 277–78

Migratory Bird Treaty Act of 1918 (U.S.), 278

military installations, U.S., in Puerto Rico, 192, 195–96

military secret service (SIM), Dominican, 306

milk: animal-protection societies on alteration of, 231; medicinal use of, 137

mining communities, in colonial Mexico, 128

Mintz, Sidney W., 311, 321n13

Miranda Ribeiro, Alípio de, 285, 289

Mirounga leonine. See sea elephants

Misericordia, La, 232, 241n95

missionaries, Jesuit, in Andes, 127–28, 133, 146n19

mistreatment of animals. *See* cruelty to animals, in Cuba

Mixco (Guatemala), locusts in, 73

Mixcoatl, Andrés, 36–37, 55n21, 56n23, 59n59

mizquitl tree (mesquite), 48, 60n70

miztli, 59n58

MK-Ultra experiments, 343n13

Moctezuma II, 37, 38

modernity: insect eradication in establishment of, 168, 331–32; Maya as barrier to, 153

Molina, Andrés de, 57n30, 57n33; *Vocabulario*, 47

Molina, Antonio, 69, 70, 74, 77, 88n46

Monaghan, John, 34

monkeys: in creation stories, 48; human kinship with, 334–35; in polio vaccine development, 180, 182–83; universality of, 186

monkeys, in Puerto Rico, 13, 180–202; animal-rights movement on, 194, 197; ban proposed in 2007 on select species of, 180; chupacabras legend and, 194–96; controlled breeding of, 182, 183, 197; disease among, 189, 196; free-ranging populations of, 191, 192–96; government capture and culling of, 193–94, 197; importation of, 180–81, 183–88, 192, 197; as invasive species, 181, 192, 194; native

monkeys, in Puerto Rico (*cont.*)
species of, lack of, 180; population of, 189, 192, 193; progress associated with, 181, 187, 194, 198, 201, 202, 335; public opinion on, 187, 192–93; in U.S. imperialism, 181, 182–88, 199. *See also* Cayo Santiago monkey colony

Montalvo y Castillo, José, *Tratado general de escuela teórico-práctica para el gobierno de los ingenios de la Isla de Cuba en todos sus ramos*, 216–17

monteros, 311–13

Montt, Jorge, 259

morality. *See* ethics and morality

Morejón y Gato, Antonio, 213

Morel, Antonio, "Mataron al Chivo," 302, 303

Moreno Fraginals, Manuel, 218, 325n60

mosquitoes: chemical-resistant strains of, 170; lack of response in, to human attacks, 26n52; as vector of dengue fever, 170; as vector of yellow fever, 23n21, 164–69

mosquitoes, in Mexico, 164–69; DDT in 1950s campaigns against, 62; Rockefeller campaigns to eradicate, 151, 165–68

mosquito nets, 167, 343n12

Motolinia (Franciscan friar), 43

Movimiento 14 de Julio, in Dominican Republic, 306

Moya, Antonio de, 316–17, 318

Moziño, José Mariano, 81–82, 91n106

muermo, 233

mules, in Cuba: animal-protection societies on, 228, 230; labor of, in sugar plantations, 212

multiculturalism, 333, 336

multinaturalism, 333, 336

Mundo, El, 191

municipal ordinances, in Cuba, on protection of animals, 224, 226–31

Muñoz Marín, Luis, 189

mupollo, medicinal use of, 138–39

Museu Nacional (Brazil), 279, 282, 285, 287–88

Museu Paraense (Pará), 280, 281, 288

Museu Paranaense (Paraná), 288

Museu Paulista (São Paulo), 280, 284, 286, 287, 288

mushrooms, 51

musical instruments, in locust-extermination campaigns, 74–75

musullo, medicinal use of, 138–39

Myocastor coypus. See nutrias, hunting of

Na, Pedro, 1–2, 19, 20, 338

Nación, La (newspaper), 240n63

NAFTA, 195, 196

Nagel, Carlos, 192

nagualismo, in Mesoamerican cosmovision, 34, 50, 54n9. *See also* human-animal transformation

naguals (animal companions), 46, 59n50

nahuallis: definitions of, 34, 35; vs. sorcerers, 34, 55n15; Teton as, 34, 35

Nahuas: creation story of, 45–46; diet of, 12, 47; as linguistic group, 55n22; ritual lustral baths of, 57n36; on soul, 35, 54n13; on turkeys and nobles, 47–48

Nahuatl language: names for exotic species in, 9, 24n32, 57n32; personal names in, 36, 55n20; words for humans and animals in, 34, 45, 54n12, 58n49; writings in, 32, 53n4

National Association of Audubon Societies, 276

National Audubon Society, 276

National Institutes of Health (NIH): in Cayo Santiago monkey colony, 191–92, 196, 197; National Primate Research Centers of, 183, 191; origins of, 183

nationalism, Puerto Rican, 181, 189–91, 196

national parks, in Brazil, 294

National Park Service, U.S., 332

National Primate Research Centers, origins of, 183, 191

national reserves, for fur seals, 260

national symbols, birds as, 289

National Union of Jurists of Cuba, 210

National Zoo of Iraq, 194

Natural History of the Kingdom of Guatemala (Ximénez), 66, 67, 75

natural order, humans in, ix, x

Natural Resources and Environment Department (DRNA), Puerto Rican, 194

nature: Brazilian views on, 293–94; vs. culture, problems with classic distinction between, 333; in multiculturalism, 333

Nature and Society (Descola and Pálsson), 126–27

Nazarea, Virginia D., 157

Neiva, Artur, 281–82, 285

neocolonialism, in call for animal rights, 330

neoshamanism, 337

nests, medicinal use of, 137, 148n49

New England: colonial agriculture in, x; fur seal hunters based in, 244, 250–58, 264n7

New England's Creatures (Benes), 6

New Fire, 38, 56n26

New Spain (colonial Mexico): administrative units of, 91n91; ancestral deities in, as devils, 40, 57n35; anthropocentrism in, 105, 109; archives of, limitations of, 4; bestiality cases in, 1–3, 14; calendar of (*see* calendar, traditional Mexican); dogs in (*see* dogs, in colonial Mexico); elimination of indigenous peoples in, as goal, 17–18; insect-extermination campaigns in, 62; introduction and spread of European livestock in, 41–44; Teton case in (*see* Teton, Juan)

New State (Brazil), 271, 292

New Testament. *See* Bible

New World of Animals, A (Asúa and French), 7

New York (city), animal-protection societies in, 220

New York Times, 168

Nicaragua, locusts in, 72

nicknames, in Dominican culture, 315–19. *See also* Trujillo, Rafael

Nicolas of Tolentino, Saint, 73, 89nn61–62

NIH. *See* National Institutes of Health

nobility, turkeys associated with, 47–48, 60nn67–68. *See also* elite

Noguchi, Hideyo, 166, 167, 179n87, 179n89

Nogueira, José, 243, 251, 252

noise, in locust-extermination campaigns, 74–75

nonnative animals. *See* exotic species; invasive species

Nora, Pierre, 321n9

North American Indians, on boundaries between humans and animals, 58n49

North Atlantic Free Trade Agreement (NAFTA), 195, 196

North Atlantic sealers, in Patagonia, 244, 250–58, 251, 261–62

Norton, Marcy, 3, 316

Nueva Era, La, 224

Nunn, Charles, 114n7

nutrias, hunting of, 260

ocelotl, 59n58

Ocelotl, Martín, 36–37, 55n21, 56n23

O'Farril, José Ricardo, 212–13

Ogilby, John, *America*, 8

Ohlsen, Theodor, 254, 254

Ohnuki-Tierney, Emiko, 335

ojoy balche' (evil eye), 158

Old Testament. *See* Bible

oligarchies, of Brazil, 271

Olivier, Guilhem, 47

Olmos, Andrés de, 46

Ometochtzin, Carlos, 55n18

1-Rabbit years, 31, 37–38, 39, 52, 56n24, 56n26

Operation Bootstrap, 189

opossums, 8, 24n31, 140

orientalism, simian, 185–86

ornithology. *See* birds

Orquera, Luís, 249

Ortiz de Montellano, Bernard R., 44, 133

Otaria flavescens. *See* sea lions

otariids, 246–47, 265n18. *See also specific types*

otherness: of animals, 205n52; of dogs, 105

Otomi: definition of, 54n6, 55n22; as linguistic group, 55n22; Teton as, 36; tlaciuhqueh (diviners) among, 32, 33, 35, 36

overgrazing, 44

Oviedo y Valdés, Gonzalo Fernández de, 324n55

owls, 229, 240n77

ox-drivers, 214, 215–16, 217 18, 235

oxen, in Cuba: concern for care of, 215–17; labor of, in sugar plantations, 210–20, 215, 234–35; laws protecting, 224; number of, 209, 210, 214–15, *216*

Ozaeta y Oro, Pedro de, 79

Ozma, Miguel de, 80

Ozomatli (deity), 48

pack animals, benefits of, 42

Palestinians, 342n4

Palmer, T. S., 283

Pálsson, Gisli, 155; *Nature and Society*, 126–27

Pamela (dog), 102–5, 106–7, 112

pandemics, cholera, of 1830s, 151, 154–56, 171n7

papa pahspa, medicinal use of, 141–42

Pará (Brazil): Museu Paraense in, 280, 281, 288; opposition to killing of birds in, 280–81

parakeets, demand for feathers of, 276

Paraná (Brazil), Museu Paranaense in, 288

parasites, on untended animals, in cholera epidemics, 155–56

Paris (France): animal-protection societies in, 220; Exposition of 1889 in, 131; pet-keeping practices in, 105, 107

parks, national, in Brazil, 294

Parrish, Susan Scott, 24n31

parrots, demand for feathers of, 276

Parteiro, María Agustina, 99

partridges, hunting of, 274

Pássaros do Brasil (Santos), 292

passenger pigeon, extinction of, 278

Pasteur, Louis, 164

Patagonia, southern: definition of, 263n2; fur seals in (*see* fur seal hunting, in Patagonia)

Patamuna, 334, 338, 344n24

patas monkey, in Puerto Rico. *See* monkeys, in Puerto Rico

patrimony, Brazilian, 294

Paul III, Pope, 119n63

Pauw, Cornelius, 325n61

Pazón (Guatemala), locusts in, 75–76

Pearson, Susan J., "Does 'the Animal' Exist?," 10

peccaries, 334

Pedro II (emperor of Brazil), 271, 282

Peguero, Valentina, 316

pelagic hunting, of fur seals, 249, 259

People for the Ethical Treatment of Animals, 194

Peralte, Charlemagne, 326n76

Pérez, Juan B., 325n64

personhood, Mesoamerican vs. European notions of, 34. *See also* human(s)

perspectivism, 333–34

Peru, colonial, medicinal remedies of. *See* medicine, Andean

pests: context-specific categorization of, 158; insects as, 17, 151, 164; travelers' perception of, 149–50; untended animals as, 157, 158; vs. vectors, 158–59; as vectors of disease, shift in thinking about, 17, 151, 158–59; vs. vermin, 177n74

pets: in colonial Mexico, social meaning of, 105–9; origins of category of, 99; social class and ownership of, 106. *See also specific types*

Petapa (Guatemala), locusts in, 79

petits chiens de dames, Des (Bonnardot), 99

Pezet, José, 146n27

pharmaceutical research, on animals. *See* vivisection

Philo, Chris, x

Physicians Committee for Responsible Medicine, 197

piaii, 344n24

Piana, Ernesto, 249

pictographic writing, 32, 53n4

Piedra Buena, Luís, 251

Pigafetta, Antonio, 309

pigeons, hunting of, 278

pigs: indigenous consumption of, 44; indigenous names for, 57n33; sacraments defiled by, 96; transformation of humans into, 39

pigs, in Dominican Republic, 305; arrival of, in colonial era, 311; consumption of, 305–6, 311; creole, 305, 321n7, 321n15; in economy, 304–6, 312–13; eroticization of, 325n70; hunters of, types of, 311, 324n55;

mass slaughter of (1979), 321n7, 321n15; as symbol of freedom, 313, *314*

pine nuts, 46

pinnipeds, 265n18

Pinto, Olivério, 284–85, 286, 288

Pinula (Guatemala), locusts in, 73

Pinzón, Vincent Yáñez, 8

pipillo, 47. *See also* turkeys

pipilpipil, 47

pipiltin, 47

Pires, Fernão, 97

Pitt-Rivers, Julian, 55n15

Pizarro, Francisco, 6

Plague of Sheep, A (Melville), 6, 332–33

plant(s): in animal-protection societies, 222, 231; for Cayo Santiago monkey colony, 186; in creation stories, 46–47, 49; Delgar's cataloging of, vs. animals, 136–42; in indigenous diets, 46–47, 49; in Mayan medicine, 153; as scientific curiosities to Europeans, 130, 146n29

plumes. *See* feathers

Poey, Felipe, 228–29

Point Four Program, of U.S. State Department, 189

polio vaccine, monkeys in development of, 180, 182–83

political power, in Brazil, 271

political resistance, 330, 342n4

politics: of bird conservation in Brazil, 280, 282–85; of experience vs. identity, 336–37

Poma, Guaman, 127

Popol Vuh, 45, 48, 59n51, 60n72

Portugal, Brazilian independence from, 271

Portuguese Inquisition, 97

positivism, in ideology of elite, 162

postcolonial studies, gaps in field of, 11

postdomestic societies, 26n59

posthumanism, 9, 13, 14

post-traumatic stress disorder (PTSD), 342n4

Potosí (Bolivia), Delgar in, 128, 129

poverty: animal-protection societies' work on, 232; in Brazil, 281, 282, 295; in Dominican Republic, 318

pozole, for treatment of disease, 153

Pozos Dulces, count of (José Francisco de Frías y Jacott), 213, 214; *Memoria sobre la industria pecuaria en la Isla de Cuba*, 222–23

prayer, in locust-extermination campaigns, 72–74

Presas, Manuel, 223

preservation movement, on birds of Brazil, 270–71

press coverage. *See* media coverage

Pribilof Islands (Alaska), fur seal hunting on, 244

Primeira Conferência Brasileira de Proteção à Natureza, 293

prisons, of Mexican Inquisition, 94, 114n7

processions. *See* public processions

progress: Maya as barrier to, 153, 162; medical, monkeys as sign of, 181, 187, 194, 198, 201, 202, 335

Property and Environment Research Center, 342n7

protection of animals: change in sensibilities and, 110, 277; European laws on, 222, 277; societies promoting (*see* societies, animal-protection). *See also* protection of animals, in Cuba

protection of animals, in Cuba, 220–34; advocates' arguments for, 222–24, 226–29; concern for laboring animals and, 215–17, 223–24, 228, 235; international societies' collaboration with, 225–26; international societies' rise and, 210, 220–22, 223; laws on, 210, 223, 224, 226, 228, 233; modern movement for, 235; origins of movement for, 15, 210, 211, 220; under republic, 233–34, 241n100. *See also* Sociedad Cubana Protectora de Animales y Plantas

"Protegei os passarinhos" (Ihering), 287

protein, precontact sources of, 44

Protestant Reformation, 97, 109

protopeasantry, 304, 321n13

Prussia: animal-protection laws in, 222; animal-protection societies in, 220

PTSD, 342n4

Puar, Jasbir K., 5, 17

Reyes García, Luis, 53n3
rheas, demand for feathers of, 276
rhesus monkey. *See* monkeys, in Puerto Rico
rice, sociocultural meaning of, 306
rifles, fur seal hunting with, 254, 254–56
rights, history of expansion of, 222, 329–30
Riley Locust Catcher, 90n89
Rio de Janeiro (Brazil): demand for feathers in, 272–73; remodeling of city center, 272–73
Ritvo, Harriet, 262
roaches, 62
Rockefeller Foundation, International Health Board (IHB) of, 332; challenges facing, 151; international work of, 165, 168; in malaria eradication, 151, 165, 168; Mexican government's relationship with, 151, 165, 166, 178n79; mosquito eradication by, in Mexico, 151, 165–68; in yellow fever eradication, 151, 165–68
Rodríguez, María, 100
Rodríguez Ferrer, Miguel, 219–20
Roig, Antonio, 185
Romero Cuyás, José, 228, 230
Rondon, Cândido, 278–79, 285
Roosevelt, Theodore, 278–79
Roquette-Pinto, Edgar, 279, 289
Rosenberg, Charles, 171n11
Rostros culturales de la fauna (Ulloa and Baptiste-Ballera), 7
Rothfels, Nigel, 9, 10
Royal Australasian Ornithologists Union, 277
Royal Society for the Prevention of Cruelty to Animals (England), 220, 222. *See also* Society for the Prevention of Cruelty to Animals
rufous-bellied thrush, hunting of, 274–75
rufous-collared sparrow (tico-tico), agricultural role of, 289, *290*, 294
rufous hornero, moral qualities of, 292
Ruiz de Alarcón, Juan, 59n56

Sacatepéquez (Guatemala), locusts in, 72–73, 74
sacraments: animals in desecration of, 96–97, 114n12; in boundaries between humans and animals, 97, 108. *See also specific sacraments*
sacrifice: of chickens, in Dominican Republic, 321n13; of goats, in Dominican Republic, 304, 314, 321n13; of humans, turkeys as substitute for, 47
Saguah, Cristobal, 80
Sahagún, Bernardino de: *Florentine Codex*, 8, 35, 56n27; on human-animal transformation, 50; on nahuallis, 34, 35; on tlaciuhqueh (diviners), 33, 35; on turkeys, consumption of, 60n68
Said, Edward, 186
saints, in locust-extermination campaigns, 72–74. *See also specific saints*
Salisbury, Joyce E., 114n12
salt, in baptisms, 100, 101, 116n32
saltón, 68. *See also* locusts
San Cristobal Amatitán (Guatemala), locusts in, 68, 75, 76, 79, 80, 81
San Juan (Puerto Rico), School of Tropical Medicine, 180, 183–85, *184*, 190–91, 192
San Juan Amatitán (Guatemala), locusts in, 68, 79, 80, 82
Santería, goats in, 314
Santiago de Guatemala: as capital, 84n3; locusts in, 68, 70, 79
Santos, Eurico, 271, 291–92; *Pássaros do Brasil*, 292
Santos Fernández, Juan, 234
São Paulo (Brazil): Bird Day in, 286–87; demand for feathers in, 273; killing of birds in, 274, 281, 282; Museu Paulista in, 280, 284, 286, 287, 288
Satan, 307–10, 314
satyrs, 314
Scammon, Charles, 253, 254
scapegoating, 322n29
scarlet ibis: demand for feathers of, 276; opposition to killing of, 280, 281, 282
Schiavini, Adrian, 249, 266n24
Schomburgk, Robert, 312
School of Tropical Medicine (San Juan), 180, 183–85, *184*, 190–91, 192
Schwartz, Marion, 118n53

Schwartz, Stuart, 95

Science magazine, 197

scientists, on killing of birds in Brazil, 270–71, 279–85, 287–88, 294–95

Scott, James, 317

sea elephants, 250

sea lions: fish farming and, 261; fur of, 247, 265n19; hunting of, 249, 250, 260; in indigenous diets, 249; reproduction of, 266n21; taxonomic classification of, 246–47

seals. *See* fur seal, South American; fur seal hunting, in Patagonia

"sea wolves," 261

sensibilities, changes in: in movement for protection of animals, 110, 277; toward birds, 277, 285–86, 295; toward dogs, 98–99, 109–13

Serra dos Órgãos (Brazil), 294

sertão (backcountry), of Brazil, 275, 278–79

Serviço de Proteção ao Índio (Brazil), 279

Seville (Spain): animal-protection societies in, 220; locusts in, 89n67, 89n71, 91n97

sexuality: female, in Puerto Rico, 182; goats in Dominican Republic and, 307, 313; pigs in Dominican Republic and, 325n70

shamanism, 46, 335, 337, 344n24, 344n29

shape-shifting: in Dominican culture, 309–10, 314; by jaguars, 46; by nahuallis, 34, 35

Shapiro, Kenneth, 9

sheep: indigenous consumption of, 44, 60n67; indigenous names for, 9, 57n32; introduction and spread of European, 41–44, 58n45; land destroyed by, 42; in Patagonia, 243; transformation of humans into, 39, 41

shellfish, in indigenous diets, 248, 249, 262

shigella, 189

Shubert, Adrian, 241n85

SIM (military secret service), Dominican, 306

Simalo (goat), 315

simian orientalism, 185–86

Simon, Antoine, 314–15

Singer, Peter, 205n52; *Animal Liberation*, 9, 329; *Animal Rights and Human Obligations*, 329

Sixtus IV, Pope, 97

skunks, medicinal use of, 128–29, 146n25

slavery, in Brazil, abolition of, 291

slavery, in Cuba: abolition of (1886), 212, 218, 235; animal labor and, 210–11, 212, 215–16, 218; animal-protection societies and, 222, 226, 234, 235

slaves, runaway, in Dominican Republic, 305, 311

slave ships, importation of monkeys on, 180

Sleigh, Charlotte, 84n2

smallpox: as historical agent, 69; locusts associated with, 70

smoke, from animal parts, medicinal use of, 141–42

smooth-billed anis, agricultural role of, 289–90

smuggling, of livestock products, 44

social behaviors, of fur seals, 255

social class. *See* class, social

social Darwinism, 162

socialism: in Eastern Europe, collapse of, 209; in Mexico, 166

Sociedad Barcelonesa Protectora de Animales y Plantas, 226

Sociedad Cubana Protectora de Animales y Plantas, 15, 220–34; agenda of, 224–25, 230–33; anthropocentrism of, 226–28; asylum for beggars created by, 232, 241n95; board of directors of, 225, 231–32, 240n63; *Boletín* of, 225, 226–29, 240n66; bylaws of, 224, 232; demise of (1890), 232; in enforcement of animal-protection laws, 226, 228, 230–31; establishment of, 211, 220, 223–24, 277; first calls for, 222–23; goals of, 211, 220, 222, 224, 232; logo of, 221; name change of, 232; plants in work of, 222, 231; reforms of 1885 in, 232; slavery and, 222

Sociedade Brasileira Protetora dos Animais, 285–86

Sociedade Rural Brasileira, 291

Sociedad Humanitaria, Protectora de los Niños y contra la Crueldad con los Animales (Cuba), 233–34

Sociedad Madrileña Protectora de los Animales y de los Plantas (Spain), 226, 232

Sociedad Protectora de Animales y Plantas de Cádiz (Spain), 220, 241n85

Sociedad Protectora de los Niños de la Isla de Cuba, 234

societies, animal-protection: on birds, 277, 285–86, 291; international collaboration of, 225–26; international rise of, 210, 220–22, 223; plants in agendas of, 222, 231; slavery and, 222, 226, 234, 235; on vivisection, 222, 224, 226, 229. *See also specific countries and groups*

Society for the Prevention of Cruelty to Animals (England), 15, 210, 220

Society for the Protection of Birds (England), 277

Society for the Protection of Children on the Island of Cuba, 234

Solorzano, Armando, 179n87

Soluri, John, 10, 11, 15–16, 26n52, 330, 336, 337

sorcery and sorcerers: Catholic accusations of, 310; vs. nahuallis, 34, 55n15; Spanish use of term, 34; tlaciuhqueh as, 33, 34

sorrino, medicinal use of, 128–29, 146n25

soul(s): Catholic Church on, 97–98, 108–9; indigenous notions of, 35, 45, 50, 54n13, 58n50

sources. *See* historical sources

South American fur seal. *See* fur seal, South American

South Shetland Islands, fur seal hunting on, 256, 264n9

Soviet Union, collapse of, 209

Spain: animal-protection movement in, nineteenth-century lack of, 297n16; animal-protection societies in, 220, 226, 230, 232, 241n85; bullfighting in, 230, 231, 241n85; locusts in, 89n67, 89n71, 91n97; shipment of new-world animals to, 123, 144n3

Spanish colonists: on ancestral Mexican deities as devils, 40, 57n35; European ecology transferred to colonies by, 6–7; fur seal hunting by, 244; introduction and spread of livestock by, 41–44; in locust-extermination campaigns, 74, 78–79;

royal land grants to, 42; on tlaciuhqueh (diviners), 34; as tzitzimimeh deities, 36–37, 39, 41, 50, 53

Spanish Conquest: decimation of indigenous population after, 39, 42; indigenous eschatology in decades after, 36–39

Spanish Crown: fur seal hunting in territories of, 250; land grants by, 42

Spanish Inquisition, 97

sparrows, in Brazil, 282

species, hierarchies of, in Mexico, 152, 153, 161–62

speciesism, 205n52

spectacles. *See* public spectacles

squirrel monkey, in Puerto Rico. *See* monkeys, in Puerto Rico

State Department, U.S., Point Four Program of, 189

steam power, in Cuban agriculture, 210, 212, 213–14, 234, 237n11

Stegomyia, 167, 178n86

Stephenson, Carl, "Leiningen versus the Ants," 340

sterilizations, in Puerto Rico, 182

stray dogs, in Cuba, 233

Suárez de Peralta, Juan, 42

suches, medicinal use of, 139

suffocation, in locust-extermination campaigns, 75

sugarcane cultivation: in Cuba, 210–20, 215, 234–35; in Dominican Republic, 304, 312, 325n60; in Guatemala, 69; in Haiti, 211, 213; in Jamaica, 218; in Puerto Rico, 185

suicide bombings, 342n4

supernatural beings, locusts as, 74, 83

swallows: agricultural role of, 289; moral qualities of, 292

swine-flu epidemic, 321n7

Switzerland, animal-protection societies in, 220

syphilis, 343n13

Tambiah, Stanley, 318

Taunay, Afonso de Escragnolle, *Zoologia fantástica do Brasil*, 7

technology: in insect extermination, 77–78,

technology (*cont.*)
90n89; in sugarcane cultivation, 212, 234–
35. *See also specific types*
Tecol, Juan, 40
tecuani (people-eaters), 58n49, 59n58
telegraph lines, in Brazil, 278–79
Teneek, on turkeys and nobles, 48
Tenochtitlan (Mexico), fall of, 41
Ten Years War (1868–78), 220
Teocaltiche (Mexico), bestiality in, 14
teocintle, 47
Teotlixcans, on human-animal transformation, 50
Tepaneca, 36, 37, 39, 55n22
Teton, Juan, 12, 31–53; *Anales de Juan
Bautista* on, 32–33, 35–36, 39–40, 53n1,
53n3, 55n20; arrest of (1558), 31, 37, 55n18;
baptism of, 36; baptisms undone by, 31,
36, 40, 51; as commoner, 35–36; cosmovision of, 32; creation stories and, 45–49;
on eating of European livestock, 12, 31, 32,
39–41, 50–53; eschatological beliefs of, 32,
33, 35, 36–39, 49–53; introduction of
European livestock and, 41–44; as nahualli, 34, 35; origin of name, 36, 55n20;
prosecution of, lack of record of, 55n18;
religious office of, 32–33; as tlaciuhqui,
32, 33, 35; on transformation of humans
into livestock, 31, 39–41, 50–51
teuih, 49
teyolia, 50, 54n13, 59n50
Tezcatlipoca (deity), 38, 46, 47, 49, 56n26
Thomas, Keith, 6, 297n16, 336; *Man and the
Natural World*, 98–99
Thomas Aquinas, Saint, 98
Thomas Hunt (schooner), 251–52, 255–57
Thompson, Edward, 163
threatened species, of birds, in Brazil, 295
thrushes: agricultural role of, 289; hunting
of, 274–75
ticks, 62, 149–50
tico-tico, agricultural role of, 289, 290, 294
Tierra del Fuego: fur seals in (*see* fur seal,
South American; fur seal hunting, in
Patagonia); land grants in, 243; Yámana
in, 248–49

tigers: in creation stories, 46; in Dominican
culture, 310, 316–17, 318, 319; in India,
conservation of, 342n8; indigenous
names for, 59n58; as jaguars, 59n58; Tony
the Tiger, 310, 324n48
tlaciuhqueh (diviners), 33–36; definition of,
33; human-animal transformation by, 34,
54n9; social functions of, 33–34; Spanish
responses to, 34; Teton as, 32, 33, 35
Tlaltipaque (deity), 52
Tlantepozilama (deity), 52
Tomilin, Michael, *190*
tonalli soul, 35, 45, 54n13, 58n50
Tony the Tiger, 310, 324n48
töpöra, 249
Toro, Emilia, 243
Torres, Alberto, 278, 284, 285, 289
Tortorici, Zeb, 11, 18, 182, 200, 334, 335, 338,
339, 341
toucans, demand for feathers of, 276
tourism: in free-market environmentalism,
330–31; in Patagonia, 260; in Puerto Rico,
197
Townsend, Eben, 258
Tractado breve de anothomía y chirugía (Farfán), 147n34
trade, fur sealskin. *See* fur seal hunting, in
Patagonia
transportation: alternative modes of, in
Cuba, 210; of animals, from Europe to
colonies, 6–7
transspecies ethnography, 7
transspecies violence, 338
*Tratado general de escuela teórico-práctica
para el gobierno de los ingenios de la Isla de
Cuba en todos sus ramos* (Montalvo y Castillo), 216–17
treatment of animals. *See* cruelty to animals,
in Cuba; protection of animals
Trent, Council of, 97, 116n32
tributary labor system: in locust-
extermination campaigns, 74–76, 83;
Spanish uses of, 76
Triunfo, El (newspaper), 226, 231
Trujillo, Rafael, 18–19, 302–20; anniversaries
of death of, 302–3, 306, 307–10, 320n2;

characteristics of goats shared by, 306–7; media coverage after death of, 306; origins of goat nickname, 303, 320n4; other nicknames of, 306; reasons for goat nickname, 303–4, 306–7, 314, 319–20, 320n4

Trujillo, Ramfis, 306

Tsing, Anna, 200

tuberculosis, 189

Tugwell, Rex, 189

Tula (Mexico), animal husbandry in, 42, 57n38

tungui tungui, medicinal use of, 141

turkey buzzard, 145n18

turkeys: bestiality with, 1–2; in creation stories, 47–48, 59n62, 60n73; indigenous consumption of, 47–48, 51, 60nn67–68; indigenous names for, 47, 59n63; interspecies grazing with chickens, 1, 2–3; nobility associated with, 47–48, 60nn67–68

2-Reed years, 37–38, 56n26

Tying of the Years festival (xiuhmolpilli), 37, 38, 56n27

Tyler, Tom, 109

typhoid fever, 343n11

typhus fever, 168, 178n79

tzitzimimeh deities: pregnant women as, 38; Spaniards as, 36–37, 39, 41, 50, 53

Ulloa, Astrid, Rostros culturales de la fauna, 7

Unanue, Hipólito, 129

unemployment, in Dominican Republic, 318

ungulates, absence of native, 44. See also livestock; specific types

Unión Cívica (newspaper), 306

Unión Cívica movement (Dominican Republic), 306

Unión Nacional de Juristas de Cuba, 210

United Front for the Defense of the Lajas Valley, 193, 195, 196

United States: animal-protection societies in, 220, 222, 225–26; biology in national identity of, 270; birds in (see birds, in United States); cows in colonial history of, x; on creole pigs of Dominican Republic, 321n7; Dominican Republic

invaded by, 317; imperialism of, in Puerto Rico, 181, 182–88, 195–96, 199; insect-extermination campaigns in, 62; invasive species in, 332; locusts in, 86n17; medical research in (see monkeys, in Puerto Rico); Puerto Rican relations with, changes of 1940s in, 189–92

University of Puerto Rico (UPR), 183, 196

urbanization, in international movement for protection of animals, 221

urination problems, 139, 140

vaccination, human: of Mexican children, 166; monkeys killed for use in production of, 183; for polio, 180, 182–83; for yellow fever, 166, 167, 179n87, 179n89

vaccination, livestock, in Mexico, 160–61

Valle, José de, 82

Valle, Pedro de, 119n61

Vandenbergh, John, 192–93, 198–99

Vargas, Getúlio, 271, 294

Vargas Llosa, Mario, 319

Varner, Jeannette Johnson, Dogs of the Conquest, 5–6

Varner, John Grier, Dogs of the Conquest, 5–6

Vázquez, María Manuela, 100

vectors of disease, 17, 149–70; insects as, 17, 151, 164; livestock as, 155–56, 157–58, 161; mosquitoes as, 23n21, 164–69; vs. pests, 158–59; pests as, shift in thinking about, 17, 151, 158–59; vs. vermin, 177n74

vegetarianism: animal-protection societies on, 222; in early Christian sects, 98

vegetation. See plant(s)

Venezuela, bird hunting in, 278

Veríssimo, José, 286

vermin: European conceptions of, 83; vs. pests and vectors, 177n74

vervet monkey, in Caribbean, 180

Vibreo cholerae, 154. See also cholera

Vicious: Wolves and Men in America (Coleman), 6

victims, historical, animals as, 330, 338

vicuña, medicinal use of, 139–40

Vidas, Ariel de, 48

Viel, Oscar, 259

Vieques Island (Puerto Rico), U.S. military installations on, 195–96
vinco vinco, medicinal use of, 140
vinik (person), 34, 54n12
violaceous euphonia, demand for feathers of, 276
violence: among humans, protection of animals and, 221; in political resistance, 342n4; transspecies, 338
Virgin Mary, in locust-extermination campaigns, 73
viscacha, medicinal use of, 141
visibility, of animals in history, 3, 4, 5, 19
Viveiros de Castro, Eduardo, 333–34
vivisection: animal-protection societies on, 222, 224, 226, 229; history of, 123–24; laws against, 222. *See also* monkeys, in Puerto Rico
Vocabulario (Molina), 47
vocalizations, of marine mammals, 247
vodú, 310, 314
Voekel, Pamela, 117n41, 175n43
vomiting, of Eucharist, 96, 114n12

Walton, William, 312
war: chemical weapons in, and development of insecticides, 90n89; vs. hunting, weapons of, 334
Warren, Adam, 11, 12–13, 336–37
water: cholera spread through, 154; as energy source, in Cuba, 213, 237n11; in ritual lustral baths, 40, 57n36; in Yucatán Peninsula, 151–52. *See also* baptisms
wealth, in Brazil, from exportation, 274
weapons, of hunting vs. war, 334. *See also* specific types
weasels, medicinal use of, 140–41
weddings, dog, in colonial Mexico, 18, 93–113; anthropocentrism and, 109–13; descriptions of, 93–96; Inquisition on, 18, 93–96, 108–9, 111; prevalence of, 99; social meanings of, 105–9
Weismantel, Mary, 322n18; "Does 'the Animal' Exist?," 10
welfare, animal, Indian regulations on, 187. *See also* protection of animals

werewolves, 309
Whalemen's Shipping List and Merchant's Transcript, 252, 253
whales: hunting of, 250, 269n84; in indigenous diets, 248
wheat, locust attacks on, 69
White, Richard, 262
Whitehead, Neil L., 10, 11
white-tailed doves, hunting of, 274
Wilbert, Chris, x, 13
wild-domestic dualism: fur seals in, 258; goats in, 313; insects in, 64
wildlife conservation, free-market environmentalist approach to, 330–31
Wilson, Woodrow, 168, 276
wind energy, in Cuba, 213, 237n11
Windle, William, 198–99
winds, in origins of disease, Maya on, 158
wisdom, of tlaciuhqueh (diviners), 33
Wolfe, Cary, 13, 14
wolves, 6, 261
women: pregnant, as tzitzimimeh deities, 38; sexuality of, in Puerto Rico, 182
Wood, Carlos, 251
wood-rails, demand for feathers of, 276
wool: exportation of, 44; indigenous use of, opposition to, 39, 41
work, as virtue in Brazil, 290
workers, indigenous. *See* labor, human
"Workers Revolt" (Darnton), 106–7
worms, medicinal use of, 138–39
worship, of animals, Bible on, 98, 115n18

Xalatlauhco (Mexico): Catholicism in, 53n1; location of, 31; Teton in, 31; Teton on downfall of, 40–41, 50
Xico, Pedro, 40
Xilotepec (Mexico), animal husbandry in, 42
Ximénez, Francisco, *Natural History of the Kingdom of Guatemala*, 66, 67, 75
xiuhmolpilli (Tying of the Years festival), 37, 38, 56n27
xiuhpohualli (year count), 32
xkantumbú, for treatment of cholera and yellow fever, 153
xoloitzcuintli, 49

Yámana: diet of, 248–49; fur seal hunting and, 244–45, 248–50, 262; nutria and sea lion hunting by, 260; population decline of, 244–45; population densities of, 248–49, 266n24

year count (xiuhpohualli), 32

yellow fever: mosquitoes as vector of, 23n21, 164–69; vaccination for, 166, 167, 179n87, 179n89

yellow fever, in Yucatán Peninsula, 164–69; in Caste War period, 151; Maya treatments for, 153; Rockefeller campaigns to eradicate, 151, 165–68

Yelvington, Kevin, 316

yerba de guinea, 218

Yucatán Peninsula (Mexico): bestiality cases in, 1–3; Caste War in (*see* Caste War of Yucatán); climate and physical environment of, 151–52; disease in (*see* disease, in Yucatán Peninsula; public health, in Yucatán Peninsula); Mexican Revolution in, 166; travelers' perception of pests in, 149–50

yuris pájaro, medicinal use of, 141

Zamudio, Juan, 57n38

Zapotecs, diet of, 47

Zayas, Andrés de, 215–16

Zimbabwe, CAMPFIRE program in, 343n9

Zoologia fantástica do Brasil (Taunay and Matos), 7

Zumárraga, Juan de, 55n18